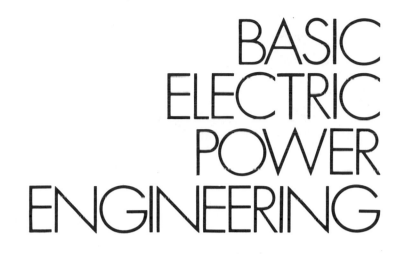

BASIC
ELECTRIC
POWER
ENGINEERING

BASIC ELECTRIC POWER ENGINEERING

OLLE I. ELGERD
University of Florida

ADDISON-WESLEY
PUBLISHING COMPANY

Reading, Massachusetts
Menlo Park, California
London · Amsterdam · Don Mills, Ontario · Sydney

This book is in
ADDISON-WESLEY SERIES IN ELECTRICAL ENGINEERING
Consulting Editors
David Cheng, Leonard Gould, Fred Manasse

ISBN 0-201-01717-2
JKLMNOP-MA-8987654

PREFACE

The oil embargo of 1973 and the subsequent drastic price increases of all prime energy resources have had worldwide repercussion in economical, political, and, particularly, technological areas. The United States is embarking upon a course that, hopefully, will lead to energy independence, that is, an eventual state when the nation's own energy resources will satisfy the total domestic demand. To reach this goal, it will be necessary to shift the prime energy use away from scarce oil and natural gas to those solid fuels, uranium and coal, which are available in abundant domestic supply. New exotic energy resources must be explored; but realistically assessed, they will not make much of an impact in this century due to techno-economical time lags.

The dramatic developments in the energy field have had profound effects on educational institutions throughout the United States. Energy research now occupies top priority on most campuses, and energy-related courses are mushrooming. As the energy content of solid fuels can be utilized only after its transformation into electric form, electric power technology will occupy a key position in the nation's energy future. A strong reemphasis on electric power is seen in most engineering colleges.

This text is tailored to help meet the new challenges that will be faced by the engineering teaching profession in the electric power field in the years ahead. Specifically, the text is of introductory level; it is assumed, however, that the student has had a course in electrical engineering covering the basic circuit concepts. At the University of Florida, where this text was developed, the total text material is covered in about sixty one-hour lectures in a junior-level course sequence. The students come from various engineering disciplines.

The introductory chapter gives a general background to the role played by energy in our society and explains in particular how electric power fits into the picture. Chapters 2 and 3 summarize the basic "energy physics" upon which the electric power technology is based.

Chapters 4 through 8 constitute the core of the text. We have used the novel idea of following the flow of energy from the generators, via the step-up transformers, through the power grid, and finally ending up with the most important consumer devices—the electric motors. Chapter 9 discusses probable future developments in the electric power area. The unique feature of this text is that it addresses itself to the *total* bulk electric power area; generation, transformation, transmission, and the final conversion processes in electric motors are all covered.

A major effort has been made to present the material in the simplest possible terms, by use of mathematical models with proven validity but stripped of unnecessary complexity. Prolific use of text examples and analogs characterize the book. Although the chapters follow in a logical order, an individual treatment of the various topics permits a teacher to make an arbitrary chapter selection.

The United States is going metric and, consequently, the SI unit system is used throughout the book. Because the old units will linger during the conversion period, an effort is made in examples and a separate appendix to demonstrate unit conversion techniques. The very unique "per unit" system, used by electric power engineers throughout the world, is given ample coverage. Worldwide adoption of SI units will most probably not make obsolete this special unit system.

In preparation of the text, I have been greatly assisted by many useful discussions and interchanges with my power colleagues at the Department of Electrical Engineering, University of Florida. In particular, my thanks go to Professor Erv Priem who has helped to test the early versions in class.

I am also greatly indebted to Professor Leonard A. Gould of the Massachusetts Institute of Technology for his thorough review of the entire manuscript and his many valuable suggestions throughout the project. Professors S. A. Sebo and N. A. Smith of Ohio State University read the manuscript, checked for errors, and contributed improvements. The enthusiasm of the editorial staff of Addison-Wesley has been a source of inspiration.

The assistance extended over the years by the Florida investor-owned electric utilities, Florida Power Corporation, Florida Power and Light Company, Gulf Power Company, and Tampa Electric Company, has greatly aided the University of Florida's electric energy engineering program, including the development of this teaching material.

Finally, I acknowledge my debt to Miss Edwina Huggins for her valuable work with the constantly changing classnotes and to my wife Margaret for the typing of the final manuscript.

Gainesville, Florida   O. I. E.
October 1976

CONTENTS

ENERGY THE BASIS OF CIVILIZATION

chapter 1

Man has no Body distinct from his Soul;
for that called Body is a portion
of Soul
discern'd by the five Senses,
the chief
inlets of Soul in this age.
Energy is the only life and is
from the Body;
and Reason is the bound or outward
circumference of Energy.
Energy is Eternal Delight.
—WILLIAM BLAKE
The Marriage of Heaven and Hell, 1793

Overhead high-voltage transmission lines in a Brazilian "energy corridor." (Courtesy ASEA)

This text offers an introductory treatise on electric energy, its generation, transmission, and conversion to and from other forms of energy. Electric energy is unquestionably the most versatile and universally useful form of energy available. Its demand is growing at a rate faster than that of any other energy form. In the overall energy picture, however, electricity constitutes only a fraction of the total energy demand of our society. Before we turn our attention to our main topic we find it useful to view the role of electricity from both a historical angle and an overall energy point of view.

1.1 HISTORICAL PERSPECTIVE

Since his emergence on this planet man has learned to master with increased skill the use of the various forms of energy that nature has provided. In the evolutionary model of man's development, *homo erectus* populated the temperate zones about a million years ago. Archaeological evidence supports the view that for most of that period our forebears knew how to control and use fire. By releasing the energy stored in chemical form in wood into heat energy he was able to improve the comfort of his habitat and also the quality of his diet. This gave man a distinct advantage over other species and contributed immeasurably to his survival and further development.

The first ancient civilizations began rising about ten thousand years ago. It is interesting to note that our present civilization thus represents the combined work of only the latest—about 500 human generations. In terms of one human lifetime, this is a long period but in relation to the total life span of our specie, our civilization developed at the very last "moment."

Why did our civilization develop with such relative suddenness? Let us identify a few important factors.

First, there was the invention of the bow and arrow about 15–20 thousand years ago. We may classify this device as a "weapon," "implement," or "tool," but "energy converter" is probably the most accurate technical description. It enabled the hunter to transform a small portion of his muscle energy into highly controllable kinetic energy (see Chapter 2) of a deadly missile. To an enormous degree it simplified the food-gathering task of early man and offered him "leisure time" that he could now use for more ennobling ventures into art, religion, etc.

The abandonment of the nomadic life of the hunter and settlement into the ancient city-states represented an important milestone in the civilizing process of man. Agriculture and animal husbandry were the prime prerequisites for this revolutionary change of lifestyle. Plants and animals have always provided the life-sustaining links between the energy of the sun and man's food energy. By their domestication, man was now able to control and also improve those essential energy transformation processes.

Pottery making and metallurgical skills increased the demand for man's early energy source—wood. At the same time forests were burned to leave space for agriculture. In many areas, this resulted in the first recorded "energy crises."

Three millennia before the birth of Christ, man harnessed the wind energy. The pre-Christian era saw fleets of Phoenician, Egyptian, Greek, and Roman sailing vessels plying the Mediterranean. The added mobility of wind-driven ships contributed immeasurably to trade and communications. Domestication of horses added "horsepower" to man's means of land travel. The water wheel was invented at about the beginning of the Christian era.

It is noteworthy that, up to the time of the Renaissance, man had not to any great extent begun tapping the energy resources which today are of greatest importance—the fossil fuels. When he finally realized the potential of coal, oil, and natural gas, he was, in effect, ready for that last and still ongoing hectic period—the industrial revolution. When we contemplate to what extent science and technology have changed (and also prolonged) human life in ten short generations, the term "revolution" is most appropriate.

The quality of life that our modern technological civilization offers can be sustained and further improved only if we are able to keep the wheels of industry turning. As this is being written, the most severe energy crisis in man's recorded history is bluntly demonstrating that the era of fossil energy abundance is passed. One of the most important future tasks of technology and engineering must be to develop new energy sources to maintain the continued high standard of living in the industrialized countries and to help meet the aspirations of the developing nations of the world. These new energy sources must be safe and reliable and characterized by acceptable economic and environmental cost. As all technological evidence seems to indicate that some of the most important of these new energy sources can be utilized only after first being transformed into electric form (see Chapter 9), it is reasonable to predict that this particular energy form will be of dominant future importance.

1.2   ENERGY FLOW IN OUR INDUSTRIALIZED SOCIETY

Let us now study in some detail the present energy situation. Figure 1.1 depicts in flow-diagram form the energy picture in the United States. The flow of energy is coarsely traced from its various *primary sources* to the ultimate conversion into various end uses and waste heat. The percentage figures refer to the year 1976. The figure illustrates the following interesting facts.

■ Hydrocarbon fossils, that is, oil, natural gas, and coal in that order, constitute ninety-four percent of our primary energy sources. We are in fact "hooked" on these fuels at the present time.

Hydropower, the energy obtained from impounded rivers, amounts to an unimpressive four percent. About half of this country's rivers (Alaska not included) are already developed, and grave doubts must be raised

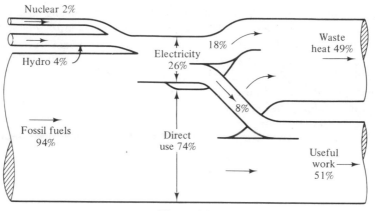

Figure 1.1

about damaging the relatively few remaining free-flowing rivers in order to capture their energy. Realistically appraised, this energy source is thus fairly developed. (In many countries of the world, hydropower plays a much more vital part in the overall energy picture than it does in the United States of America.)

Nuclear energy, although presently amounting to only about two percent, in the opinion of many energy experts is our greatest hope for the future (see Chapter 9).

- Our overall energy use efficiency is only about fifty percent. Many of our most important energy-conversion devices, for instance the internal combustion engine, are characterized by relatively low efficiency. The energy loss always appears as "waste heat"; for example, the radiator heat loss in a car engine. By "useful work" we mean the intended end use of an energy conversion process. For example, in driving an automobile the useful work would represent the energy needed to overcome the friction forces.

  In this connection it is worth pointing out that *energy never is lost, only transformed into new forms.* When these new forms of energy appear in the form of "low-grade" heat, then we talk about "waste energy." The immutable laws of thermodynamics prevent us from "recycling" this low-grade heat energy back to a high-grade form.

- In Fig. 1.1, the various energy flows are given in *percentage* values. It is of special interest to look also at the *absolute* values. Consider the end stage referred to as "useful work" in the figure. For the year 1976, this energy component amounted to about 40×10^{18} joule per annum.† This is so

---

† Joule (J) is the basic energy unit, and J/s or watt (W) is the basic power unit in the *metric* or SI system (*Système Internationale d'Unités*). The system is sometimes referred to as MKSA for the basic units, meter, kilogram, second, and ampere. Energy and power units will be discussed further in Chapter 2 and Appendix C.

large a number as to be totally meaningless. We divide it, therefore, first by 2.15×10^8, the 1976 United States population figure, to obtain a *per capita* figure. Then we divide it further into 31.5×10^6, which is the number of seconds per year. We obtain a value of about 5900 which thus represents an *average per capita energy use rate*, or *power*, measured in joules per second or *watts*.

The reader probably has still only a vague feel for the magnitudes just computed. Is 5900 W a large or small value? If we convert this power into horsepower (hp), a more familiar power unit, we obtain the figure 7.9 hp. The average reader may have a better feel for this unit. Actually, a strong person working at a sweat-driving tempo can with difficulty sustain an output of about 0.1 hp.

The above figure of 7.9 hp thus means that in 1976 each one of us in the United States was assisted by 79 "energy slaves" in our daily lives. These slaves power our cars and airplanes, keep our homes cool or warm, and in general help us sustain our affluent life styles.

No other country is so energy dependent as ours. Citizens in some countries have only a couple of energy slaves each. The United States, with only five percent of the total world population, actually accounts for about one third of the world's energy use.

1.3 ENERGY CONSUMPTION GROWTH

Figure 1.1 gives a *static* picture as it depicts the United States energy situation for a specific year. It is of particular interest to study the *dynamic* aspects of our energy use. Figure 1.2 plots the growth of electric energy consumption in Gainesville, Florida during the period 1940–1975. The graph indicates that the consumption growth during this period was *exponential†* or *geometric†* in nature, with a doubling time of about five years. This corresponds to a compounded growth rate of about fifteen percent annually.

The United States overall consumption of electric energy during the same period doubled every ten years, corresponding to an annual compounded growth rate of about seven percent. During the same period, the *total* U.S. energy consumption grew at the slower annual rate of 4.5%, corresponding to a doubling time of about fifteen years. We make the following additional

---

† A geometric or exponential growth results when constant percentage increments are added periodically in a *compounded* fashion. For example, put \$100 in a bank that gives 7% interest compounded annually. After one year you obviously have accumulated \$107, after two years \$114.49, after three years \$122.50, etc. In about ten years you have *doubled* the initial investment, in twenty years you have *quadrupled* it, etc. A geometrically growing process thus is characterized by a doubling at regular intervals, referred to as the "doubling time" of the process. If shown in a semilogarithmic scale (Fig. 1.2) a geometric process plots as a straight sloping line.

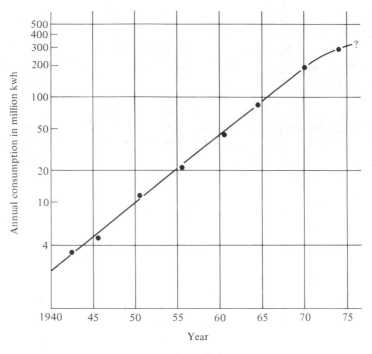

Figure 1.2

observations:

■ The growth of energy demand is much faster than the population growth itself, indicating a growing per capita energy consumption. (The world population growth is unfortunately also geometric. It took over a million years to reach the 1976 figure of about four billion people. If present trends persist, this figure will double in only thirty more years. Geometric growth, for better or worse, seems to be inherent in our life processes— one cell divides into two, two into four, etc.)

■ A geometric growth process is held in great respect by most engineers and scientists since it signifies a situation that is *uncontrolled, unstable,* or *runaway.*

■ A geometric growth process, when the growth limits are approached, has an "explosive" result.† The geometrically growing energy demand on a

---

† Consider the following example: About one percent of the surface of a Florida lake is covered by water hyacinths. This pesky plant, like a cancer, grows at a geometric rate. Assuming that the doubling time is one year, the hyacinths would then grow to cover two percent of the surface in one year, four percent in two years, etc. The reader can easily confirm that it will take about six years to cover half the lake's surface but then, very suddenly, the total lake will be choked in only one additional year.

worldwide basis caught up with the supply in the early 1970s. The tumultuous events that followed on the world energy market created shockwaves that have not abated as of this writing. The resulting energy crisis will have far-reaching consequences for the future U.S. energy supply. (In Chapter 9, when we hopefully have gained some better insights into electric energy technology, we shall venture some predictions of future energy trends.)

1.4 ELECTRIC ENERGY

As shown in Fig. 1.1, electric energy constitutes only about one fifth (eight percent out of fifty-one percent) of all "end uses." But this is an important one fifth which in a multitude of ways influences our daily lives. When on rare occasions we experience electric "blackouts" we are being reminded of our total dependency on this energy form.

The extreme versatility and usefulness of electricity stem from these unique features:

- Instant availability
- Easy transmittability
- Easy controllability

Electromagnetic waves carry intelligent commands to and experimental data from man-made robots exploring the outer reaches of our planetary system. The robots themselves can operate unaided for years via finely tuned electromechanical control systems powered by solar-electric cells. Reliable electric pacemakers give new leases on life to crippled heart patients. By "direct distance dialing" we can establish instantaneous voice communication with parties on the other side of the globe.

The student solving the end-of-chapter exercises in this book is freed from the numerical computational drudgery by an electronic calculator that fits in the palm of his hand. Modern large-scale electronic computers perform computations in a few seconds that would take the average unaided human brain a lifetime to complete—and unlike man the computer will do the job with unerring accuracy.

Electric motors varying in size from fractions of a watt to tens of megawatts turn the vast majority of the wheels in our complex industrial society. In all these and a multitude of other uses of electricity, electric energy is involved in one way or another.

Figure 1.1 indicates that electricity in bulk quantities is generated either as "thermal power" in fossil- or nuclear-powered stations or as "hydropower." Figure 1.3 gives a somewhat more detailed picture (the technical details will be explored in later chapters).

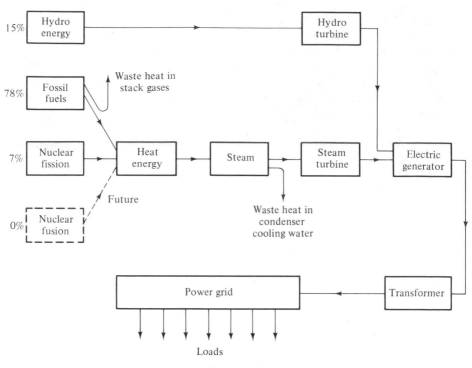

Figure 1.3

1.4.1 Hydroelectric Power

About fifteen percent of the electricity in the United States is derived from hydropower, that is, water stored behind river dams. The potential energy of the water is released in turbines which drive electric generators, whence the energy, via electric transformers, is fed into the national power grid.

The simplicity of the hydroelectric process makes it by far the most reliable electric bulk generation process known. Its efficiency is also very high, of the order of ninety percent. Because the initial cost can be very high, dam construction is often combined with water irrigation projects. In the absence of any fuel costs, hydroelectric power usually is economically very attractive. Hydroelectricity is pollution-free per se, but dam construction has vast and often negative impact on natural river systems.

An attractive feature of hydro power is its high *generation controllability*, meaning that the generation can be varied conveniently and fast. In a hydro plant, the electric power level is changed by simply opening or closing the turbine water gates. One basic feature of electric bulk energy is that it cannot be stored in any large quantities. It must, therefore, be generated the instant the customer demands its use. As a power company has no control over the

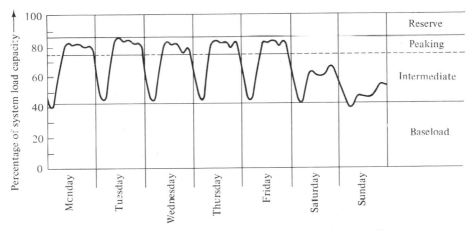

Fig. 1.4 Weekly power demand variation of an electric utility.

timing of the customer demand, it must, therefore, instantaneously match its generation to the demand. The electric power demand typically varies widely throughout the day and week as shown in Fig. 1.4. The ease with which a hydro plant can be turned on or off makes it very suitable as a "peaking unit" (see Fig. 1.4). The hydroelectric generation process is also finding increased use in *pumped hydro storage*, an energy storage method to be described in Chapter 2.

1.4.2 Fossil-Fuel Steam Power

About eighty percent of the electric power generated in the United States is presently obtained from burning fossil fuel. The energy is released in the form of high-grade heat in the boiler. The heat is used to produce steam of high temperature and pressure in a complex system of heat exchangers (see also Fig. 2.15). Via the steam drum (which serves as a low-capacity buffer steam storage device) the steam is led to the turbine, where part of its thermal energy is transformed into mechanical form. The steam turbine drives an electric generator from which energy in electric form is fed into the power grid.

The expanded steam is cooled in the condenser and condensed into water. The water is pumped back to the boiler, thus completing a closed "steam cycle." The steam–electric generation process is a very complex and roundabout way of obtaining electric energy—but it is the best one that technology offers when fossil fuels must be used as primary energy source. The process, furthermore, has poor efficiency—about thirty-five to forty percent at best. The efficiency can be increased by raising the pressure and temperature of the steam, but metallurgical constraints set limits here.

Most of the energy is lost as low grade waste heat in the condenser cooling water. This water, when it exits from the condenser, usually has an elevated temperature in the approximate range of 10°–20°C above ambient. Due to the large quantities needed it may cause environmental impact ("thermal pollution") when issued into relatively small bodies of water. For that reason, closed cooling ponds are often used, or the cooling may take place in cooling towers. Heat energy is also lost to the atmosphere via the stack gases.

The latter contain, in addition to their waste energy, the chemical air pollutants which constitute the greatest problem associated with generation of fossil power, particularly when coal is used as fuel. Due to its complexity, a steam plant has much lower reliability than a hydro plant and its power level cannot be varied as conveniently and as fast. Changes in its power outputs must be accomplished by changes in the combustion rates. If those changes take place too rapidly, they result in unacceptable thermal stresses in the complex of heat exchangers. As a consequence, once a steamplant is "on line" one tries to keep its power level fixed by letting it carry the power system "baseload" (Fig. 1.4).

1.4.3   Nuclear Electric Power

Nuclear electric power plants ("nukes") are playing an increasingly important role in our electric generation picture. In Chapter 9, we shall discuss to greater length the probable future role of nuclear power in our nation.

A nuke is a thermal-type power plant. It differs from a fossil-fueled plant by the absence of a combustion chamber. The heat source is now a nuclear reactor. The nuclear reactor fuel in a controlled fission process transforms some of its mass into energy according to Einstein's formula,

$$E = mc^2.$$

The energy appears in the form of high-grade heat which is then used to produce steam. The steam drives a turbine generator in a conventional way. A nuke requires condenser-cooling water like any thermal power plant and will therefore have the potential for thermal pollution. It does, however, produce zero air pollution, a feature that greatly enhances its attractiveness in comparison to fossil–thermal units. Due to the energy density of mass (compare Section 2.12), a nuke requires a very minute amount of fuel.

Great care must be taken in the design of a nuke to prevent radiation leaks. This adds greatly to the initial cost and construction time lags of present-day nuclear plants. By a far-reaching program of component standardization, this problem will hopefully be solved.

When all the possible future energy candidates are scrutinized in the cold light of technological reality (and we shall make a comparison attempt in Chapter 9) nuclear energy comes out as the probable winner. Nuclear energy can be utilized *only via the medium of electricity*. It seems, therefore, reasonable to predict that we will in future years increasingly turn an electric switch when we want a job done that is beyond the ability of our own muscle power.

1.5 SUMMARY

Electricity like no other single factor helps sustain our modern technological civilization. All signs indicate that it will assume still greater importance in our future. In particular, it may within a couple of generations assume the role as our major energy form, being the only viable link between the inexhaustible supplies of nuclear energy and the end uses.

In view of this probable development, it seems reasonable to foresee the need for better understanding by both electrical and nonelectrical engineers of the basic characteristics of electric energy technology. The objective of this book is to provide this knowledge.

EXERCISES

1.1 It was stated that the 1976 total U.S. "end use" of energy amounted to 40×10^{18} joule. Assume that you could obtain all this energy by means of 100% efficient nuclear processes, the energy thus obtained from transformation of mass according to the Einstein formula

$$E = mc^2.$$

How many kilograms of mass would be required?

1.2 Exponential growth of all human processes lies at the heart of many of the resource problems that we are now facing. To demonstrate the speed with which an exponential process "takes off" consider the following problem:

In 1626 the governor of Dutch West India Co. bought from the Manhattan Indians the island that now carries their name for about $24. If this amount had been invested in a bank yielding 6% compounded annual interest, what would be the value of the investment in the year 1977?

REFERENCES

Energy and Power: A Scientific American book. Ed. by Scientific American Editors. San Francisco: W. H. Freeman, 1971.

Energy and Man: Technical and Social Aspects of Energy. M. G. Morgan (ed.). New York: IEEE, 1975.

Fisher, John C. *Energy Crises in Perspective.* New York: Wiley, 1974.

Krenz, H. J. *Energy—Conversion and Utilization.* Boston: Allyn and Bacon, 1976.

"Megawatts from Municipal Waste," *IEEE Spectrum,* Vol. 12, No. 11 (November 1975), pp. 46–50.

Odum, Howard T. *Environment, Power, and Society.* New York: Wiley (Interscience), 1971.

Penner, S. S., and L. Icerman. *Energy*, Vols. I and II. Reading, Massachusetts: Addison-Wesley, 1974.

Putnam, P. C. *Power from the Wind*, New York: Van Nostrand Reinhold, 1975.

Reynolds, William C. *Energy: From Nature to Man.* New York: McGraw-Hill, 1974.

Spencer, D. F. *The Spectrum of Future Electric Generation Alternatives.* Electric Power Research Institute (EPRI), 3412 Hillview Avenue, Palo Alto, California, 94304. (Many excellent and relevant articles on electric power technology are being published in the EPRI Journal.)

Williams, J. R. *Solar Energy Technology and Applications.* Ann Arbor, Michigan: Ann Arbor Science, 1974.

chapter 2

ENERGY FUNDAMENTALS

*Artist's concept of how the sun's energy could be
used for large-scale production of electricity. Solar
heat is reflected from movable mirrors and focused
on a boiler at the top of the tower. Water in the
boiler is converted to high-pressure steam, which is
led to a conventional steam-turbine generator
at the base of the tower. (Courtesy Honeywell,
Systems and Research Center)*

13

In this chapter we shall introduce the reader to some fundamental physical characteristics of energy. Many of the devices to be discussed in later chapters, like motors, generators, transducers, etc., transform energy from electric to mechanical form or vice versa. In electric heaters, energy is being transformed from electric to *caloric* (thermal) form. In a storage battery, a transformation takes place between electric and chemical energy forms.

Clearly, a proper understanding of electric energy technology is facilitated by a broad knowledge of energy in its many varied forms. The objective of this chapter is to tie together seemingly unrelated pieces of energy-related topics which the reader has picked up in several basic science and engineering courses like physics, statics, dynamics, etc. In fact, we shall concern ourselves in this chapter mostly with nonelectrical forms of energy.

The units used here and throughout the book are those of the *SI unit system*. The reader will find a brief description of this important unit system in Appendix C.

2.1 ENERGY AND GRAVITATION

It is man's fate to spend his whole life under the constant influence of earth's gravity forces. No other factor has so deeply affected our lives and, in fact, the very nature of what type of creatures we are. Gravity has had basic influence on all our sciences and our technology. It is particularly strongly intertwined with the energy concept and we find it natural, therefore, to choose it as the takeoff point in our energy story.

Man learned early the effects of gravity. He found it easier to walk downhill than to climb a mountain. Long before he invented bow and arrow, he had learned to chase animal herds to the edge of a precipice and then let the gravity forces relieve him from the dangerous job of killing his prey.

Quite probably—though we will never know for sure—the thunderous fall of a round boulder inspired one of the unknown early inventors to the design of the wheel. Only the taming of the horse and the wind antedate man's use of falling water in his age-old and continuing process of adding "energy slaves" to his employ. The gravity-powered waterwheel was the most important power source for many centuries and in later decades played a dominant role in the process of generating electric energy.

Up to the present generation, gravity forces confined man to his earthly habitat. The historic trip to the moon marks the beginning of a new era when man has learned to employ and control powers of such magnitude as to free him from the gravity grip of his own planet.

2.2 GRAVITATIONAL FORCE FIELD

Our familiarity with the gravitational pull makes "force" one of the easily accepted and best-understood concepts of physics. As we as engineers make daily use of this concept in our designs, creations, and analyses, we rarely dwell

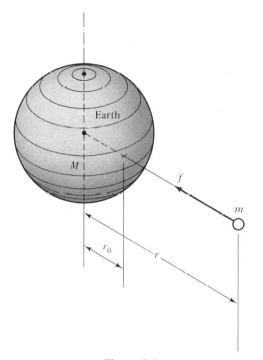

Figure 2.1

on the fact that we know as little now as did our distant ancestors about what "really does the pulling." This is a philosophical question that most probably will never be fully and adequately answered.

However, what makes us different from Cro-Magnon man is our ability to describe, measure, utilize, and control the forces of nature, even though we don't understand their innermost working. Because gravitational forces serve as a logical and easy way of defining *energy* and *power* and *because we can by analogs extend those concepts later to electrical forms of energy and power*, we find it useful to briefly dwell on the physical character and mathematical modeling of the earth's *gravitational field*.

Careful measurements of the gravitational force f acting† on a mass m placed in the vicinity of earth (Fig. 2.1) or some large spherical body of mass M reveal that the force *outside*‡ the large body is everywhere radially directed, is attractive and of magnitude

$$f = k \frac{Mm}{r^2} \quad \text{N.} \tag{2.1}$$

---

† An equal but opposite force is felt by earth.

‡ The gravity force inside earth decreases approximately linearly with distance to earth's center.

The symbols are defined as follows:

f = force magnitude, newton, N;

r = distance between mass centers, meter, m (Fig. 2.1);

m, M = masses, kilograms, kg.

Because the force is characterized by magnitude *and* direction it is a *vector.*

The "universal gravitational constant" k has the numerical value

$$k = 6.670 \times 10^{-11} \qquad \mathrm{N \cdot m^2/kg^2}. \qquad (2.2)$$

It is useful to write the formula (2.1) in the following alternate form:

$$f = m\left(k\frac{M}{r^2}\right) \qquad \mathrm{N}. \qquad (2.3)$$

The expression inside parenthesis is a vector with physical dimension of "force per mass," having the same direction as the force but a magnitude different from the force and *independent of m.* We call it the *gravitation vector,* or *gravity* for short, and define it thus:

$$g \equiv k\frac{M}{r^2} \qquad \mathrm{N/kg}. \qquad (2.4)$$

In terms of this new vector we can write the formula (2.1) in the following alternate form:

$$f = mg \qquad \mathrm{N}. \qquad (2.5)$$

The force has now been expressed as the product between mass m and a new entity, the gravitation vector g, which now embodies the gravitation character of the earth. The two formulas (2.1) and (2.5) are, of course, mathematically identical. But the latter permits the following physical interpretation:

The gravity g, being independent of m and solely depending upon the presence and mass of earth, is an "earth fixture" that exists around our earth *independent of the presence of the mass m.* We refer to it as the *gravitational field.* This field is, of course, invisible and Fig. 2.2 represents a means to "visualize" something that otherwise cannot be sensed with our best sense, sight.

Most of man's activities take place in a limited region close to the surface of earth (Fig. 2.3). In such a limited region, when we can set $r \approx r_0$, where r_0

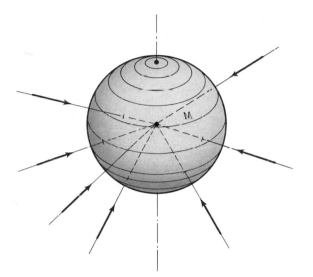

Fig. 2.2 The earth's gravity field is everywhere radially directed.

equals the earth radius, the field lines are essentially parallel and the gravitational field, and thus the force, are of *constant* magnitudes:

$$g \approx k \frac{M}{r_0^2} \equiv g_0 = \text{constant}, \tag{2.6}$$

$$f \approx mg_0 = \text{constant}. \tag{2.7}$$

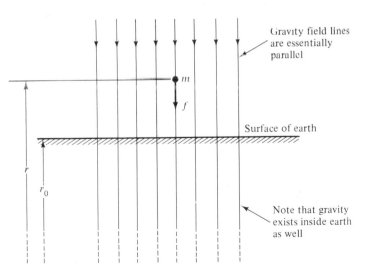

Figure 2.3

The *surface gravity* g_0 varies slightly around the earth, which is not perfectly spherical. (Furthermore, the centrifugal force, which adds an indistinguishable component to the gravity force, varies with latitude.) One has chosen the *standard surface gravity* to equal internationally

$$g_0 \equiv 9.80665 \text{ N/kg.}† \tag{2.8}$$

or 9.81, for short.

In terms of the surface gravity the force magnitude can also be written in the alternate form:

$$f = k\frac{Mm}{r^2} = m \cdot \frac{kM}{r_0^2} \cdot \frac{r_0^2}{r^2} = mg_0\frac{r_0^2}{r^2} \quad \text{N.} \tag{2.9}$$

2.3 GRAVITATIONAL ENERGY EXCHANGE—DEFINITION OF ENERGY

One dictionary defines energy (from the Greek *energos* meaning active) as "capacity for doing work." This seems a reasonably clear definition until one finds that the same dictionary defines "work" as "transference of energy from one body to another."

It is a well-known experience that the work or energy required to climb a hill increases both in proportion to the person's weight and the elevation gain. Similarly, the energy that can be released from an impounded river increases both with the water flow and the height of the dam. *Energy*, certainly in a gravitational sense, thus *is a quantity that depends upon the product of force and distance.*

In elevating a mass by moving it *against*‡ the gravity force we need to *expend* energy. This energy is being *stored*. (Whether the storage in actuality takes place in the mass or in the field is a philosophical question.) As the mass is being lowered the stored energy is being released.

For our engineering purposes, we need to look at this well-known "energy exchange" process from a more quantitative point of view. Consider the energy expended in elevating the mass m in Fig. 2.4 the vertical distance h.

Guided by the "product rule" stated above, we define the energy or work increment dw needed to move the mass the incremental distance dx as *the product of force and distance* according to

$$dw \equiv \text{force} \cdot dx. \tag{2.10}$$

---

† The physical unit for gravity is N/kg. If the mass m is released, the gravitational force will impart upon it an acceleration a which, according to Newton, follows the law $f = ma$. By comparison with formula (2.5) we thus have $a = g$. This means that g also has the physical unit for acceleration, m/s^2.

‡ In the case of the river water above, this job is being done for us by atmospheric forces originally emanating from the sun.

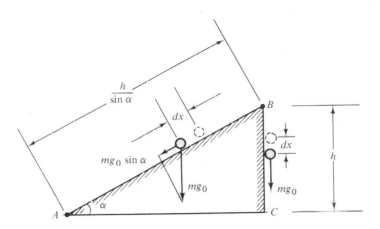

Figure 2.4

By "force" we must, of course, mean that component that we in reality encounter. In moving vertically from $C \rightarrow B$ we must buck the full strength of the gravity force mg_0. In moving up the incline from $A \rightarrow B$ we encounter the smaller component

$$mg_0 \sin \alpha.$$

We thus obtain the two alternate energy expressions

$$dw = mg_0 \, dx \qquad \text{(along } C \rightarrow B\text{),}$$
$$dw = mg_0 \sin \alpha \, dx \quad \text{(along } A \rightarrow B\text{).}$$
(2.11)

The *total* energy expended in the two cases is obtained by integration:

$$w = \int_0^h dw = \int_0^h mg_0 \, dx = mg_0 h \qquad\qquad \text{(along } C \rightarrow B\text{),}$$
$$w = \int_0^{h/\sin \alpha} dw = \int_0^{h/\sin \alpha} mg_0 \sin \alpha \, dx = mg_0 h \quad \text{(along } A \rightarrow B\text{).}$$
(2.12)

The energy is clearly independent of the path chosen and depends only upon the elevation gain, h. (One can extend the proof to include any arbitrarily chosen path.)

The physical dimension for energy from the above definition will be "force times distance." In the SI system, this energy unit will be given the special name *joule* (J). The joule is thus a "derived" unit with the dimension $J = N \cdot m$ (or newton-meter). It also has the dimension *watt-second* (W $\cdot$ s) (often used by electrical engineers).

Example 2.1   What energy amount must be delivered by an elevator motor in lifting a load of five metric tons a vertical distance of 200 m (1 metric ton = 1000 kg)?

SOLUTION: Formula (2.12) gives directly

$$w = 5000 \cdot 9.81 \cdot 200 = 9{,}810{,}000 \text{ J} = 9.81 \text{ MJ}.$$

···

2.4 GRAVITATIONAL POTENTIAL—POTENTIAL ENERGY

If we express the energy in formula (2.12) in terms of joules per kilogram (of the mass m) we obtain a measure v_g, defined as follows:

$$v_g \equiv \frac{w}{m} = \int g_0 \, dx = g_0 h \qquad \text{J/kg.} \tag{2.13}$$

The new physical quantity is referred to as the *gravitational potential*. The increase in potential thus represents the added *per unit* increment in energy that we must impart to the mass to lift it from the lower to the higher level. A mass at a higher level is said to possess a higher *potential gravitational energy* than the same mass at a lower level. According to formula (2.13) each meter of altitude gain adds 9.81 J/kg to the potential of the mass. Differently expressed, for every 10.2 cm of altitude gain, its potential increases by 1 J/kg.

Should we move the mass in a horizontal direction, perpendicularly to the gravity force, no energy change is involved. That is, the potential energy of the mass will be constant on horizontal levels, and these levels would then represent *equipotential surfaces*. In a building, the floors thus constitute equipotential surfaces, whereas the walls coincide with the gravitational field lines.

The integration of formula (2.12) was made simple by the fact that we considered the gravity force to be constant. If we wish to compute the gravitational potential at greater distances from the earth's surface, then we must use the *general* force formula (2.9). We now get

$$v_g = \frac{w}{m} = \int_{r_0}^{r} g \, dr = \int_{r_0}^{r} g_0 \frac{r_0^2}{r^2} \, dr = g_0 \frac{r_0}{r}(r - r_0) \qquad \text{J/kg.} \tag{2.14}$$

(Note that if we set $r \approx r_0$ and $r - r_0 = h$ we get the relation (2.13).)

The equipotential surfaces will now consist of concentric spherical shells. (An earth satellite travelling a circular earth orbit retains a constant velocity because of this fact.)

The reader should note that in our definition of potential we have really talked only about potential *differences*. The zero level can be chosen quite arbitrarily—just as we can arbitrarily choose the zero point of a temperature scale. In our definition we arbitrarily set the gravitational potential at the surface of the earth equal to zero.

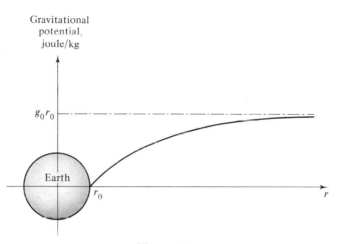

Figure 2.5

If we plot the potential (2.14) as a function of distance from earth we obtain the graph in Fig. 2.5. Note that the potential reaches an asymptotic value of $g_0 r_0$. The reader can visualize this graph as a "potential energy hill" to be climbed should one wish to leave the surface of the earth.

We make this final observation: The potential as we have defined it equals the *line integral* of the gravitation vector. It *increases* in value in negative g direction. It does not matter along what path we integrate—only the end points count. In most cases the integrations are most easily performed if we choose the path (as we did) to coincide with the gravity vector.

Example 2.2 How much energy is needed to rocket a 150-lb astronaut to an altitude of 100 miles above the surface of earth?

SOLUTION: We first convert to SI units:

$$150 \text{ lbs} = 68.0 \text{ kg},$$
$$100 \text{ miles} = 1.609 \cdot 10^5 \text{ m}.$$

Earth radius equals

$$r_0 = 6.38 \cdot 10^6 \text{ m}.$$

The exact formula (2.14) then gives

$$v_g = 9.81 \cdot \frac{6.38 \cdot 10^6}{6.5409 \cdot 10^6} \cdot 1.609 \cdot 10^5 = 1.54 \cdot 10^6 \text{ J/kg}.$$

(The approximate formula (2.13) gives

$$v_g \cong 9.81 \cdot 1.609 \cdot 10^5 = 1.58 \cdot 10^6 \text{ J/kg},$$

that is, an error of 2.5%.) Thus, for each kg of "payload" we need to expend $1.54 \cdot 10^6$ joules of energy. For our 68-kg astronaut we must "pay" an energy price of $105 \cdot 10^6$ joules. (Note that this is the energy needed to *get* him to the required altitude. It is not the energy needed to *keep* him there. Compare Exercise 2.12.)

2.5 GENERAL ENERGY EXPRESSIONS

In our previous discussions all forces involved were of gravity origin. In the practical world in which we live, forces emanate from a number of sources, for example, springs, pressurized gases, friction, etc. In this book we will be particularly concerned with forces of *electrical* and *magnetic* origin.

In most technical applications, forces perform work in either *translational* or *rotational* sense. Rocket propulsion is an example of the former, an electrical motor exemplifies the latter. We develop appropriate formulas for both cases, as shown in Fig. 2.6(a) and (b).

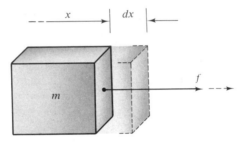

(a) Translational motion

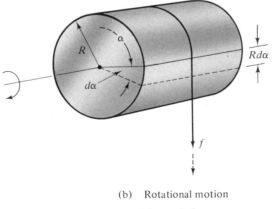

(b) Rotational motion

Figure 2.6

Guided by formula 2.10 we define the incremental energy or work performed by the force to be the product of the force and the incremental distances in each case:

$$dw = f\,dx \quad \text{(translational motion)}, \tag{2.15}$$
$$dw - fR\,d\alpha \quad \text{(rotational motion)}. \tag{2.16}$$

These formulas give the *incremental* energies performed by the forces. If the forces are permitted to act over *finite* distances, then we obtain the total energies by summation or integration:

$$w = \int f\,dx \qquad\qquad \text{J}, \tag{2.17}$$

$$w = \int fR\,d\alpha = R\int f\,d\alpha \quad \text{J}. \tag{2.18}$$

Should the forces be *constant* then the integrations are very simple:

$$w = f\int dx = fx \qquad \text{J}, \tag{2.19}$$

$$w = Rf\int d\alpha = Rf\alpha \qquad \text{J}, \tag{2.20}$$

where x and α represent the total translational and rotational movements, measured in meters and radians, respectively.

2.6   ENERGY RATE OR POWER

Assume that the forces f in Fig. 2.6 perform the incremental works (2.15) and (2.16) in the incremental time dt. We now say that the forces are capable of an *energy* or *work rate* or *power*, p, expressed by the ratios

$$p - \frac{dw}{dt} - \frac{f\,dx}{dt} \quad f\frac{dx}{dt} \qquad \text{W}, \tag{2.21}$$

$$p \equiv \frac{dw}{dt} = \frac{fR\,d\alpha}{dt} = fR\frac{d\alpha}{dt} \qquad \text{W}. \tag{2.22}$$

We identify the ratios dx/dt and $d\alpha/dt$ as the translational and rotational velocities s (m/s) and ω (rad/s) respectively. Thus we have

$$p = fs \qquad \text{(translational case)}, \tag{2.23}$$
$$p = fR\omega = T\omega \quad \text{(rotational case)}, \tag{2.24}$$

where $T = fR$ represents the moment or torque exerted by the force f.

The unit of power is newton-meter per second (N · m/s), joule per second (J/s) or watt (W). Larger power units often used by electric power engineers are kilowatt (kW) or megawatt (MW). They are defined as follows:

$$1 \text{ MW} = 1000 \text{ kW} = 1{,}000{,}000 \text{ W}.$$

Another popular power unit that still lingers among engineers is horsepower (hp), defined as $1 \text{ hp} \equiv 0.746 \text{ kW}$. The metric horsepower (*cheval vapeur*) equals 0.736 kW.

Example 2.3   The power rating of an electric drive motor for a mine elevator must be determined. The motor must be capable of elevating a five-ton load up the 200-meter vertical mine shaft at a velocity of 5 m/s. The acceleration and deceleration periods shall be five seconds each.

SOLUTION: From the given specifications we compute the acceleration a:

$$a = \frac{5[\text{m/s}]}{5[\text{s}]} = 1 \text{ m/s}^2 \qquad \text{(acceleration period)},$$

$$a = -\frac{5[\text{m/s}]}{5[\text{s}]} = -1 \text{ m/s}^2 \qquad \text{(deceleration period)}.$$

During the acceleration period, the lift distance covered will be $\frac{1}{2} \cdot 1 \cdot 5^2 = 12.5$ m. Deceleration will require equal distance. The constant velocity (steady state) lift distance is thus 175 m, which will be covered in 35 s. In Fig. 2.7, we have plotted the velocity during the entire 45 s lift cycle. We next compute the *lift force f.*

During the steady-state period the force f must equal the gravity force:

$$f = f_{ss} = mg_0 = 5000 \cdot 9.81 = 49050 \text{ N} = 49.05 \text{ kN}.$$

During the acceleration and deceleration periods, the total force acting on the load follows from Newton's law of acceleration:

$$f_{acc} - mg_0 = m \cdot 1,$$
$$f_{dec} - mg_0 = m \cdot (-1).$$

Thus

$$f_{acc} = m(g_0 + 1) = 5000 \cdot 10.81 = 54050 \text{ N},$$
$$f_{dec} = m(g_0 - 1) = 5000 \cdot 8.81 = 44050 \text{ N}.$$

The lift force is plotted in Fig. 2.7 for the total lift cycle.
Finally we obtain the *lift power p* from Eq. (2.23).

During steady-state period:

$$p = p_{ss} = 49050 \cdot 5 = 245{,}250 \text{ W} = 245.3 \text{ kW}.$$

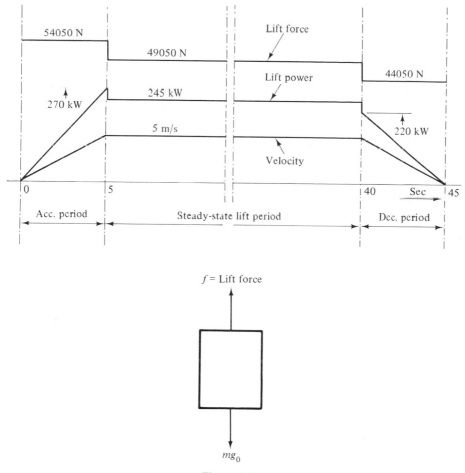

Figure 2.7

During acceleration period: The lift force is constant and the velocity increases linearly from zero to the max value 5 m/s. The lift power must therefore also increase linearly from zero to the maximum value:

$$p_{acc\,max} = 54.05 \cdot 5 = 270.3 \text{ kW}.$$

During deceleration period: A similar reasoning tells us that the power must decrease linearly from a maximum value:

$$p_{dec\,max} = 44.05 \cdot 5 = 220.3 \text{ kW}.$$

The lift power is plotted in Fig. 2.7 for the entire lift cycle.

   This example shows that the lift power will vary throughout a considerable range during the total lift cycle. The *motor control system* must provide this varying power in order to achieve desired lift cycle (compare Sect. 7.4.9).

One final note: In our above calculations we did not consider friction, wind resistance, etc. In a practical situation, these must be included as they result in unavoidable energy and power losses. The size of the motor must be chosen so as to provide a corresponding power margin.

· ·

Example 2.4   Figure 2.8 shows a so-called Pelton turbine used to transform the potential energy of water when a large "head," *h*, is involved. By means of the control valve the water flow is controlled to a ·desirable flow rate, *i* kg/s. The energy of the high-pressure, high-velocity water is "caught" by the turbine blades. The blades must have such a velocity that when the water is leaving the blades at the trailing edges the water velocity must be zero. Only then is all its potential energy fully utilized.

Find a formula for the the turbine power!

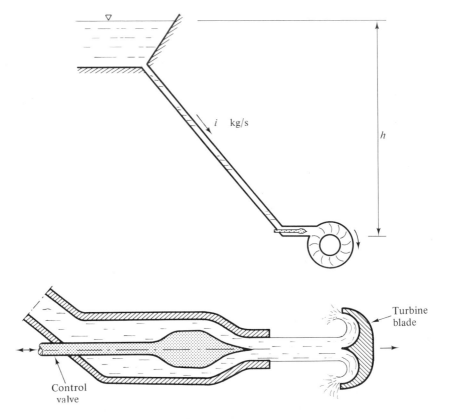

Figure 2.8

SOLUTION: Every second the potential energy of i kg of water is being released. Since the water falls through a potential difference (Eq. 2.13) of

$$v_g = g_0 h \qquad \text{J/kg},$$

a total energy amount of

$$p = v_g i = g_0 h i \qquad \text{J/s} \tag{2.25}$$

is released per second.

Numerical example:

$$h = 700 \text{ m} \qquad (2297 \text{ ft}),$$
$$i = 10,000 \text{ kg/s} \quad (353 \text{ ft}^3/\text{s}).$$

From formula (2.25) we get $p = 9.81 \cdot 700 \cdot 10,000 = 68,670,000$ W or 68.67 MW. (Due to the unavoidable losses associated with the total process the turbine power will be less than this ideal figure.)

Where does this power (or energy) go? Were the turbine not constrained it would accelerate at a high rate and the final results would be destructive. In reality the turbine is driving an electric generator. Through a mechanism (to be discussed in Chapter 4) the currents in the generator windings will create an electromechanical torque which exactly counteracts the turbine torque. The result is zero acceleration, that is, *constant* speed. At the same time the generator feeds 69 megawatts (minus losses) out to the electric grid where it is distributed among millions of electric customers.

2.7   THE LAW OF ENERGY CONSERVATION—FIRST LAW OF THERMODYNAMICS

We can compute the lift energy delivered by the lift motor in Example 2.3 by two methods. The simpler method is to apply the formula (2.19) to the three separate lift intervals. We can also use the formula (2.21) that gives for the energy

$$w = \int p \, dt \qquad \text{J.} \tag{2.26}$$

(This integral obviously equals the area under the power curve in Fig. 2.7.) The reader can easily perform either one of these calculations and obtain the lift energy delivered by the motor:

$$w = 9,810,000 \text{ J} = 9.81 \text{ MJ}.$$

This energy has not been lost! In fact, we never "lose" a single joule of energy anywhere. We simply transform energy from one form to another. This is the "First Law of Thermodynamics" which states that energy can be converted from one form to another but never destroyed.

In the above case, the electric energy delivered by the motor is used to increase the gravitational potential energy of the elevator load. This potential energy is computed (compare Example 2.1) from formula (2.12):

$$w = 200 \cdot 5000 \cdot 9.81 = 9,810,000 \text{ J},$$

which exactly equals the lift energy delivered by the motor.

Now we understand better what really took place when we elevated the five-ton load a distance of 200 meters. The energy that originally was used to produce the electricity simply has been *transformed* into the potential energy of the elevator load. For example, assume that the electricity was produced in a hydro plant where falling water is used as the primary source. If we would neglect all energy losses associated with the generation and transmission plus also motor losses, we would find that we would need exactly five tons of water falling through 200 meters (or 10 tons falling 100 meters) to produce the lift energy in previous example.

Were we to lower the five-ton load we can, in fact, by a procedure called "regenerative braking" (see Chap. 7) recapture the potential energy. The electric motor will now be driven as a *generator* feeding energy back into the electric network.

Were we to lower the load by the use of friction brakes, the energy would be transformed into heat (see Sect. 2.10) but, again, it would not be "lost."

Finally, were we to permit the load to fall freely, the energy would be transformed into *kinetic* energy of the accelerating load (to be discussed in Sect. 2.9).

Potential energy provides a means for energy *storage* in electric power systems. The curve in Fig. 2.9 (compare also Fig. 1.4) tells how the electric

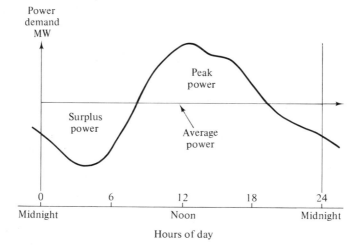

Figure 2.9

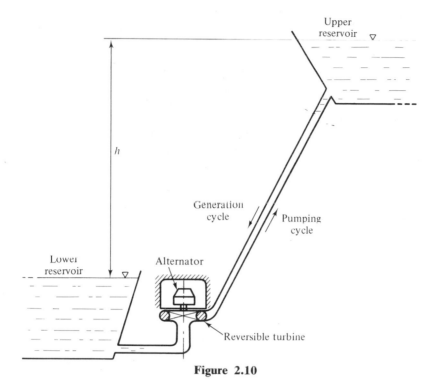

Figure 2.10

energy demand in such a system varies throughout a typical day, with a peak usually in the early afternoon and a minimum just before dawn. As it is essentially impossible to store large amounts of electric energy (see Chap. 3), in an electric power system one must vary the generation so that it matches the demand at each instant.

This varying generation requirement conflicts, however, with the strong desire to run steam-driven generator stations at a *constant* energy output (see Sect. 1.4.2). It would be nice if the steam plant could be run at the constant *average* power output shown in Fig. 2.9. This can be achieved by means of a so-called pumped hydro storage generation plant sketched in Fig. 2.10. An *alternator* is connected to a hydro turbine through which water can flow both ways.† If the water flows from the upper to the lower reservoir it will drive the alternator as a generator, feeding electric energy *into* the network. If electric energy is fed into the alternator *from* the network, then the alternator will run as a motor driving the turbine as a pump, pumping water back up to the upper reservoir.

---

† Why would not a Pelton wheel (Fig. 2.8) do the job? How would you design a reversible turbine?

The proper operating strategy would be to pump water, thus increasing the potential energy in the upper reservoir, during the night hours when surplus energy is available in the power system (see Fig. 2.9). In the afternoon hours when the steam plant needs support power we would reverse the flow and operate the facility as a "peaking generating unit."

Of course, as the peaking unit must deliver the difference between the actual varying demand and the constant steam generator output, its power output will not be constant. However, (as was pointed out in Sect. 1.4.1) a hydro plant unlike a steam plant can easily be operated with a greatly varying output.

Example 2.5   Assume that the average head (Fig. 2.10), or level difference between the two reservoirs, is $h = 300$ m. How much water capacity must the reservoirs have to store the equivalent of 1200 megawatt hours (MWh) of energy? (For example, this energy amount corresponds to a power output of 400 MW during a period of three hours.) In our analysis let us assume the overall efficiency of a full pump-generation cycle to be $\eta = 60\%$.

SOLUTION: First we express the stored energy in joules:

$$1200\text{ MWh} = 1200 \cdot 3600\text{ MWs} = 1200 \cdot 3600 \cdot 10^6\text{ Ws}$$
$$= 4.32 \cdot 10^{12}\text{ J} = 4.32\text{ TJ (terajoules)}.$$

If the required water quantity is m kg, we then get from formula (2.12) the equality

$$\eta \cdot mg_0h = 4.32 \cdot 10^{12}; \tag{2.27}$$
$$\therefore m = \frac{4.32 \cdot 10^{12}}{0.6 \cdot 9.81 \cdot 300} = 2.45 \cdot 10^9\text{ kg} = 2.45 \cdot 10^6\text{ m}^3\text{ H}_2\text{O}.$$

If each reservoir has a surface area of 1 km^2 (about 0.39 sq mi) the water level at each reservoir would thus change 2.45 meters during the total pump-generation cycle.

2.8 OTHER FORMS OF POTENTIAL ENERGY

As we have seen, earth's gravity field offers a means for potential energy storage. Other possibilities exist. One can, for example, fill earth cavities with compressed air which, upon expansion, can power air turbines. Springs, tension bars, and other elastic media can likewise be used to store energy but in less quantities. Let us demonstrate with a couple of examples.

Example 2.6   Consider the rock cavity in Fig. 2.11(a). It has the volume v_c and is filled with air under pressure p_c. Compute the energy stored in the system which potentially can be used for generating electric energy in the power plant.

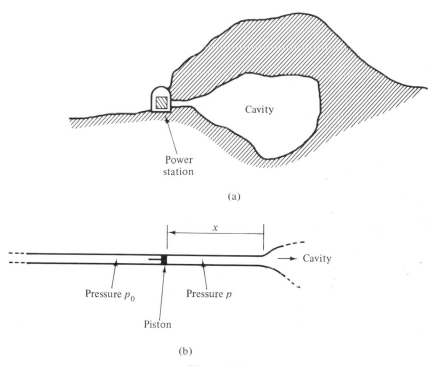

(a)

(b)

Figure 2.11

SOLUTION: The energy storage may be computed in the following way: Let the compressed air push an imaginary piston (Fig. 2.11b) along a tube. The total energy stored in the cavity will be given up to the piston when the inside pressure p equals the outside pressure p_0 at which time the piston position is $x = x_{max}$. To obtain the total stored energy we use formula (2.17) and integrate between $x = 0$ and $x = x_{max}$.

If the piston area is A m^2 then the force on the piston is $(p - p_0)A$ N. From Eq. (2.17) we thus get

$$W_{stored} = \int_0^{x_{max}} (p - p_0)A \, dx \qquad \text{N} \cdot \text{m}. \tag{2.28}$$

To perform the integration we must obviously find a relationship between p and x. For this purpose we borrow Boyle's law† from physics:

$$pv = p_c v_c, \tag{2.29}$$

_____

† Boyle's law stated in this form strictly applies to gases in *static* equilibrium and of constant temperature. We assume that the cavity is emptied at such a slow rate that we essentially have a "pseudo static" situation.

where the total gas volume v is

$$v = v_c + Ax \qquad m^3. \tag{2.30}$$

Combination of (2.29) and (2.30) gives the sought relationship between p and x:

$$p = \frac{p_c v_c}{v_c + Ax} \qquad N/m^2, \text{ or } pascal \text{ (Pa)}. \tag{2.31}$$

The maximum piston stroke $x = x_{max}$ is obtained by setting $p = p_0$ in Eq. (2.31). We get

$$x_{max} = \frac{v_c}{A} \cdot \left(\frac{p_c}{p_0} - 1\right) \qquad m. \tag{2.32}$$

By substituting these expressions for p and x_{max} into the integral (2.28) it can be readily integrated. Integration gives the simple energy storage expression:

$$W_{stored} = p_c v_c \left[\ln\left(\frac{p_c}{p_0}\right) - \left(1 - \frac{p_0}{p_c}\right)\right] \qquad J. \tag{2.33}$$

Let us consider the following numerical case:

$$v_c = 10^6 m^3 \quad (35.3 \text{ million } ft^3),$$
$$p_0 = 100 \text{ kPa} \quad (\approx 14.5 \text{ psi}),$$
$$p_c = 1000 \text{ kPa} \quad (\approx 145 \text{ psi}).$$

We get upon substitution into formula (2.33)

$$W_{stored} = 1.40 \cdot 10^{12} J = 1.40 \text{ TJ}.$$

Compare this energy storage capacity[†] with that of the pump storage facility in Example 2.5. (Note that the cavity volume is about equal to the pumped water volume in Example 2.5.)

....

Example 2.7 In the introductory chapter, we suggested that the bow and arrow was an "energy converter." Compute how much potential energy can be stored in a drawn bow (Fig. 2.12).

SOLUTION: As the bow string is pulled, a force f must be applied which grows approximately linearly with pull distance x, that is,

$$f = kx \qquad N \tag{2.34}$$

[†] In a practical situation one would not, of course, reduce the air pressure down to 1 atm during the generation cycle, for the same reason that one would not reduce the waterhead h to zero in the pump storage plant.

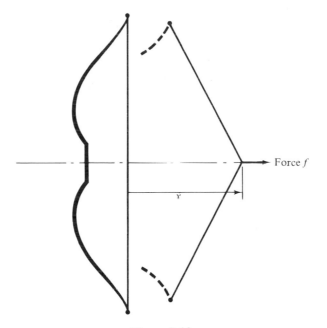

Force f

Figure 2.12

where k is the "spring constant" which determines the bow stiffness. It has the dimension N/m.

From Eq. (2.17) we get now directly

$$w_{\text{stored}} = \int_0^x kx\, dx = \tfrac{1}{2}kx^2 \qquad \text{J}. \qquad (2.35)$$

This *quadratic* energy expression is typical for all linear springs.

Numerical example: Compute the energy stored in a 40-lb bow at its full 18-in stroke.

The spring constant k is computed from the information that the force is 40 lbf (178 N) for $x = x_{\max} = 18$ in. (0.4572 m);

$$\therefore \; k = \frac{178}{0.4572} = 389.3 \text{ N/m}.$$

From Eq. (2.35) we then get

$$w_{\text{stored}} = \tfrac{1}{2} \cdot 389.3 \cdot 0.4572^2 = 40.7 \text{ J}.$$

2.9   KINETIC ENERGY FORMS

All the various potential energy forms we have discussed concern themselves with *static* systems. Later we will add potential *electric* energy forms to this list.

In many important practical applications relating to electric conversion problems, we will encounter *kinetic* forms of energy,—energy associated with systems in motion.

Consider for a moment the mass m in Fig. 2.6(a). According to Newton, the mass will experience an acceleration d^2x/dt^2 in the x direction, following the equation

$$f = m\frac{d^2x}{dt^2}. \tag{2.36}$$

We introduce the velocity

$$s = \frac{dx}{dt}$$

as a new variable and can then write Eq. (2.36) in the following form:

$$f = m\frac{d^2x}{dt^2} = m\frac{ds}{dt} = m\frac{dx}{dt}\cdot\frac{ds}{dx} = ms\frac{ds}{dx}, \tag{2.37}$$

or

$$f\,dx = ms\,ds. \tag{2.38}$$

Upon integration we obtain

$$\int_0^x f\,dx = \int_0^s ms\,ds; \tag{2.39}$$

that is,

$$\int_0^x f\,dx = \tfrac{1}{2}ms^2 \equiv w_{\text{kin}}. \tag{2.40}$$

This equation states that the work performed by the force f (left side) is absorbed by the mass in the form of a new energy form (right side) which we will refer to as *kinetic energy*. It increases as the square of the velocity.

If we perform a similar analysis of the rotational mass in Fig. 2.6(b) we would obtain a symmetric expression for its kinetic energy:

$$w_{\text{kin}} = \tfrac{1}{2}I\omega^2 \qquad \text{J} \tag{2.41}$$

where I is the moment of inertia of the rotating mass and ω equals its rotational speed expressed in rad/s.

Example 2.8   The kinetic energy of the spinning rotors of synchronous generators is of very great importance in understanding the operation of interconnected electric power systems (Chap. 6).

A turbogenerator rotor has the cylindrical form as depicted in Fig. 2.6(b).

It is spinning at the rate of 1800 rpm. The rotor dimensions are

$$\text{diameter} = 1.0\,\text{m},$$
$$\text{length} = 3.0\,\text{m}.$$

The rotor is made from steel of density $7800\,\text{kg/m}^3$. How much kinetic energy does the spinning rotor represent? (The rotor is, of course, coupled directly to a steam turbine which will add significantly to the total inertia.)

SOLUTION: The moment of inertia of a solid cylinder with radius R is obtained from the formula

$$I = \tfrac{1}{2}mR^2 \text{ kgm}^2 \tag{2.42}$$

where the mass m is computed as follows:

$$m = 7800 \cdot \pi \cdot 0.5^2 \cdot 3.0 = 18380 \text{ kg};$$

$$\therefore I = \tfrac{1}{2} \cdot 18380 \cdot (0.5)^2 = 2298 \text{ kgm}^2.$$

The rotational speed is

$$\omega = \frac{1800}{60} \cdot 2\pi = 188.5 \text{ rad/s}.$$

We finally obtain the kinetic energy from formula (2.41):

$$w_{kin} = \tfrac{1}{2} \cdot 2298 \cdot 188.5^2 = 40.8 \cdot 10^6 \text{ J} = 40.8 \text{ MJ}.$$

Compared with the energies stored in Examples 2.5 and 2.6 this is not a very impressive figure. Nevertheless, energy storage in rotating masses ("super-flywheels") is sometimes of practical usefulness.

Example 2.9   An arrow weighing 20 grams is shot from the drawn bow analyzed in Example 2.7. Compute its velocity when it leaves the string.

SOLUTION: If we disregard wind friction we can assume that the total potential energy, 40.7 J, is transformed into kinetic energy of the arrow. We thus have the energy equality:

$$\tfrac{1}{2} \cdot 0.020 \cdot s^2 = 40.7.$$

Solving for the arrow velocity yields

$$s = 63.8 \text{ m/s}.$$

2.10 CALORIC (HEAT OR THERMAL) ENERGY

2.10.1 Ordered and Disordered Energy Forms

Often in the previous discussions we have referred to energy "losses" and "efficiency" of various energy transformation processes. We noted in Chapter 1

(Fig. 1.1) that our total national energy utilization is characterized by a low 51 percent overall efficiency.

All physical energy transformation processes are inherently nonideal in the sense that *always* a certain portion of the energy "disappears" in the form of caloric energy. The temperature of a solid, liquid, or gas is a measure of the intensity of the *random motion* of its elementary particles. The more heat a body absorbs, the more intense this particle motion and the higher the body temperature.

Because caloric energy is associated with random particle motion it is referred to as a "disordered" form of energy. The disorderedness of energy is measured by its *entropy*. For example, electric energy (Chap. 3) is ordered and has zero entropy. It is this "orderedness" of certain energy forms that permits their controlled transformation into other forms and thus increases their "usefulness."

Nature has an inherent tendency to "go disordered." It is said that the *entropy of the world tends towards a maximum.* Nature itself therefore sets limits to the effectiveness of all energy transformation processes.

2.10.2 Reversible and Nonreversible Energy Transformations—Second Law of Thermodynamics

Consider for a moment the bow and arrow example discussed earlier. As the bowstring is released, the string and arrow will accelerate air molecules and thus increase slightly the temperature of the surrounding air. A certain small amount of the stored potential energy will thus be "lost" for this purpose, resulting in a somewhat smaller arrow velocity than what we earlier computed, and in an energy transformation efficiency less than 100 percent.

If this lost energy cannot be recaptured, the energy transformation process is called *nonreversible.* Were it not for this heat loss the energy transformation process of the bow and arrow would be perfectly *reversible.* (Note that we can easily conceive of the possibility of "catching" the flying arrow by the means of an identical bow, thus stopping its flight and recapturing its kinetic energy.)

The usefulness of caloric energy increases generally with its temperature level. Thus the heat in an acetylene flame is more valuable (high-grade heat) than the heat obtained from an automobile radiator (low-grade heat). We should not, however, understand this statement to mean that low-grade heat is useless. For example, home heating apparatus radiates heat which is more low-grade than that from an automobile radiator but nobody would suggest this energy to be "waste."

Furthermore, the fact that an energy process is nonreversible does not mean, necessarily, that the heat "lost" in the process cannot be used for other energy transformation purposes. Such transformations are, however, usually of low efficiency and may also involve long time lags between cause and effect.

For example, a *thermoelectric* generator based upon the Seebeck effect† will accept low-grade heat as input and deliver a low voltage electric output. Or consider the low-grade heat supplied to a greenhouse. Biologically, this energy is being transformed into chemical energy of plants, but the process takes time. The reader should contemplate how, over vast time spans, our present important fossil energy resources were formed.

An ordered energy form can always be transformed with 100 percent efficiency into disordered form (heat). For example, the flying arrow eventually will be stopped, that is, *all* its kinetic energy will have been transformed to heat. Or *all* the potential energy of an elevator can easily be transformed to smoking wreckage at the bottom of a mineshaft.

Conversely, a disordered form of energy (heat) *cannot* be transformed back into ordered form with a 100 percent conversion efficiency. This is the "Second Law of Thermodynamics."

This then is the basic reason why *all* energy transformation processes that start with heat always are fairly inefficient. For example, a thermal power plant, where the end product is highly ordered electric energy, seldom has an efficiency exceeding thirty-five percent.

2.10.3   The Caloric Energy Equivalent

If the five-ton elevator in Example 2.3 were to drop 200 m, its potential energy would first transform into kinetic energy during the accelerating descent. At the bottom of the mineshaft, this kinetic energy in an instant would be transformed into heat—as evidenced by the raised temperature of the wreckage. How much heat would be generated by the 9810-kJ of potential energy originally stored?

The answer can be found from a knowledge of the equivalence between caloric energy and mechanical energy. Before this equivalence can be established we need to define the caloric energy unit.

Definition.   *The caloric energy unit, kilocalorie (kcal), is defined as the amount of energy needed to raise the temperature of one kilogram of water by 1°K.*

Experimentally one finds the following mean value of the caloric energy equivalent:

$$1 \, \text{kcal} = 4.19 \, \text{kJ}. \quad \text{("Joule's constant")} \quad (2.43)$$

---

† Briefly, this effect is manifested thusly: Form an electric loop of two different materials. If the two junctions are kept at different temperatures, a current will flow in the loop.

Example 2.10   How much energy, Q, is required to increase the temperature of one pound of water by 1°F?

SOLUTION: We have

$$1 \text{ lb} = 0.4536 \text{ kg}, \qquad 1°F = 0.5556°K.$$

Thus

$$Q = 0.4536 \cdot 0.5556 \text{ kcal} = 0.252 \text{ kcal} \quad \text{(or } 0.252 \cdot 4.19 = 1.06 \text{ kJ)}.$$

This heat quantity (252 cal) is referred to as a "British thermal unit" (Btu).

Figures 2.13 and 2.14 summarize the most important energy and power units in use today and the conversion constants between them. In going from one unit to another, you simply multiply by the conversion constant shown. For example, 1 Btu is equivalent to 778 ft · lbf.

The caloric energy equivalent is very useful when determining the relative energy values of various fuels. We demonstrate by the following example.

Example 2.11   As noted from Fig. 1.3, fossil fuels account for the major portion of all electric energy generated in the United States. As of this writing shortages of oil and natural gas have resulted in increased reliance upon coal as primary fuel.

How much electric energy can be derived from 1 kg of coal if it is known that this particular coal releases 13100 Btu of caloric heat per pound upon burning?

The heat is used to generate steam which powers a steam turbine which in turn drives the electric generator. Figure 2.15 depicts the process in somewhat more detail than Fig. 1.3. As was mentioned in Chapter 1, the total efficiency of this process is quite low—the majority of the heat actually being lost through stack gases and condenser cooling water. For purposes of our analysis we assume the overall energy transformation efficiency to be $\eta = 33\%$ (including transmission losses).

SOLUTION: From Fig. 2.13, we find that 1 lb of coal releases upon burning $1.055 \cdot 13100 = 13820$ kJ of energy. One kg thus releases $2.205 \cdot 13820 = 30470$ kJ.

Upon taking the heat losses into account we are thus left with

$$0.33 \cdot 30470 = 10060 \text{ kJ} = 2.79 \text{ kWh} \quad \text{(of electric energy)}.$$

A medium-sized home requires a 5-kW compressor for running its central air conditioner. The energy derived from burning 1 kg of coal will thus run this unit for slightly more than a half hour.

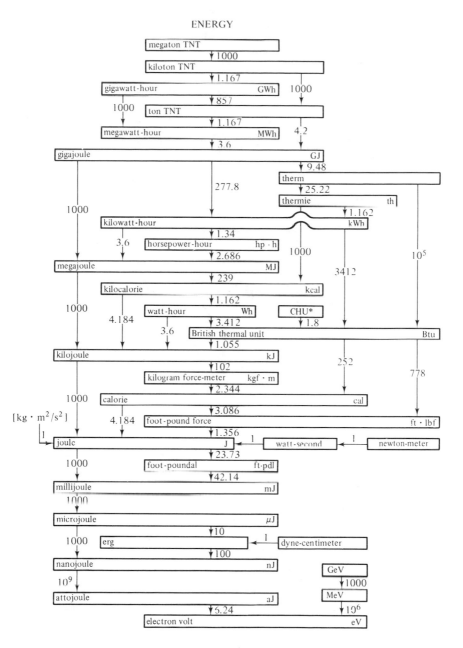

*The Centigrade Heat Unit, or pound-calorie, is the amount of heat
required to raise 1 pound of water 1°C

Fig. 2.13 From "UNITS" by Theodore Wildi. Copyright © 1971 by Volta Inc. All
rights reserved.

POWER

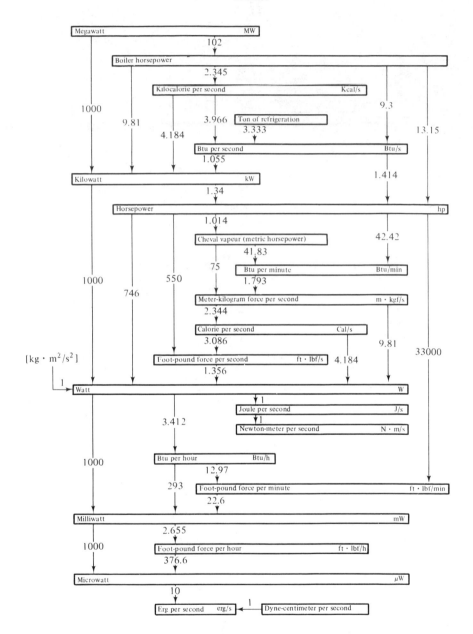

Fig. 2.14 From "UNITS" by Theodore Wildi. Copyright © 1971 by Volta Inc. All rights reserved.

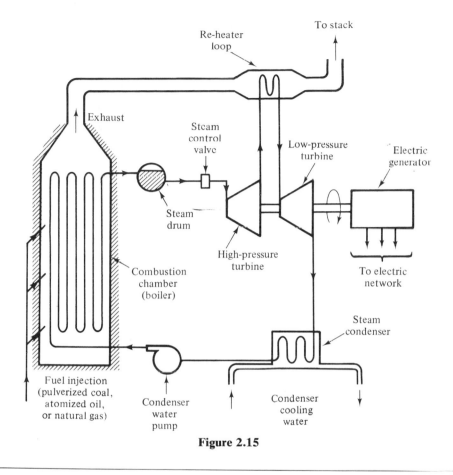

Figure 2.15

Example 2.12   A diet table states that a "normal" male, of height 6 ft 2 in. and weight 190 lbs, requires a daily food energy intake of 2900 "calories."

This energy is used, of course, to "fuel" the body for all of one's various physical activities. To get a feel for the amount of energy involved, compute the height h of the stair which our "normal" male must climb to burn his total daily food intake.

SOLUTION: When one talks about "calories" in diet matters, one means *kilo*calories. From Eq. (2.43) we compute our joule equivalence:

$$w = 4.19 \cdot 2900 = 12150 \text{ kJ}.$$

As the person's weight equals 86.2 kg, we get directly from formula (2.12)

$$h = \frac{12.15 \cdot 10^6}{86.2 \cdot 9.81} = 14{,}370 \text{ m}.$$

Clearly, a normal sedentary male does not climb the equivalent of almost two Mount Everests daily. The explanation is, of course, that the body needs energy for all vital ongoing internal processes *plus* keeping our body temperature at normal value. Also, the conversion efficiency is not 100 percent.

What is impressive in this example is the amount of energy actually contained in a normal daily food intake.

2.11 ENERGY DISSIPATION

Consider a system consisting of a mass suspended in a spring. If excited this system will perform oscillations. The energy of the system will oscillate between the spring and the mass. At one instant the total energy is stored in form of kinetic energy in the mass moving at its maximum velocity. A quarter cycle later the velocity is zero, and the total system energy is now found stored as potential energy in the spring. If the mass oscillates at a frequency of f Hz the energy oscillates at $2f$ Hz.

In many such oscillatory systems it is desirable to stop or damp the oscillations. This can be accomplished if the energy is dissipated in form of heat. A viscous damper (or shock absorber), as sketched in Fig. 2.16, will do

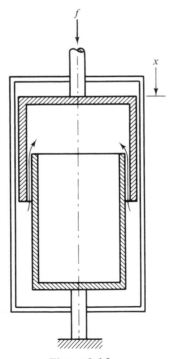

Figure 2.16

the job. As the damper is compressed, oil will flow as shown by the arrows. The force f increases with the rate of oil flow. With a certain degree of approximation and for not too high velocities we find that the force is proportional to the velocity, that is,

$$f = k_d \frac{dx}{dt} \quad \text{N} \tag{2.44}$$

where k_d is the damper coefficient. If the velocity is too high the oil flow becomes turbulent and the force changes character completely.

From Eq. (2.17) we thus get for the energy absorbed in the oil as heat:

$$w = \int f \, dx = \int k_d \frac{dx}{dt} \, dx = k_d \int \left(\frac{dx}{dt}\right)^2 dt \quad \text{J.} \tag{2.45}$$

The integrand is always positive (that is, energy is dissipated into heat independent of direction of motion) and the energy dissipation rate equals

$$\frac{dw}{dt} = k_d \left(\frac{dx}{dt}\right)^2 \quad \text{W.} \tag{2.46}$$

Electric energy can be dissipated in *resistors* (see Sect. 3.14).

Example 2.13   The oscillating mass mentioned above is mechanically coupled to the moving member of the shock absorber shown in Fig. 2.16. Describe its motion.

SOLUTION: Total kinetic plus potential energy of the mass-spring system equals, according to Eqs. (2.35) and (2.40),

$$w_{\text{tot}} = \frac{1}{2} m \left(\frac{dx}{dt}\right)^2 + \frac{1}{2} kx^2 \quad \text{J} \tag{2.47}$$

where x represents the spring elongation and k the spring constant.

This energy will dissipate in heat according to formula (2.46), and we can express this as

$$-\frac{d}{dt} \left[\frac{1}{2} m \left(\frac{dx}{dt}\right)^2 + \frac{1}{2} kx^2 \right] = k_d \left(\frac{dx}{dt}\right)^2 \quad \text{W.} \tag{2.48}$$

By performing the differentiation we obtain the following differential equation for the motion of the mass:

$$m\ddot{x} + k_d \dot{x} + kx = 0. \tag{2.49}$$

Integration of this equation produces the solutions depicted in Fig. 2.17. The reader is encouraged to confirm this.

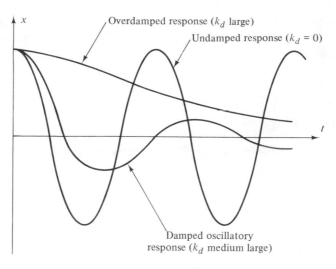

Figure 2.17

2.12   NUCLEAR ENERGY

We prophesized in Chapter 1 that nuclear energy in coming decades will become one of our most prominent sources of energy. No other form of energy can match the present growth rate of nuclear electric energy.

At the turn of the century Einstein postulated the famous energy-mass relationship

$$E = mc^2, \tag{2.50}$$

where m = mass in kg, c = velocity of light in vacuum in m/s and E = energy in J. The formula indicates that mass—any mass—is equivalent with energy in enormous quantities. For example, 1 kg of mass if completely converted into energy, according to Einstein, would yield the following:

$$E = 1 \cdot (3 \cdot 10^8)^2 = 9 \cdot 10^{16}\,\text{J}.$$

For comparison, we remember from Chapter 1 that the total annual U.S. energy demand in 1976 amounted to $40 \cdot 10^{18}$ joules. This energy demand thus could, in theory, be satisfied by the transformation of only 444 kg of mass.

Theory and practice are not always close, however, and the fact is that today we do not know how to accomplish a 100 percent efficient energy conversion of mass. We do know, fortunately, how to accomplish the feat *partially*.

Some heavy materials, like the uranium isotope U-235, are *fissionable*. The nucleus of an atom of U-235 under bombardment by neutrons will absorb one neutron and form the isotope U-236. Being highly unstable, this isotope fissions into two new heavy atoms of xenon and strontium, plus additional neutrons. The total mass of the fission fragments is slightly less than the mass of the original atom, the difference in mass having been transformed into

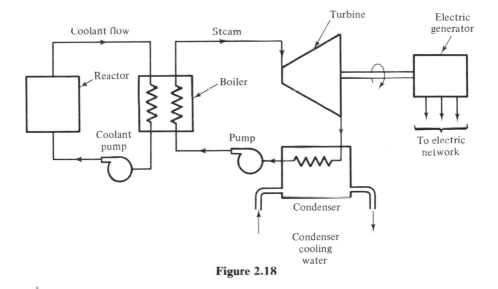

Figure 2.18

energy in quantities given by the Einstein formula. The energy appears in caloric form, most of it absorbed by the heavy fission fragments.

The particular value of the uranium fission reaction is that it can be made both self-sustaining and controllable. By simply putting together a sufficient amount (core) of U-235 a *critical mass*† is reached. The reaction rate of this *reactor* can be controlled by various means but most importantly by the positioning in the core of *control rods.*

The caloric energy released in the fission reaction will increase the core temperature and must be continuously removed. This is accomplished by the flow through the reactor of a coolant, which may be water, gas, or a molten metal. The coolant thus serves the same purpose as the steam in the fossil-fueled process shown in Fig. 2.15 it transports the energy out of the "combustion" area.

In a boiling water reactor (BWR) the water coolant exits from the reactor in the form of steam, which is led directly to the turbine. In a pressurized water reactor (PWR) the water coolant does not boil. This high-pressure, high-temperature coolant water transfers its heat energy to water in a *heat exchanger* or *boiler*, and the steam thus created drives in an ordinary manner the electric turbo generator. In a nuclear plant, the reactor thus essentially replaces the combustion part (including stack, smoke, coal, sludge, ash, etc.) of a fossil-fired unit (Fig. 2.18).

---

† A simple analogy is offered by a woods fire. It is impossible to make a fire of a single log. A sufficient number must be put together before it burns. Similarly, nuclear fuel will "burn" only if the reactor core is sufficiently large.

Although the uranium fission reaction converts only of the order 0.1% of the U-235 mass into energy, the energy "compactness" of nuclear fuel is still extremely impressive. *One kg of U-235 produces the same amount of caloric energy as about 3000 metric tons of coal.* Whereas a 1000-MW coal-fired power plant requires an almost continuous feed of coal cars, the annual fuel charge of an equal-sized nuclear station corresponds to a few truck loads.

The controlled nuclear fission process is now, after almost a third of a century of experimentation, well under technological control. We have not, however, as of this writing succeeded, even on a laboratory scale, to control nuclear *fusion.* In this energy process, which is ongoing in the sun and which, uncontrolled, takes place in a hydrogen bomb, light atoms are fused together into heavier ones under the simultaneous release of energy. (See Sect. 9.3.6.)

Man will surely one day master this process. When this happens, the energy, like in the fission process, most surely will be obtained in the caloric form. To be useful it will therefore, like now, have to be transformed into electric form. In that future day, our electric transmission networks will thus not become obsolete.

2.13   SOLAR ENERGY

The sun radiates to our planet energy amounts which far exceed the present and even the foreseeable future need of even a most technologically advanced society. This energy input fuels all our biological processes, and is, of course, the original source for hydro power, wind power and, most importantly, our fossil fuel resources. Environmentally and economically, it is our most attractive energy source. Why then don't we use solar energy directly for industrial, domestic, commercial, and transportation purposes? Three reasons account for this fact:

1) Solar energy arrives too diluted—of the order of only one kW per square meter (measured in clear weather).

2) It does not arrive at all during night hours.

3) It is very difficult to transform into any useful energy forms except low-grade heat.

Present technology knows of only two ways to capture solar energy in bulk quantities:

1) By means of parabolic or spherical collectors solar energy can be concentrated into high temperature ($> 1000°K$) caloric form. These *high-intensity* collectors work only in direct radiation—they are inoperative in cloudy weather. (See also Chap. 9 and figure on p. 13.)

2) By means of "greenhouse" type collectors, the solar energy is transformed into low temperature ($\approx 370°K$) caloric form. These *low-intensity* collectors work, with reduced effectiveness, even on cloudy days.

The above limitations put severe restrictions on the usefulness of solar energy.

Present technology permits us, however, to utilize solar energy for domestic purposes. Thus, by means of a few square meters of roof-mounted low-intensity solar collectors, we can easily satisfy the need for low-grade heat needed in every home.

A recent study of a typical Florida home revealed that 60 to 70 percent of the total domestic energy need is of this form. Heating and cooling plus, of course, all hot water needs fall in this category.

We see, therefore, a trend toward *hybrid solar electric* energy service to the domestic market and maybe also to the commercial market. In such a hybrid system, solar energy is used for those jobs it can nicely do—electricity is used for the rest. Solar energy can, conceptually at least, also be used in conjuction with pumped hydro storage for peak power generation. We shall explore those possibilities in Sect. 9.3.6.

Example 2.14   The simplest type of a low-intensity solar collector consists of a flat glass-covered box. The solar heat trapped in the box heats the water in a grid of copper tubes, which is part of a circulatory system. In order to serve the energy need during night hours, it is necessary to store enough hot water during the day to last throughout the night. This is done in an insulated hot water storage tank.

What size should this tank have if its water temperature should not drop more than $15°K$ during the night? It is assumed that the total heating load during the night period amounts to 30 kWh.

SOLUTION: If the tank capacity is x liters (1 liter H_2O equals 1 kg) it can store x kcal per °K, From Fig. 2.13 we also conclude that 1 kWh is equivalent to 860 kcal. We thus have the energy equality

$$x \cdot 15 = 30 \cdot 860;$$

$$\therefore x = 1720 \, kg \, H_2O \quad (\approx 454 \text{ gallons}).$$

2.14 SUMMARY

In this chapter, we have reviewed the physical concepts of energy and power—all of the nonelectrical variety. Emphasis has been placed upon the various forms in which energy can appear and the possibilities for transformation from one form to another. In electric power engineering, the need for bulk energy storage is of great importance. We have discussed pumped hydro and compressed gas storage possibilities.

Although energy never will be lost, it will eventually after various transformations degrade into low-grade waste heat that has little practical usefulness. An energy engineer's skill should be directed towards the objective of "squeezing" maximum use out of the energy resources as they are being transformed from their most high-grade to their inevitable low-grade state.

EXERCISES

2.1 The moon has a diameter that is 27.2% of the earth's diameter. Its mass is 1.22% of that of the earth.

a) Find the surface gravity of the moon in percent of that of earth.

b) On earth, each meter of altitude gain adds 9.81 J/kg to the gravitational potential of a mass. What would be the corresponding figure on moon?

2.2 A rocket performs a vertical takeoff under the power of its thrust force. The latter equals three million lbs and is assumed constant.

a) What power does the thrust engine deliver at takeoff?

b) What power does it deliver at the instant the rocket velocity equals 3000 mph?

Express all quantities in SI units and give your answer in both MW and hp.

2.3 Falling water exerts a torque of 9000 N · m on the turbine blades of a hydro turbine, which runs at a speed of 720 rpm. Compute the turbine power!

2.4 Consider the mine elevator discussed in Example 2.3 in the text. During the acceleration period, the lift distance is 12.5 m. At the end of the acceleration period we have thus increased the potential energy of the elevator load by

$$5000 \cdot 9.81 \cdot 12.5 = 613100 \, \text{J}.$$

The lift force during acceleration was computed to 54050 N. According to formula (2.19), the lift motor thus delivers

$$54050 \cdot 12.5 = 675600 \, \text{J}.$$

Why the discrepancy between the two values? Can you balance the energy equation?

2.5 A car has a "gas mileage" of 15 miles per gallon at a speed of 55 mph. Compute the gas mileage of the same auto when traveling at 70 mph!
Make the following realistic assumptions:

1) The windage and friction force acting upon the car increases with the square of velocity

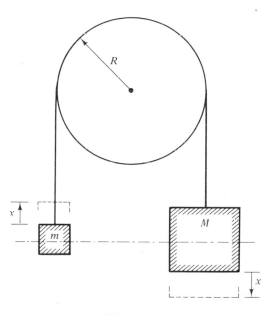

Figure 2.19

2) The gas consumption of the car as measured in *gallons per hour* is directly proportional to engine power. (The efficiency of an auto engine varies somewhat with engine speed so this assumption is somewhat erroneous. We disregard this fact in order to simplify the analysis.)

2.6 Figure 2.5 indicates that you need to impart $g_0 r_0 = 9.81 \cdot 6.38 \cdot 10^6 = 6.26 \cdot 10^7$ J/kg to a mass to free it from the gravity grip of earth.

  a) What vertical muzzle velocity ("escape velocity") must a bullet have never to return to earth? For simplicity, neglect the air resistance in your analysis. (In reality the air friction is *very* important and will in a major way change your answer.)

  b) Use the moon data given (and computed) in Exercise 2.1 to find the escape velocity from the moon's surface!

2.7 Consider the pulley arrangement in Fig. 2.19. Upon release the system will obviously accelerate in clockwise direction. Find the vertical acceleration, $\ddot{x}$, of the masses!
Assumptions: (1) zero friction, (2) inertia of pulley $= I$ kg m^2, (3) no slippage between pulley and string, (4) string is inelastic.

  [*Hint:* Write an expression for the total energy of the system, w_{tot}, as a function of x. As this energy must be constant you have $(d/dt)(w_{tot}) = 0$.]

2.8 Figure 2.20 depicts the drive system for an elevator. The elevator load is 10 metric tons (10,000 kg). The pulley radius is 1 m. The gear ratio is 25 : 1. The

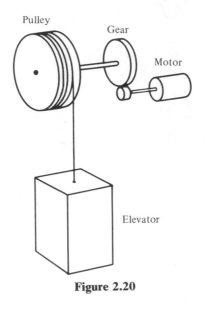

Pulley

Gear

Motor

Elevator

Figure 2.20

load is hoisted at the rate of 10 m/s.

Compute:

a) torque on pulley shaft, $N \cdot m$,
b) motor torque, $N \cdot m$,
c) pulley speed, rad/s,
d) motor speed, rad/s and rpm,
e) motor power, kW.

Neglect all losses.

2.9 In Example 2.8 in the text the kinetic energy of a spinning rotor was computed. If the energy price is five cents per kWh, how much did it cost to accelerate the rotor to full speed?

2.10 If the five-ton elevator in Example 2.3 in the text were to drop 200 m and crash, what would be the temperature rise, ΔT, of the wreckage? The specific heat, c_T, of the elevator plus load is assumed to be 0.3 (meaning that it takes 0.3 kcal to raise the temperature by 1°K of 1 kg).

2.11 The total drop of a river from the mountain glaciers to the ocean is 2000 m. As the water tumbles down its rapids, by how many degrees will its temperature rise due to transformation of its potential energy into heat?

2.12 In Example 2.2 in the text, we computed the potential energy needed to propel a payload to an altitude of 100 miles. Assume now that we wish to keep it in a circular orbit, 100 miles above earth.

a) How much additional (kinetic) energy must be imparted to the payload to make this possible!

b) What respective shares do the potential and kinetic energies constitute of the total?

[*Hint:*   By giving the payload an orbital velocity s of such a magnitude as to make the centrifugal force ms^2/r exactly equal to the gravity pull mg, the payload will circle the earth without "falling down."]

REFERENCES

Aimone, M. A. *The Hybrid Solar-Electric Service Concept.* Master's thesis, Department of Electrical Engineering, University of Florida, Gainesville, Florida, 1974.

Duffie, J. A., and W. A. Beckman. *Solar Energy Thermal Processes.* New York: Wiley, 1974.

Feynman, R. P., et al. *The Feynman Lectures on Physics*, Vol. I. Reading, Massachusetts: Addison-Wesley, 1963.

Holman, J. P. *Thermodynamics.* New York: McGraw-Hill, 1974.

Hubbert, M. K. "The Energy Resources of Earth." *Scientific American*, 224, September 1971, pp. 60–70.

NSF/NASA Solar Energy Panel. "An Assessment of Solar Energy As a National Energy Source." University of Maryland, 1972.

Rose, D. J. "Nuclear Electric Power." *Science*, 184, April 19, 1974, pp. 351–359.

ELECTRIC ENERGY FUNDAMENTALS

chapter 3

Steel pressure vessel being lifted by crane up the side of the nuclear reactor building at Brown's Ferry site in Decatur, Alabama. (Courtesy General Electric Company, Nuclear Energy Programs Division)

3.1 ELECTRIC ENERGY ENGINEERING

In the previous chapter, we reviewed the basic definitions of energy and power. The more important nonelectrical forms of energy and energy transformations were discussed. Now we will focus our attention on energy of the electrical variety and the various means of its transformation to and from other forms of energy. At a first glance it would seem unavoidable that a study of electric energy would involve us in all the varied aspects of electrical engineering. However, it is possible to divide, in broad sense, electrical engineering into two major subareas: (1) Communication and data technology, (2) Energy conversion and control.

In the first area we include all applications that have as their collective primary function the processing, transmission, and general handling of "information." To most electrical engineers the single term "electronics" suffices to adequately describe this entire spectrum of important electrical engineering activities. As a general statement it can be said that electronic components operate at relatively low electric power levels, usually in the range 10^{-6} to 10^2 watts.

We shall concentrate in this book on the second area of electrical engineering—energy conversion and means for its control. Generally the electric "energy" or "power" engineer is concerned with electricity in the power range of 10^3 to 10^9 watts. He designs and operates the electric power stations where electric energy is generated in bulk quantities, the transmission systems over which it is routed, and the distribution systems from which it is retailed to the users. Generators, transformers, and transmission and distribution lines constitute a *power system* or *grid*, the proper operation of which requires a thorough knowledge of systems and control theory. The electric energy engineer's interest area also includes design and operation of motors and transducers which transform electric energy into mechanical energy and other energy forms.

Our goal is to present the material in this book in a manner as to make it digestible to *all* engineering students. Matters electrical have a tendency to appear mysterious to most nonelectrical engineers, the reason being that "electricity" itself is often poorly understood.

In nonelectrical engineering the most common concept is *mass*, which we first encounter in the cradle. In addition, it also excites every one of our five senses. Early in life, everyone therefore feels quite at ease with this concept that forms the cornerstone of the nonelectrical world.

Electricity is something quite different. Our first contact with it can be quite shocking and we never really lose the respect for this strange medium that we cannot see but the effects of which can be quite dramatic.

Some of the mystery of electricity can be removed by utilization of mechanical analogs. This technique is often possible and shall be used prolifically in this text. However, the reader is cautioned that this is not always possible. After all, electricity is a unique medium with very unique features.

The reader has in all probability obtained a basic knowledge of electricity from a physics course. Maybe he has also taken a basic circuits-oriented or survey-type course in electrical engineering. An electric power engineer is interested in electric and magnetic phenomena for reasons quite different from those of an electronics engineer. In this chapter, phenomena shall be stressed that are of particular import for our later discussions. Although we run a certain risk of duplicating material from other courses, we think it is worthwhile having all electric *energy* theory summarized in one chapter.

3.2 PHYSICAL NATURE OF ELECTRICITY—ELECTRIC CHARGE

Electrical science and engineering are based upon the physical characteristics and behavior of that medium we call "electricity." Already the ancients knew that the fossil resin called amber (Gr. *elektron*) upon rubbing would exert attraction on certain light materials. By the end of the eighteenth century it was established that there were actually two kinds of electricity in nature. Benjamin Franklin introduced the names "positive" and "negative" which are still in use.

Today we know that electricity is closely interrelated with the microstructure of all matter, the basic building blocks of which, the atoms, obtain their identity to a great extent from the amount of electricity they hold. The modern view depicts the atom as consisting of a core or a *nucleus* surrounded by a "cloud" of orbiting *electrons*. The nucleus of a stable atom is made up of *protons* and *neutrons*. To each proton is ascribed an elementary positive *charge* of electricity, to each electron an elementary negative charge of equal magnitude. The neutrons are electrically neutral. These elementary charges are the smallest *electricity quanta* found in nature.

An undisturbed atom contains exactly as many electrons as protons, and the positive and negative charges thus balance or neutralize each other. If, due to external influences, atoms are robbed of some of their electrons, they will hold a surplus of positive charge and are referred to as *positive ions*. An atom may also acquire more than its normal complement of electrons, thus becoming a *negative ion.*

Particular combinations of protons, neutrons, and electrons form the atoms of the about 100 known *chemical elements*. The number of protons in the nucleus identifies, in general, the element. Elements with equal numbers of protons, but different numbers of neutrons, constitute *isotopes* of the element in question. Protons, neutrons, and electrons have physical mass. Furthermore, they all perform enormously complicated spin- and orbital-motion patterns.

In all engineering applications, we shall be concerned with macrostructures of matter. Even the tiniest microcircuit used in today's electronics will contain trillions of atoms. A typical solid contains of the order of 10^{20} atoms per cubic millimeter. Should we negatively *charge*, or *electrify*, this material by adding an electron to only every billionth atom, the net charge of this single small cube will still contain the incredible number of 10^{11} charge quanta.

Electricity as we encounter it in most engineering applications thus involves myriads of elementary charge quanta. To picture electricity, we may thus think of it as an extremely fine-grained, or fluidic, substance. Maybe a useful definition would be "an invisible fluid that shocks on touch."

The fine-grained character of electricity is best demonstrated in the units in which it is measured. The SI unit of charge, coulomb (C), is defined as

$$1 \text{ C} = 6.242 \cdot 10^{18} \text{ charge quanta.}$$

Obviously, the charge quantum of a proton is thus $+1.602 \cdot 10^{-19}$ C; of an electron, $-1.602 \cdot 10^{-19}$ C.

It may seem impractical to define the basic charge unit in terms of an uneven number of quanta, but that choice was made long before we knew the quantized character of electricity.

3.3   COULOMB'S LAW—THE GRAVITY ANALOG

The natural takeoff point in introducing electric energy concepts is Coulomb's law which describes electrostatic force action between *static*† charges.

The electrostatic forces have far-reaching similarities with gravity forces between masses. We can make better use of analogies between the two if we present Coulomb's law for a charge system that has the same spherical symmetry features as the mass system that we discussed in Chapter 2.

Consider therefore the large spherical ball shown in Fig. 3.1 containing the electric charge Q coulombs.‡ In its vicinity, at the distance of r meters, we

---

† Although as we have said electricity in a microsense is never at rest, it is still possible to ascribe to macro-conglomerations of electricity static characteristics. Consider the following analogy. All gases are made up of individual gas molecules, each of which is in a state of perpetual and complex motion. As we consider a tank filled with gas under certain pressure, we can still derive certain static relations between volume, pressure, and temperature of this gas. For example, what we measure as a constant "pressure" is a result of a statistical averaging of countless collisions of billions of molecules against the tank walls.

‡ Where in the ball would this charge reside? If the sphere were made of metal or some other good *conductor*, that is, material in which electricity can travel or move freely, then we would find the total charge Q spread out uniformly over the surface of the sphere in the form of a *surface charge*. The reason for this is that the charges repel each other (as this experiment demonstrates) and being free to move they would thus travel to the surface and distribute themselves uniformly around it. (Compare the shape a balloon will take when subject to uniform internal pressure.)

If the ball were made of insulating material, such as amber, the internal charge distribution would be more complex. In general charges would now be found throughout the interior as well as on its surface.

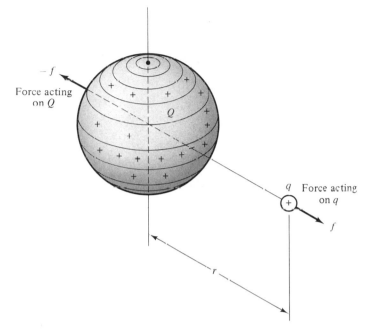

Figure 3.1

place a small test charge, q C. We can now observe an *electrostatic* force on the q-charge. (An equal and opposite force is also felt by the charge Q.) Careful experiments establish the following formula for the magnitude of the force which is everywhere radially directed:

$$f = \frac{1}{4\pi\epsilon\epsilon_0} \cdot \frac{Qq}{r^2} \quad \text{N,} \tag{3.1}$$

where

$\epsilon_0 = 8.854 \cdot 10^{-12}$ (the *dielectric constant of vacuum*),

ϵ = constant that depends upon the material surrounding the charges. For air (or vacuum), $\epsilon = 1$.

We note the great similarity with the gravity formulas in Chapter 2. However, whereas the gravity force between two positive masses is attractive, the electrostatic force between two positive (or two negative) charges is repulsive. If the charges are of unequal signs the force is attractive.

3.4   THE ELECTRIC FIELD

We can write the formula (3.1) in the alternate form:

$$f = q\left(\frac{1}{4\pi\epsilon\epsilon_0} \cdot \frac{Q}{r^2}\right) \quad \text{N.} \tag{3.2}$$

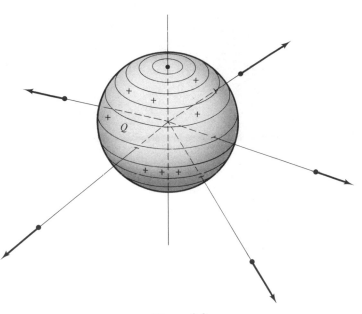

Figure 3.2

In a manner very analogous to the introduction of the gravity vector, g, in the previous chapter, we now identify the expression within parenthesis as a new vector, the *electric field vector*. It represents the force-per-unit charge q and has the magnitude

$$E \equiv \frac{1}{4\pi\epsilon_0} \cdot \frac{Q}{r^2} \qquad \text{N/C.} \tag{3.3}$$

Coulomb's law can thus be written

$$f = qE \qquad \text{N.} \tag{3.4}$$

The electric field is a vector that evidently "radiates" from the Q-charge in a manner analogous to the g-vector around earth. They are of opposite directions, however.

We can view the electric field as a fixture associated with the Q-charge, just as the gravity field around earth can be said to be associated with the earth mass M. The radial nature of E is shown in Fig. 3.2 (compare Fig. 2.2).

3.5 ELECTROSTATIC ENERGY

As we move a mass m against the gravity field, we store in it a potential gravitational energy. Similarly, as we move the charge q against the electric field, we expend energy, which will be stored in the charge or in the field (depending upon what philosophical view we take).

As we move the charge radially from the initial radial distance r_1 to the final position r (Fig. 3.3) we obviously must impart the energy amount

$$w_e = \int_{r_1}^{r} qE(-dr) = q\int_{r_1}^{r} E(-dr) \qquad J \tag{3.5}$$

upon the charge. (Note the negative sign for dr resulting from our definition of r to be positive in an outward sense.)

By making use of Eq. (3.3) we get the following energy expression:

$$w_e = -\int_{r_1}^{r} q\frac{1}{4\pi\epsilon_o} \cdot \frac{Q}{r^2}\, dr \qquad J \tag{3.6}$$

which upon integration yields

$$w_e = \frac{qQ}{4\pi\epsilon_0}\left(\frac{1}{r} - \frac{1}{r_1}\right) \qquad J. \tag{3.7}$$

In the gravitational field the stored energy increases as we move *away* from the earth. In the electric case the stored energy increases as we move *toward* the Q-charge (assuming, of course, that q is positive).

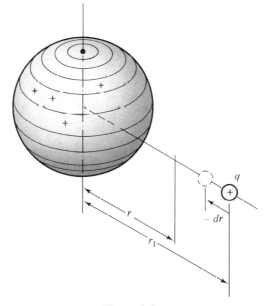

Figure 3.3

Example 3.1 Find the energy stored in a charge which is moved from a very large distance!

SOLUTION: If we set $r_1 = \infty$ in formula (3.7) we obtain the energy required to move a charge from "infinity":

$$w_{e\infty} = \frac{qQ}{4\pi\epsilon_0} \cdot \frac{1}{r} \qquad \text{J.} \tag{3.8}$$

3.6 ELECTRIC POTENTIAL

Analogous to the concept of gravitational potential, we consider now the energy storage per unit charge:

$$v \equiv \frac{w_e}{q} = -\int_{r_1}^{r} E \, dr \qquad \text{J/C.} \tag{3.9}$$

In analogy with the gravity case we refer to this new "per unit energy" as the *electric potential*. It obviously has the unit of joules per coulomb. In electrical engineering this unit is of such great importance that we give it a very special name—volt (V). The unit for electric field intensity was earlier given as newton/coulomb. In view of Eq. (3.9) we note that its unit also can be expressed as volts/meters. This unit is more popular among electrical engineers.

Like in the case of gravitational potential we are free to choose our zero potential point wherever we wish. In this case we shall find it practical to set the potential at great distance equal to zero.† Combination of expressions (3.8) and (3.9) thus gives the potential around our spherical Q-charge:

$$v = \frac{1}{4\pi\epsilon_0} \cdot \frac{Q}{r} \qquad \text{V.} \tag{3.10}$$

Figure 3.4 depicts the hyperbolic variation of the electric potential outside the Q-charge. Note that the "potential hill" now slopes in the opposite direction as compared with the gravitational case (compare Fig. 2.5).

The potential is equal at all radially equidistant points and the *equipotential* surfaces are thus spherical shells like in the gravitational case. The formulas (3.1) and thus (3.10) are valid *outside* the large sphere. The variation of the potential *inside* the large sphere depends upon the inside charge distribution. If the sphere is made of conducting material, and the Q-charge

---

† In most practical applications we define "ground" to have zero potential. Electrical apparatus usually have their chassis connected to ground (grounded) to avoid the possibility of charge buildup and resulting shock.

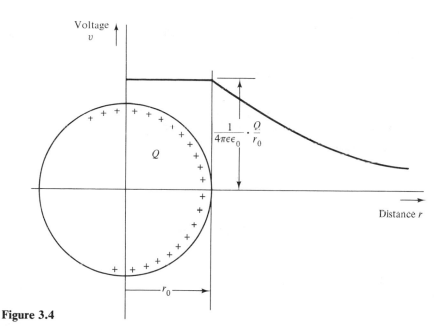

Figure 3.4

thus is a surface charge, then the inside potential is constant and equals the surface value (this case has been assumed in Fig. 3.4).

***Example* 3.2**   Electric power engineers often work with very high voltages. There is a physical limit to how much voltage a certain conductor can withstand. When the field strength E reaches a value of about 3 kV per mm ($3 \cdot 10^6$ V/m) in *dry air* the air molecules become ionized and *corona discharges* will occur, accompanied by a hissing sound and a bluish glow around the conductor. (In vacuum or for gases under high pressure this dielectric break-down will take place at much higher field strengths. See also Sect. 9.5.2B).

Find the highest voltage that can be tolerated on a 20-cm spherical conductor in air!

SOLUTION: The E-field has the largest value according to Eq. (3.3) at the surface of the ball, that is, for $r = r_0 = 0.1$ m. The largest possible charge, Q_{max}, is thus obtained from

$$E_{max} = 3 \cdot 10^6 = \frac{1}{4\pi\epsilon_0} \cdot \frac{Q_{max}}{(0.1)^2} .$$

Solving for Q_{max} yields

$$Q_{max} = 3.338 \cdot 10^{-6} \, C, \quad \text{or} \quad 3.338 \; \mu C.$$

From Eq. (3.10) we then solve for the corresponding voltage:

$$v_{max} = \frac{1}{4\pi\epsilon_0} \cdot \frac{Q_{max}}{0.1} = 300{,}000 \text{ V}.$$

....

Example 3.3   What would be the electrostatic force between two 20-cm balls placed with centers 1 m apart and charged to their maximum charge?

SOLUTION: Eq. (3.1) yields the repulsive force

$$f = \frac{1}{4\pi\epsilon_0} \cdot \frac{Q_{max}^2}{1.0^2} = 0.100 \text{ N} \qquad (\approx 0.02 \text{ lbf}).$$

....

The above examples teach us a couple of important facts:

1) Even very small static charges result in very high voltages.
2) The highest obtainable electrostatic forces are small. Motors based upon electrostatic force actions are therefore very weak and of little practical interest.

3.7 GENERAL FIELD CONFIGURATIONS

The electric field in the vicinity of a spherical charged conductor, as we have seen, takes on simple geometrical forms. A small charge q placed in this field will distort the field picture, but only slightly.

The field picture around more complex conductor configurations is more complex. In Fig. 3.5 we have sketched the field picture around two irregularly shaped conductors, of equal charges but opposite signs.

For more general conductor geometries, finding the field and potential distributions can be very difficult. In fact, if the conductors have no simple geometric character, the field picture can be obtained only after tedious computational processes. We give some general results here:

1) The electric field lines will always penetrate the equipotential surfaces and terminate upon the conductor surfaces at right angles.
2) Field concentrations occur at sharp conductor corners.†

The effect of the electric field in the vicinity of electric overhead transmission line conductors is of very great importance in electric power engineering. In Fig. 3.6(a) is shown the electric field in the vicinity of a cylindrical single conductor. The electric field is radial and most intense at the conductor surface. (Note that the closeness of the E-lines is a measure of the field intensity.)

---

† This explains why electrostatic discharges ("corona") always start at perturbations. In high-voltage technology all conductors must have rounded forms.

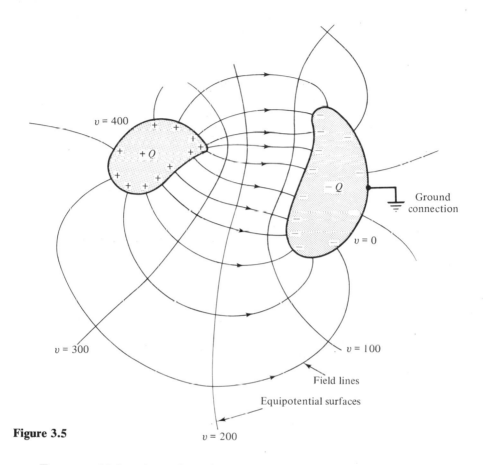

Figure 3.5

For extra high voltage (EHV) lines, usually in excess of 250 kV, the field concentration becomes so intense that the single conductor must be replaced by *bundled conductors*. Figure 3.6(b) depicts "triple bundling."† The three conductors are kept apart by spacers about 20–30 cm long. We also have sketched the field picture around the bundled conductors. It is obvious that the field intensity at each conductor has been reduced—the reason being, of course, that the total charge is now divided equally among the three conductors. (Note that in order to make a meaningful comparison, the two cases have equal *total* cross-sectional conductor area.)

3.8 ELECTROSTATIC ENERGY STORAGE—CAPACITANCE

From Eq. (3.10) we note that there is a proportionality between the charge and the voltage of the spherical conductor. By setting $r = r_0$ we can write the

---

† The lines shown on p. 1 have "double bundling."

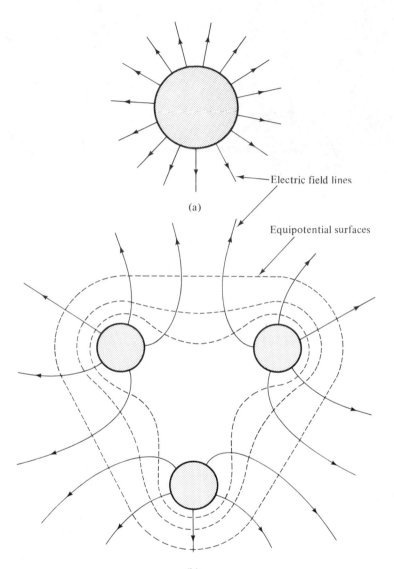

(a)

Electric field lines

Equipotential surfaces

(b)

Figure 3.6

equation as follows:

$$Q = 4\pi\epsilon\epsilon_0 r_0 v \qquad \text{C.} \qquad (3.11)$$

More compactly, we can express this proportionality as

$$Q = Cv \qquad \text{C} \qquad (3.12)$$

where

$$C \equiv 4\pi\epsilon\epsilon_0 r_0$$

is referred to as the *electric capacitance* of the sphere. It has the physical unit coulomb/volt and is named *farad* (F).

As we move a small charge dQ "from infinity" to the sphere, we obviously add to its charge and also to its potential. The potential increase dv follows from Eq. (3.12) and equals

$$dv = \frac{1}{C} dQ \quad \text{V.} \tag{3.13}$$

The addition of the charge dO adds, according to Eq. (3.9), the energy increment

$$dw_e = v \, dQ \quad \text{J} \tag{3.14}$$

to the sphere.

By combining formulas (3.13) and (3.14) we thus have

$$dw_e = Cv \, dv \quad \text{J.} \tag{3.15}$$

As we "charge" the sphere, thus increasing its potential from $0 \to v$, we increase its stored electric energy in the amount

$$w_e = \int dw_e = \int_0^v Cv \, dv = \tfrac{1}{2} Cv^2 \quad \text{J.} \tag{3.16}$$

Example 3.4   Compute the stored energy of the charged sphere in Example 3.2.

SOLUTION: We have:

$$C = 4\pi \cdot 8.854 \cdot 10^{-12} \cdot 0.1 = 11.13 \cdot 10^{-12} \, \text{F}$$

or

$$11.13 \text{ picofarad} \quad \text{(pF).}$$

As the voltage equals 300 kV we get for the stored energy

$$w_e = \tfrac{1}{2} \cdot 11.13 \cdot 10^{-12} \cdot (300{,}000)^2 = 0.501 \text{ J.}$$

Not a very impressive energy amount! Capacitor storage of large amounts of electric energy is not very promising.

Example 3.5   The two communicating vessels shown in Fig. 3.7 form a *hydraulic capacitor*. By means of the pump, fluid can be redistributed between the vessels. The figure shows a certain fluid mass Q "stored" in the left tank. Find the capacitance of the storage system and also the energy stored!

SOLUTION: If A represents the vessel area and ρ the density of the fluid we evidently have for the stored mass

$$Q = \rho A h \quad \text{kg.} \tag{3.17}$$

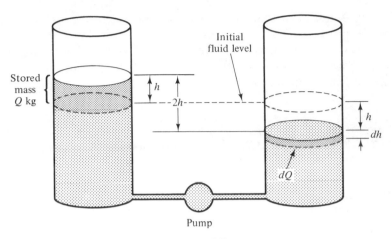

Figure 3.7

The head $2h$ represents a gravitational potential increase of

$$v_g = g_0 2h \qquad \text{J/kg.} \qquad (3.18)$$

In terms of this potential difference we can write for the stored mass:

$$Q = \frac{\rho A}{2 g_0} v_g \qquad \text{kg.} \qquad (3.19)$$

We introduce now the *hydraulic capacitance,*

$$C_h \equiv \frac{\rho A}{2 g_0}, \qquad (3.20)$$

and can then write for the stored fluid mass

$$Q = C_h v_g \qquad \text{kg.} \qquad (3.21)$$

The stored mass Q evidently represents a certain hydraulic energy w_h. If we were to release it, it could drive a water turbine and perform work. Let us compute how much energy is in fact stored.

As the fluid is being pumped into the left vessel clearly the hydraulic head is increasing. We can therefore expect that the pump effort, or *power*, will have to increase as more fluid is being transferred.

Consider the energy dw_h required to move the mass element dQ (see Fig. 3.7) against the head $2h$.

Using Eq. (2.12) we have

$$dw_h = g_0 2h \, dQ = g_0 2h \rho A \, dh \qquad \text{J.} \qquad (3.22)$$

The total energy need is obtained by integration:

$$w_h = 2g_0\rho A \int_0^h h\, dh = g_0\rho Ah^2$$

$$= \frac{1}{4}\frac{\rho A}{g_0} v_g^2 \quad \text{J.} \tag{3.23}$$

In view of Eq. (3.20) we finally get

$$w_h = \tfrac{1}{2}C_h v_g^2 \quad \text{J.} \tag{3.24}$$

Note the analogy with formula (3.16)!

Consider the following numerical case:

$$A = 1\,\text{m}^2,$$
$$\rho = 1000\,\text{kg/m}^3,$$
$$h = 1\,\text{m}.$$

We have

$$C_h = \frac{1000 \cdot 1}{2 \cdot 9.81} = 50.97$$

$$v_g = 9.81 \cdot 2 = 19.62$$

From Eq. (3.24) we then compute the hydraulic energy storage:

$$w_h = \tfrac{1}{2} \cdot 50.97 \cdot 19.62^2 = 9810\,\text{J.}$$

Note the magnitude difference with that of Example 3.4. It is much easier to store large energy quantities hydraulically than electrically. In practice we see examples on this. (Recall the pump storage facilities discussed in Chapter 2.)

3.9 PRACTICAL ELECTRIC CAPACITORS

A single conductor, like the sphere in Section 3.8, the charge of which had to be taken "from infinity," is not a very practical storage arrangement. In practice an electric capacitor is made up of two adjacent conductors separated by an insulator or dielectric. For example, the charge $+Q$ on the positive conductor in Fig. 3.5 can be "pumped" from the opposite conductor which now ends up with the charge deficiency $-Q$.

The potential difference

$$v = v_1 - v_2 \quad \text{V}$$

between the two conductors increases linearly with the charge transfer Q and again we can express this in the formula

$$Q = C(v_1 - v_2) = Cv.$$

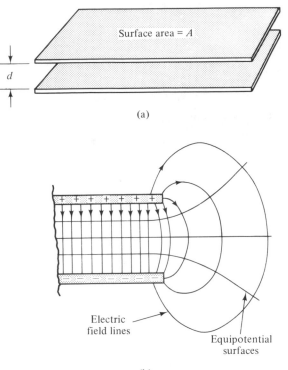

(a)

(b)

Figure 3.8

The capacitance C depends upon the conductor geometry, size, and dielectric constant of the insulating material. For example, we give without proof the following capacitance formula for a *plate capacitor* (with d and A defined as in Fig. 3.8a):

$$C = \epsilon\epsilon_0 \frac{A}{d} \quad \text{F.} \tag{3.25}$$

The electric field is uniform between the plates (except for the "fringe effect" shown in Fig. 3.8b) and equals

$$E = \frac{v}{d} \quad \text{V/m,} \tag{3.26}$$

where v is the voltage between plates. The equipotential surfaces are equidistant planes.

The capacitance increases with increased area and decreased plate distance. Commercial capacitors are made of layered foils (Fig. 3.9) which are rolled into bundles.

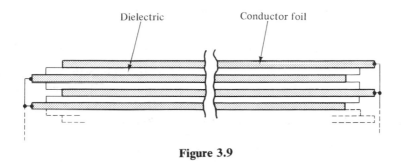

Dielectric   Conductor foil

Figure 3.9

Example 3.6   The plate capacitor shown in Fig. 3.8 is connected to a 12-volt auto battery, that is, a voltage difference of 12 volts exists between the two plates. The plates are parallel and 1 mm apart, and the electric field between the plates therefore equals

$$E = \frac{12}{0.001} = 1.2 \cdot 10^4 \quad \text{V/m.}$$

For all practical purposes the field is of uniform strength. An electron, the smallest quantum of electric charge, is placed at the negative plate. Under the influence of the electrostatic force the electron will accelerate towards the positive plate. Disregard any collisions (assume vacuum) and describe the electron motion.

SOLUTION: The electron mass is $m = 9.11 \cdot 10^{-31}$ kg and its charge $q = -1.602 \cdot 10^{-19}$ C.

If we neglect all relativistic effects (which we are permitted to do because the velocity will be much less than that of light) the electron acceleration a will be

$$a = \frac{\text{force}}{\text{mass}} = \frac{qE}{m} = \frac{(-1.602 \cdot 10^{-19}) \cdot 1.2 \cdot 10^4}{9.11 \cdot 10^{-31}} = 2.11 \cdot 10^{15} \text{ m/s}^2.$$

Considering that the electron acceleration due to earth's gravity is only 9.81 m/s² we can thus conclude that *the electric force is overwhelmingly dominant.*

The distance d between the plates will be covered in t_0 seconds and because the acceleration is constant we find t_0 from the formula

$$d = \frac{a}{2} t_0^2. \tag{3.27}$$

Thus

$$t_0 = \sqrt{\frac{2d}{a}} = \sqrt{\frac{0.002}{2.11 \cdot 10^{15}}} = 0.974 \cdot 10^{-9} \text{ s} = 0.974 \text{ ns.}$$

The electron, when it hits the positive plate, will have attained the velocity

$$s = at_0 = 2.11 \cdot 10^{15} \cdot 0.974 \cdot 10^{-9} = 2.06 \cdot 10^6 \text{ m/s}.$$

....

In words: The electron, under the influence of the relatively modest electric field caused by a car battery, will in the short distance of 1 mm accelerate to a velocity of about 1300 miles per second in the time span of about one nanosecond. *Electricity certainly is a volatile medium!*

Example 3.7   In our next example we shall estimate the energies involved in a typical thunderstorm. We hasten to explain that our modeling of the storm cloud is a highly simplistic one but one that should give us "ball park" figures.

Figure 3.10 depicts a thundercloud, having its negatively charged base at an altitude of 1500 meters above the earth. The base area of the cloud is 90 km^2. The cloud is about to discharge a lightning bolt. Experimental measurements indicate that discharge occurs when the electric field reaches a value of about $4 \cdot 10^5$ volts/meter. (Note that this is considerably lower than the dry air breakdown in Example 3.2.)

a) Compute the voltage between cloud and earth before discharge.

b) If you model the cloud–earth system as a plate capacitor what is its capacitance?

c) Compute the negative charge on the cloud just before discharge.

d) Compute total stored electric energy before strike.

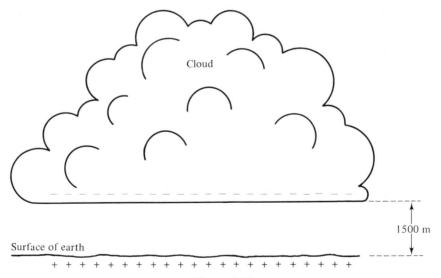

Figure 3.10

e) If the discharge lasts about 100 milliseconds, compute the average power developed during the stroke.

f) The cloud has a lifetime of about 1 hour and delivers 100 strokes during this time. What average power does this correspond to?

SOLUTION:

a) If we consider the electric field uniform between the "plates" then we obtain for the cloud base voltage

$$v = -1500 \cdot 4 \cdot 10^5 = -6 \cdot 10^8 \text{ V}.$$

b) From the formula for capacitance we get

$$C = 8.854 \cdot 10^{-12} \cdot \frac{90 \cdot 10^6}{1500} = 0.531 \cdot 10^{-6} \text{ F} = 0.531 \ \mu\text{F}.$$

c) Formula (3.12) gives for the cloud charge

$$Q = -0.531 \cdot 10^{-6} \cdot 6 \cdot 10^8 = -319 \text{ C}.$$

d) The energy formula (3.16) yields for the stored cloud energy

$$w_e = \tfrac{1}{2} \cdot 0.531 \cdot 10^{-6} \cdot (-6 \cdot 10^8)^2 = 9.56 \cdot 10^{10} \text{ J}.$$

e) If this energy is being discharged in 10^{-1} seconds then the energy release rate will be

$$p_{\text{stroke}} = \frac{9.56 \cdot 10^{10}}{0.1} = 9.56 \cdot 10^{11} \text{ W}.$$

Expressed in the large power unit, gigawatt, the discharge power equals 956 GW. (For comparison, the total electric generating capacity of the United States is about 400 GW.)

f) If the above stroke is of average size then the total energy discharge during the lifetime of the cloud will be

$$w_{\text{tot}} = 100 \cdot 9.56 \cdot 10^{10} = 9.56 \cdot 10^{12} \text{ J}$$

The average discharge power is obtained by dividing this energy into the cloud lifetime (3600 sec.)

$$p_{\text{ave}} = \frac{9.56 \cdot 10^{12}}{3600} = 2.6 \cdot 10^9 \text{ W}$$

or

$$2600 \text{ MW}.$$

This power would supply a city of about two million people, *but, we do not know how to harness it.* The example demonstrates the vast energies in play in our atmosphere.

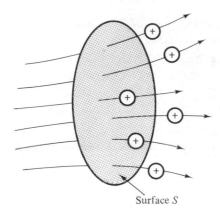

Surface S

Figure 3.11

3.10   ELECTRODYNAMICS—ELECTRIC CURRENT

So far we have concerned ourselves exclusively with static electricity and associated energy forms. The most important electric energy conversion phenomena involve electricity in motion or *electric current*.

If ΔQ coulombs pass through a certain surface S (Fig. 3.11) in the time interval Δt there exists an electric current of magnitude

$$i \equiv \frac{\Delta Q}{\Delta t} \qquad \text{C/s or amperes, A,} \qquad (3.28)$$

through the surface in question.

Example 3.8   The cloud charge of $-319\,$C in Example 3.7 discharges to ground in 100 ms. What is the corresponding (average) current between the cloud and ground?

SOLUTION: From the definition we get directly

$$i_{\text{ave}} = \frac{-319}{0.1} = -3190\,\text{A}\dagger \qquad \text{or} -3.19 \text{ kiloamps, (kA).}$$

Example 3.9   A positive charge of $10^{-6}\,$C is placed on a toroidal wheel, the spokes of which are insulators (Fig. 3.12). This charge is the maximum that the conductor can withstand (compare Example 3.2).

The wheel is spun at the speed of 1200 rpm. What is the current through the surface S?

---

† The negative sign means the current is directed from ground to the cloud. Why?

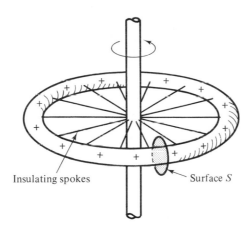

Figure 3.12

SOLUTION: In 1 second the wheel makes 20 full revolutions. The current will thus be

$$i = 20 \cdot 10^{-6} = 2 \cdot 10^{-5} \text{ A} \qquad \text{(or 20 microamps, } \mu\text{A).}$$

The above two examples demonstrate very large and very small electric currents. Neither represents a very useful application.

3.11   CURRENTS IN ELECTRIC CONDUCTORS

Electric currents take on practical significance when occurring in *electric conductors.*

The basic feature of an electric conductor is the appearance within its atomic lattice of so-called free electrons. In a nonconducting material all electrons are bound to their "home" atoms. In a conducting material, however, the free electrons are not bound to any particular atom but are free to drift throughout the material.

Consider copper as an example. This metal has $0.849 \cdot 10^{20}$ atoms per cubic millimeter. If we assume the existence of at least one free electron per atom then the total negative charge of all the free electrons in one cubic millimeter equals

$$(0.849 \cdot 10^{20}) \cdot (-1.602 \cdot 10^{-19}) = -13.6 \text{ C.}$$

Each copper atom contains 29 electrons and therefore every cubic millimeter contains a total negative charge of about -394 C. The protons carry an equal positive charge.

Example 3.10   The above charges are enormously large. For example, the total negative charge in one single cubic millimeter of copper exceeds the total cloud charge in a thunderhead (compare Example 3.7).

Assume that you could somehow separate the positive and negative charges in one cubic millimeter of copper and place these "blobs" of charge one mile (1609 m) apart. With what force would they attract each other?

SOLUTION: Equation (3.1) gives directly

$$f = \frac{1}{4\pi\epsilon\epsilon_0} \cdot \frac{(394)^2}{(1609)^2} = 5.39 \cdot 10^8 \, \text{N}$$

or 55,000 tons.

....

These figures give us some idea of the cohesive forces that hold matter together.

Because within each atom there are as many positive charge quanta as there are negative charge quanta, *on the average we have complete charge neutrality.* The free electrons can drift around "freely" within the atomic lattice but only so as to preserve this charge neutrality. If for example, a local "lumping" of free electrons would occur then there would be a negative charge concentration. Due to the repulsive force between charges of equal sign this concentration would instantly dissipate.

The free electrons can thus be thought of as a "cloud" or a "fluid" drifting around within the conductor, having on the average a uniform distribution so as to preserve overall charge neutrality. As the internal electric forces prevent any deviation from this uniformity, this *electron fluid is in effect completely incompressible.* Also, because the electric forces completely dominate over mass forces, the *electric fluid is essentially inertialess.* (Compare Example 3.6.)

Each individual free electron performs a random motion as it bounces around within the atom lattice. The overall velocity of the total electron fluid is, however, zero. A gas-filled balloon serves as a good analogy.† Each individual gas molecule performs random motions but the total gas volume has zero velocity.

When an external *E*-field is now superimposed upon this fluid every electron will be subject to a force component in the negative *E*-direction, and the whole fluid starts a collective *drifting motion* against the *E*-field, but in such a manner as to preserve the previous constant and uniform electron density and thus keeping charge balance within the atomic lattice. (We assume, of

---

† In one important aspect the analogy fails. The electron cloud is incompressible: this certainly is not true about a gas volume.

course, that the conductor is part of a *closed* circuit so that the motion is not impeded.)

From the moment the E-field is applied until a steady state drift velocity is achieved a certain start-up time will elapse. However, due to the very small inertia of the electrons (compare Example 3.6) this acceleration period is of the order of nanoseconds (10^{-9} seconds). For practical purposes we can assume "instantaneous" or inertialess response. The drift velocities are normally very small. Consider a current of 1 A in a 1 mm^2 copper wire. (A current of 1 A represents a charge flow of 1 coulomb per second.) Earlier we had concluded that copper contains approximately 14 coulombs of free electrons per cubic millimeter (assuming one free electron per atom). If all these free electrons would drift at the rate of 1 mm/s in our 1-mm^2 wire, a current of 14 A would result. Consequently, a 1-A current corresponds to a drift velocity of only $\frac{1}{14}$ mm/s. Compare this slow velocity with the high charge velocities in Examples 3.8 and 3.9!

3.12   OHM'S LAW

For most conductor materials the drift velocity (and thus the current, i) can be found experimentally to be proportional to the E-field. As this field in turn is proportional to the voltage v applied across the conductor, we can thus write

$$i = Gv \quad A$$

or

$$v = \frac{1}{G} i = Ri \quad V. \tag{3.29}$$

This is *Ohm's law*. The proportional constant, R, is referred to as the conductor *resistance*, measured in ohms (Ω). Its inverse, G, is the *conductance*, measured in mhos ($\mho$).

The resistance R varies with conductor length l and cross sectional area A in accordance with the formula

$$R = \rho \frac{l}{A} \quad \Omega \tag{3.30}$$

where ρ is the resistivity. (For copper $\rho \approx 1.75 \cdot 10^{-8}$ ohm·meter.) ρ increases slightly with conductor temperature.

3.13   ELECTRIC ENERGY TRANSFORMATIONS—THE BASIC ELECTRIC POWER FORMULA

The potential gravitational energy stored in water, for example, can be transformed into other forms of energy. In a *free* fall followed by impact, the energy is first transformed into kinetic and then, suddenly, into caloric form. In a *controlled* fall, for example in a Pelton turbine (Example 2.4), most of the energy ends up as "useful work" performed by the turbine. The turbine typically will power some device like a mill, electric generator, etc.

Some of the energy never reaches the turbine but is being transformed in the penstock into heat due to friction between the water and the tube walls and also within the fluid itself. This *resistive heat loss* p_{res} increases with the fluid flow i and also with increased tube *resistance R*. The latter parameter is a function of tube dimensions and tube surface characteristics.

The potential energy stored in electric charges can likewise be transformed into other energy forms. If we let the charges "fall freely" to lower potential levels (or to higher potential levels if the charges are negative) the energy is being transformed into caloric form.

If we control the "fall" of the charges by guiding them via conductors through an electric energy converter, like for example a motor, then the potential energy of the electric charges will be transformed into mechanical energy, utilized by whatever load is being driven by the electric motor.

We can derive a simple and useful formula for the electric power involved in such transformations. Assume that the incremental charge element ΔQ is experiencing a potential drop of v volts. In view of formula 3.14 it will obviously release an energy amount equalling

$$\Delta w_e = v \, \Delta Q \qquad \text{J.} \tag{3.31}$$

Assume that this energy transfer takes place in the short time interval Δt. The rate of energy transfer, or *electric power*, will thus equal

$$p = \frac{\Delta w_e}{\Delta t} = v \frac{\Delta Q}{\Delta t} = vi \qquad \text{W.} \tag{3.32}$$

This is the *basic electric power formula*. Compare it with the hydraulic power formula (2.25).

3.14 RESISTIVE OR OHMIC POWER DISSIPATION

When an electric current i flows through a conductor of resistance R, electric energy will dissipate in the conductor in the form of heat called *ohmic heat dissipation*. If the voltage drop across the conductor is v, then according to the power formula (3.32) electric power in the amount

$$p_\Omega = vi \qquad \text{W}$$

will be lost in heat. According to Ohm's law v and i are related by the formula

$$v = Ri,$$

and we thus have for the ohmic heat loss

$$p_\Omega = Ri^2 \qquad \text{W.} \tag{3.33}$$

3.15   ELECTRIC POWER TRANSMISSION

Figure 3.13 depicts the simplest possible electric energy transmission system. A "generator" delivers energy to a "load" via a transmission line. The generator voltage v_g is only slightly higher than the load voltage v_l, due to voltage drop along the line, and so we let v represent an average voltage.

According to formula (3.32) the transmitted power equals

$$p_{trans} = vi \quad W. \tag{3.34}$$

If R represents the total line resistance (both leads) then we have for the total transmission loss, according to Eq. (3.33),

$$p_\Omega = Ri^2, \tag{3.35}$$

or, in view of (3.34)

$$p_\Omega = R\left(\frac{p_{trans}}{v}\right)^2 \quad W. \tag{3.36}$$

If we put the loss power in relation to the transmitted power we get the following expression for the *relative* loss power.

$$\frac{p_\Omega}{p_{trans}} = R\frac{p_{trans}}{v^2}. \tag{3.37}$$

This formula tells us that the transmission loss decreases inversely with the square of the transmission voltage. We thus have established the need for high transmission voltage levels. In the United States today, the highest transmission voltages are close to one million volts (to be further discussed in Chapter 6).

How fast will the energy travel in the system in Fig. 3.13? As the voltage is applied to the sending end we in effect give the electron fluid a "push" which will be felt as a "pressure wave" propagating along the "incompressible" electron cloud of the conductor. Because of the extremely low inertia of the electrons the wave will travel extremely fast, with a velocity which in fact is slightly less than the speed of light. For most practical situations this is

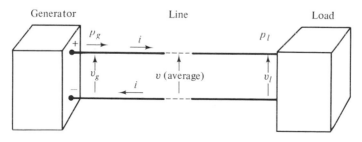

Figure 3.13

"instantaneous." (In a hydraulic transmission system, the velocity of the pressure wave would equal the sound velocity for the fluid in question.)

3.16   ELECTRIC SOURCES

We have on several occasions referred to "charge reservoirs," "sources," and "generators." In electrical engineering a variety of electric energy supply sources are available. The range varies from giant synchronous generators which can deliver, continuously, energy at rates exceeding 1000 megawatts to small capacitors which can supply "one-shot" energy pulses in the micro-joule region.

A direct current (dc) source delivers a unidirectional current and power. An electric storage battery is probably the most common and versatile dc source in use. It may power a hearing aid, start an automobile engine, or serve as an emergency hospital power source.

In an electric storage battery, positive and negative charges are "separated" from each other in an electrochemical process. The charges are forced to the two battery terminals and, if these are open-circuited, the charges will accumulate there. The result will be a buildup of terminal potential and an electrostatic field. As this field increases, the charge separation process within the battery decreases, and a balance is reached when no new charge separation takes place and a steady potential difference, the *electromotive force* (emf), exists between the terminals.

When an external load, in the form of a resistor for example, is connected across the terminals, the emf causes a current to flow. This current in effect represents a "charge leakage" between the terminals that will upset the previous balance. Immediately new charges are separated so as to maintain the emf. If the external current drain persists then we obviously have a continuous process. The process cannot go on ad infinitum, as a battery can discharge only a certain maximum charge. Usually a battery is rated in ampere-hours. A rating of one ampere-hour represents 3600 ampere-seconds or 3600 coulombs, and means that the battery can deliver 1A for 1 hour, or 2A for a half hour before the need for recharging.

During discharge the terminal voltage v_t (Fig. 3.14a) varies with the current drain in a *linear* fashion, according to the formula

$$v_t = e - R_i i \qquad \text{V} \qquad (3.38)$$

where

$$e = \text{emf in volts}$$

$$R_i = \text{internal resistance, } \Omega.$$

During battery charge, positive charges are "pumped" into the positive pole of the battery. The current is now reversed and the terminal voltage follows from the formula

$$v_t = e + R_i i \quad \text{V.} \tag{3.39}$$

During battery discharge the chemical composition of the cell changes so as to gradually "destroy" the charge separation process. By charging the battery the proper chemical conditions are restored. The electrical energy supplied to the battery during the charge process is used to build up the *chemical potential energy* of the electrolyte. This potential energy decreases during discharge.

Example 3.11 Let the transmission system in Fig. 3.13 represent the dc supply system in a mobile medical unit. The system is characterized by the following data:

Generator: Consists of a storage battery having an emf $e = 200.0$ volts and internal resistance $R_i = 0.031 \ \Omega$.
Line: Consists of 100 m (both leads) copper cable of sectional cross area $A = 20 \ \text{mm}^2$. From formula (3.30) we find its line resistance:

$$R = 1.75 \cdot 10^{-8} \frac{100}{20 \cdot 10^{-6}} = 0.0875 \ \Omega.$$

Load: An assortment of equipment drawing together a total current of $i = 95.0$ A from the battery.

Find the voltages v_g and v_l, ohmic transmission losses plus power drained from battery!

SOLUTION: We find first the battery terminal voltage by use of formula (3.38):

$$v_g = 200.0 - 0.031 \cdot 95.0 = 197.1 \ \text{V.}$$

The voltage drop across the cable is

$$\Delta v = 95.0 \cdot 0.0875 = 8.3 \ \text{V}$$

The load voltage, v_l, is found next:

$$v_l = 197.1 - 8.3 = 188.8 \ \text{V.}$$

Power supplied by battery:

$$p_g = 197.1 \cdot 95.0 = 18.72 \ \text{kW.}$$

Power lost in cable:

$$p_\Omega = 0.0875 \cdot 95^2 = 0.79 \ \text{kW.}$$

Power supplied to load:

$$p_l = 18.72 - 0.79 = 17.93 \ \text{kW.}$$

Example 3.12 A battery (Fig. 3.14a) has an emf $e = 100$ volts and an internal resistance of $R_i = 1 \, \Omega$. Determine the maximum power that can be supplied by the battery to a load resistance R_L of variable size.

SOLUTION: The current drawn from the battery is

$$i = \frac{e}{R_i + R_L} \qquad \text{A.} \tag{3.40}$$

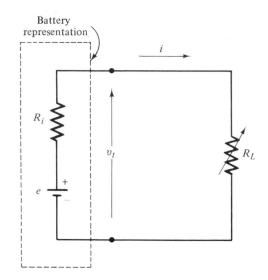

(a)

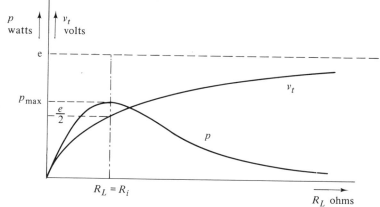

(b)

Figure 3.14

The terminal voltage will be

$$v_t = e - R_i i = e - R_i \frac{e}{R_i + R_L} = \frac{R_L}{R_i + R_L} e. \qquad (3.41)$$

The battery will thus supply the power

$$p - v_t i = \frac{R_L}{(R_i + R_L)^2} e^2 \qquad \text{W}. \qquad (3.42)$$

Figure 3.14(b) indicates that the power will approach zero for *both* $R_L = 0$ and $R_L = \infty$. Obviously, for some finite R_L-value, we must have a maximum power supply. We can find this p_{max} by setting

$$\frac{dp}{dR_L} - 0. \qquad (3.43)$$

A simple analysis reveals that power maximum will occur for $R_L = R_i$ and be equal to

$$p_{max} = \frac{1}{4} \cdot \frac{e^2}{R_i} \qquad \text{W}. \qquad (3.44)$$

In our numerical case we have:

$$p_{max} = \frac{1}{4} \cdot \frac{100^2}{1} = 2.5 \text{ kW}.$$

Note that the terminal voltage drops to half of its open-circuit value and note also that the power dissipation in the internal resistance equals the power dissipation in the load, that is, 2.5 kW. The battery would overheat in a short time, and this type of optimum battery discharge can therefore be accepted for short time periods only (like the starting of an auto engine). In normal battery operation the load resistance, R_l, typically is much larger than the internal battery resistance, R_i. In Fig. 3.14(b) this corresponds to points far to the right of p_{max}.

A fully charged battery can deliver a current in the ampere range for hours at a relatively constant voltage. As 1 ampere-hour equals 3600 coulombs, a storage battery in effect may store tens of thousands of coulombs. This charge must *not* be thought of as a *free* charge existing on the electrolytic plates within the battery somewhat like the situation in a charged capacitor. Rather the charges exist in *neutralized* fashion ready to be separated at a slow and controlled rate to match the external "leakage" from the terminals.

A capacitor, on the contrary, has much less storage capacity (Example 3.4)—usually much less than one coulomb, and often only micro coulombs. As a capacitor is being discharged it is thus to be expected that its terminal voltage will diminish rapidly.

Example 3.13   Consider the capacitor in Fig. 3.15. It is initially charged to a voltage v_0.

Study the voltage decay during discharge!

SOLUTION: Upon closing the switch S, charges will flow through the resistor R where energy will be dissipated in the form of ohmic losses.

As the loss power v^2/R must equal the negative time rate of the stored capacitor energy, we get the following power balance:

$$-\frac{d}{dt}(\tfrac{1}{2}Cv^2) = \frac{v^2}{R} \quad \text{W.} \tag{3.45}$$

Upon differentiation we obtain the differential equation

$$\frac{dv}{dt} + \frac{1}{RC}v = 0. \tag{3.46}$$

As the reader easily can verify, this differential equation has the solution

$$v = v_0 e^{-(t/T_c)} \quad \text{V,} \tag{3.47}$$

where $T_c \equiv RC$ is referred to as the *time constant* of the circuit. The voltage is plotted versus time in Fig. 3.15. We note that the capacitor is essentially discharged after about $5T_c$ seconds. (If we arbitrarily define the capacitor to be

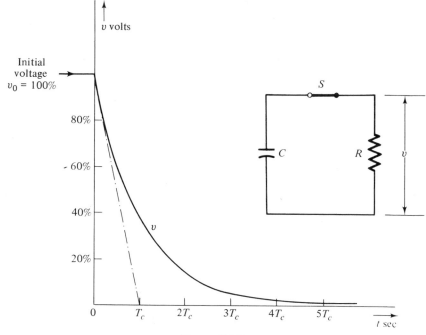

Figure 3.15

discharged when the voltage has dropped to 1% of its original value, the discharge time will amount to 4.605 T_c seconds.)

Numerical case:

$$C = 1.0 \,\mu\text{F},$$
$$R = 1000 \,\Omega,$$
$$v_0 = 100 \text{ V}.$$

We have in this case

$$T_c = 10^3 \cdot 10^{-6} = 10^{-3}\text{s} \qquad \text{or 1 ms.}$$

This capacitor circuit would discharge in about 5 milliseconds.

3.17 THE MAGNETIC FIELD

One of the most interesting features of electrical science, and certainly the most important one from a technological point of view, is the ability of moving charges to produce a so-called magnetic field. Like a gravitational or electrostatic field the magnetic field can be detected, mapped, and described in terms of its measurable effects. On these effects are based the workings of the most important electrical energy conversion apparatus.

In this section we shall present the character of the magnetic field; in sections to follow we shall describe its effects.

Consider the toroidal coil depicted in Fig. 3.16. When a current of magnitude i amps circulates in the toroid, a magnetic field is created having the circular symmetry and geometry as depicted in the figure. We summarize immediately some important characteristics of this field.

1) The field (symbol B) has vector character.
2) The field lines are *closed* (compare the open-ended character of electrostatic fields). They *always* enclose the current from which they originate.
3) The field exists both outside and inside of the conductor.
4) The magnitude B in each point in space is directly proportional to its causative current i (except in ferromagnetic materials (Section 3.26) where the relationship between B and i is nonlinear).
5) The magnitude B is greatest close to the conductor and decreases as the distance from the conductor increases (as always when we try to visualize a field on paper, the density of the field lines indicates magnitude).
6) The field direction can be obtained by the following "right-hand rule": Hold the conductor with your right hand, the thumb pointing in the current direction. The field vector is now pointing in the direction of your fingers.

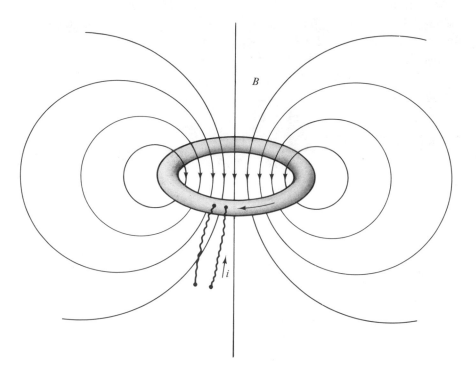

Figure 3.16

The SI unit for B is tesla (T), or weber per square meter (Wb/m^2), the latter unit revealing the nature of B as a flux density (magnetic flux density is actually another name for B).

Once the conductor geometry is known and the magnitude of current given, the magnetic field, magnitude, and direction, can be computed. The computations are, for general conductor geometries, not simple and we shall consider here only the simplest possible cases. For example, the B-field around a long straight wire (Fig. 3.17) consists of concentric circles and the magnitude of B decreases inversely with the radial distance r from the wire center in accordance with the formula

$$B = \mu_0 \frac{i}{2\pi r} \quad \text{T} \tag{3.48}$$

where $\mu_0 = 4\pi \cdot 10^{-7}$ is a constant, called the *permeability of vacuum*.

Another conductor geometry, which has great practical interest, is a *coil* (Fig. 3.18). The usefulness of this device is due to its field-amplifying feature. To explain this, consider the toroid wire in Fig. 3.16. If we double the current to $2i$ the B-field will double in every point in space. However, the same result would clearly be obtained if instead of one conductor carrying $2i$ A we would

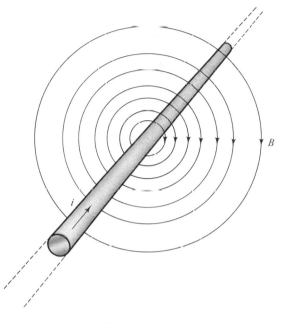

Figure 3.17

employ two adjacent conductors each carrying the current i. Using n adjacent conductors, that is an n-turn coil, evidently will amplify the field in ratio $n:1$.

3.18   MAGNETIC FLUX

Consider a magnetic field penetrating a surface S as depicted in Fig. 3.19. The field that penetrates the differential surface element dS is approximately constant over the whole element and has the component $B \cos \beta$ measured perpendicularly to the element. The differential *magnetic flux* $d\Phi$ penetrating the surface element is defined by the product

$$d\Phi \equiv B \cos \beta \, dS \qquad \text{Wb.} \qquad (3.49)$$

The total magnetic flux penetrating the finite surface S is then obtained by integrating over the total surface:

$$\Phi = \iint d\Phi = \iint B \cos \beta \, dS \qquad \text{Wb.} \qquad (3.50)$$

(We could, of course, similarly have defined gravitational and electric fluxes in terms of the vectors g and E respectively. They were not, however, needed in our elementary energy story. The magnetic flux, on the contrary, will be indispensible.)

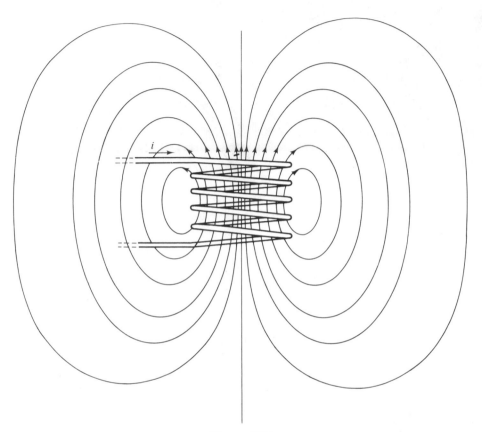

Figure 3.18

Example 3.14 Compute the total magnetic flux through the rectangular surface (Fig. 3.20) located in the same plane as a long current-carrying conductor.

SOLUTION: As shown in Fig. 3.17, the flux density is everywhere perpendicular to the surface, that is, $\cos \beta = 1$ over the total surface. As the flux density as given by Eq. (3.48) is constant along the differential surface strip of width dr, we obtain for the differential flux through this strip

$$d\Phi = BL\,dr = \mu_0 \frac{i}{2\pi r} L\,dr \qquad \text{Wb.} \qquad (3.51)$$

By integration from $r_1 \rightarrow r_2$ we thus obtain the total flux

$$\Phi = \int_{r_1}^{r_2} \mu_0 \frac{iL}{2\pi r}\,dr = \mu_0 \frac{iL}{2\pi} \ln\left(\frac{r_2}{r_1}\right) \qquad \text{Wb.} \qquad (3.52)$$

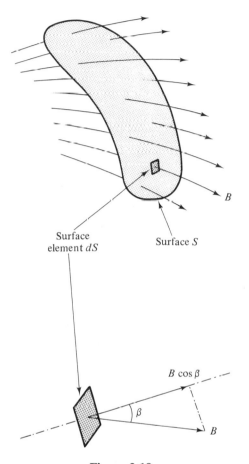

Figure 3.19

3.19 ELECTROMAGNETIC INDUCTION—FARADAY'S AND LENZ'S LAWS

An extremely important magnetic field effect is the *electromagnetic induction,* or induction for short. It can be described in the following way:

When the magnetic flux Φ through a surface S changes its magnitude an electric field is induced along the contour of S.

Faraday's law more specifically gives the quantitative nature of this electric field. It states:

If a thin N-turn coil is placed along the contour of S a voltage v can be measured across its open terminals of magnitude

$$v = N\frac{d\Phi}{dt}.$$

(3.53)

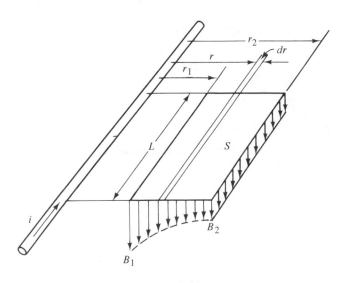

Figure 3.20

(The formula is often written in the form $v = (d/dt)(N\Phi) = (d/dt)(\Phi_c)$ where $\Phi_c \equiv N\Phi$ is referred to as the *coupled* or *linked* flux, or the flux that effectively induces voltage.)

The polarity of this voltage follows Lenz's law that states:

If the induced voltage is permitted to produce a current (by closing the coil) this current will have such a direction that the flux change is opposed.

Example 3.15   Let us examine the nature of magnetic induction by studying a specific example. Consider the coil shown in Fig. 3.21. It is placed so as to coincide with the contour of the surface S in Fig. 3.20. The flux Φ which we earlier computed (Eq. 3.52) can be changed in at least three different ways:

1) by changing the current i in the straight wire, thus changing the B-field,
2) by moving the coil radially without changing its plane (as the coil thus moves from higher to lower flux densities the flux must decrease),
3) by rotating the coil around its axis.

Let us consider the two first cases and let us also settle for the following numerical data:

$$N = 100\,\text{t,}$$
$$i = 100\,\text{A,} \qquad r_1 = 0.1\,\text{m,}$$
$$L = 1\,\text{m,} \qquad r_2 = 1\,\text{m.}$$

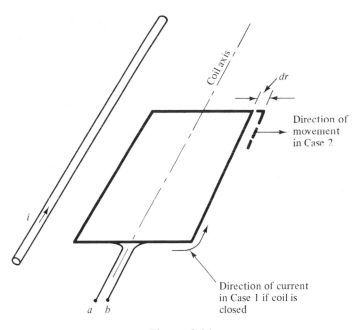

Direction of movement in Case 2

Direction of current in Case 1 if coil is closed

Figure 3.21

CASE 1. The flux change is now accomplished by keeping the coil position fixed but increasing the current i at the rate

$$\frac{di}{dt} = 10^5 \text{ A/s.}$$

(At this rate we obviously would double the current from initially 100 to 200 amperes in 1 millisecond.) The resulting flux change follows directly from Eq. (3.52):

$$\frac{d\Phi}{dt} = \mu_0 \frac{l}{2\pi} \ln\left(\frac{r_2}{r_1}\right) \frac{di}{dt} \qquad \text{Wb/s,} \qquad (3.54)$$

or numerically,

$$\frac{d\Phi}{dt} = 4\pi \cdot 10^{-7} \cdot \frac{1}{2\pi} \cdot \ln\left(\frac{1}{0.1}\right) \cdot 10^5 - 0.04605 \text{ Wb/s.}$$

The coil voltage would then follow from Faraday's law:

$$v = N\frac{d\Phi}{dt} = 100 \cdot 0.04605 = 4.605 \text{ V.} \qquad (3.55)$$

The voltage would be *constant* and prevail for as long as the current *change* prevails.

What would be the polarity of this voltage? According to Lenz, if we were to close the coil, by shorting out the terminals a and b, the voltage would give rise to a coil current having a direction such as to oppose the flux *increase*. With the given direction for the conductor current i, the coil flux is directed downwards. This means that the induced coil current would try to build up a coil flux directed upward. According to the right-hand finger rule (see page 83) the coil current would thus get the directionality indicated in Fig. 3.21, that is, the terminal a is positive relative to b.

CASE 2.   In this case we accomplish the flux change by moving the coil radially in the r-direction. Let us assume that we move the coil from its original position with a velocity of 10 m/s: that is,

$$\frac{dr}{dt} = 10 \text{ m/s}$$

(dr is depicted in Fig. 3.21).

The position change dr means that the coil flux decreases with the amount $B_1 L \, dr$ at its left side and increases with the amount $B_2 L \, dr$ at its right side. B_1 and B_2 are the flux densities at the radial distances r_1 and r_2 (see Fig. 3.20). The total flux change experienced by the coil will thus be

$$d\Phi = B_2 L \, dr - B_1 L \, dr \qquad \text{Wb.} \qquad (3.56)$$

For the flux derivative we therefore get

$$\frac{d\Phi}{dt} = L\left(B_2 \frac{dr}{dt} - B_1 \frac{dr}{dt}\right) = L \frac{dr}{dt}(B_2 - B_1) \qquad \text{Wb/s.} \qquad (3.57)$$

In our case we have (Eq. 3.48)

$$B_1 = 4\pi \cdot 10^{-7} \cdot \frac{100}{2\pi \cdot 0.1} = 2 \cdot 10^{-4} \text{ T},$$

$$B_2 = 4\pi \cdot 10^{-7} \cdot \frac{100}{2\pi \cdot 1} = 0.2 \cdot 10^{-4} \text{ T}.$$

The induced voltage† will thus be (neglecting polarity)

$$v = N\frac{d\Phi}{dt} = 100 \cdot 1 \cdot 10 \cdot (2 - 0.2) \cdot 10^{-4} = 0.18 \text{ V}.$$

What is the polarity of the voltage in this case? Evidently, coil movement *away* from the current-carrying conductor results in a flux *decrease*. The

---

† This voltage will *not* be constant as in Case 1. The computed value (0.18 V) applies *only* to the instant when the coil occupies the assumed position. As the coil travels on, the induced voltage approaches zero.

polarity of the induced voltage must thus be the opposite to the one in Case 1, that is, terminal b will be positive relative to a.

Let us make one comment concerning the result in (3.57). The two terms in the expression for flux change can be interpreted as the voltages induced in each coil side of length L. In a conductor moving perpendicular to a flux B with velocity s we thus induce a voltage of magnitude

$$v = sB \qquad \text{V/m.} \qquad (3.58)$$

(Note that the two remaining coil sides do not cut any flux and will therefore have no voltage induced in them.)

Example 3.16 A coil having a sectional area of 0.4 m^2 with $N = 10$ turns, rotates around its horizontal axis with a constant speed $n = 3600 \text{ rpm}$ in a vertical magnetic field which is uniform and of density $B = 1 \text{ T}$. The situation is depicted in Fig. 3.22(a). Find the voltage induced in the coil!

SOLUTION: If A is the coil sectional area, it will project an area equaling $A \cos \alpha$ on the horizontal plane. The total magnetic flux passing through the coil is

$$\Phi = BA \cos \alpha \qquad \text{Wb.} \qquad (3.59)$$

The coil will spin through $n/60$ full revolutions per second, that is, it will rotate at the angular speed

$$\omega = \frac{2\pi n}{60} \qquad \text{rad/s.} \qquad (3.60)$$

As the angular speed is constant, the angle α will grow in direct proportion to time t, and we can thus write

$$\alpha = \omega t \qquad \text{rad.} \qquad (3.61)$$

For the flux we therefore have

$$\Phi = BA \cos \omega t \qquad \text{Wb.} \qquad (3.62)$$

By applying Faraday's law we obtain now directly from Eq. (3.62),

$$v = N\frac{d\Phi}{dt} = N\frac{d}{dt}(BA \cos \omega t) = -N\omega BA \sin \omega t \qquad \text{V.} \qquad (3.63)$$

We thus conclude that the induced voltage in a coil spinning in a magnetic field is sinusoidal or *periodic*. A voltage of this alternating type, plotted in Fig. 3.22(b), is called an ac voltage.

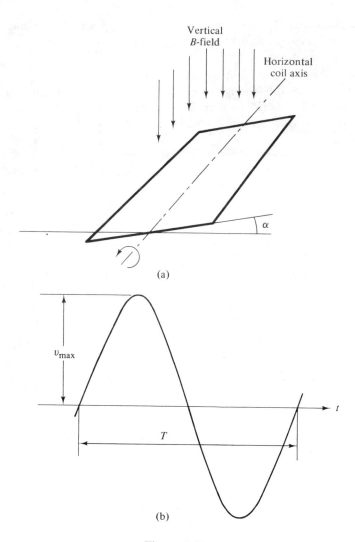

Figure 3.22

According to Eq.(3.63) the voltage wave has a *peak value*

$$v_{max} = N\omega BA \qquad \text{V.} \tag{3.64}$$

The voltage wave completes $n/60$ full cycles per second, that is, its *frequency, f,* is computed from

$$f = \frac{n}{60}\,\text{cps,} \qquad \text{or hertz (Hz).} \tag{3.65}$$

The *period time, T,* (see Fig. 3.22b) is

$$T = \frac{1}{f} \quad \text{s.} \tag{3.66}$$

In our case we have

$$f = \frac{3600}{60} = 60 \text{ Hz,}$$

$$T = \frac{1}{60} = 0.0167 \text{ s,}$$

$$\omega = 60 \cdot 2\pi = 377 \text{ rad/s,}$$

$$v_{\text{max}} = 10 \cdot 377 \cdot 1 \cdot 0.4 = 1508 \text{ V.}$$

The induction process in Example 3.15 which results in a dc emf cannot, for obvious reasons, be maintained on a continuous basis. In addition the voltage magnitude is small. The induction process described in Example 3.16, and which results in an ac emf, can be maintained indefinitely, or as long as the coil rotates. In addition the voltage magnitude is considerable. In fact, this very induction method produces the vast majority of the world's electricity. We shall make use of it repeatedly in the chapters of this book.

3.20 THE ELECTROMAGNETIC FORCE LAW

Electromagnetic induction provides the physical basis for electric *activation* of conductor terminals thus transforming a "passive wire" into an emf generator. The *electromagnetic force law* establishes the means whereby interaction between a magnetic field and a current-carrying conductor results in a mechanical force. The hundreds of millions of electric motors that are operating at this instant around the world base their functioning upon this law.

Consider the three-dimensional coordinate system shown in Fig. 3.23. A uniform magnetic field of intensity B T is oriented in the z-direction. A thin conductor carrying a current i A is contained in the x-z plane, forming an angle α with the B vector. The force law states that the conductor will be subject to an electromechanical force f directed in the y-direction and having a magnitude

$$f = BiL \sin \alpha \quad \text{N,} \tag{3.67}$$

where L is the conductor length.

The force evidently reaches a maximum

$$f_{\text{max}} = BiL \quad \text{N,} \tag{3.68}$$

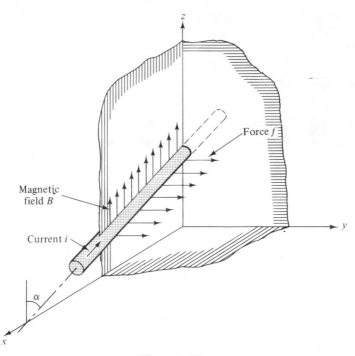

Figure 3.23

when $\alpha = 90°$. In this case f, i, and B are all orthogonal, and parallel with the y, x, and z axes, respectively.

Those readers familiar with the *vectorial cross product* realize that we can express the force as

$$f = i \times B \qquad \text{N/m.} \qquad (3.69)$$

Note that the force will always be perpendicular to both the magnetic field vector and the conductor.

Also we make the following important observation: As we place the current-carrying conductor in the field B, *the field will change due to the effect of current i.* As we use the formula (3.67) the field B must be understood to mean that magnetic field that exists *before the presence of the current i.*

Example 3.17 It is not uncommon in electrical machines to work with flux densities of magnitude 1.5 T. If we place a conductor carrying 100 A orthogonally to this field the conductor will be subject to a force of magnitude

$$f = 1.5 \cdot 100 = 150 \text{ N/m} \qquad \text{(or 10.3 lbf/ft).} \qquad (3.70)$$

Obviously we can easily obtain considerable force (and torque) action by utilizing this law. Compare the weak forces (Example 3.3) obtained by interaction between *static* charges.

Example 3.18 A thin N-turn coil carrying current i is free to rotate in a vertical B-field around its horizontal axis. Show that the coil will be subject to a torque, and find the magnitude of this torque!

SOLUTION: Figure 3.24 depicts the arrangement. Also shown are the forces acting on the four coil sides. Evidently, the forces acting on the coil sides of length b will cancel each other. The forces acting on the coil sides of length a will exert a torque ("couple") trying to rotate the coil in the direction of decreasing β-angle. The force on each side has a horizontal direction and equals $BNia$ newtons. The torque on the coil thus will be

$$T = BNiab \sin \beta \qquad \text{N} \cdot \text{m}. \tag{3.71}$$

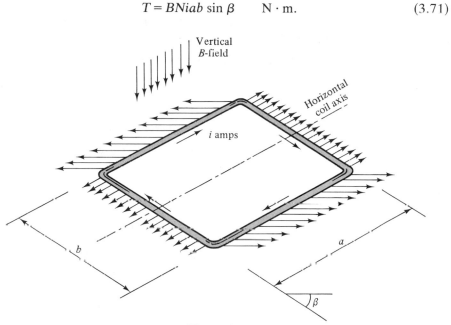

Figure 3.24

If the coil is free to move it will assume a horizontal position. In this equilibrium position all four coil sides will be subject to equal forces (expressed in N/m), all acting in the coil plane and all trying to expand the coil.

Example 3.19 Find the force between two long parallel conductors placed a meters apart and carrying equal currents i.

SOLUTION: The magnetic field caused by each conductor forms concentric circles (Fig. 3.17), with a magnitude given by Eq. (3.48). By application of the force law, we conclude that if the currents have equal directions the conductors will attract each other; if opposite directions, they will repel. We treat the latter case. By using Eqs. (3.48)† and (3.68) we find the magnitude of the repulsive force that acts on 1 m of the conductor:

$$f = Bi = \left(\mu_0 \frac{i}{2\pi a}\right)i = \mu_0 \frac{i^2}{2\pi a} \quad \text{N/m.} \tag{3.72}$$

Bus bars in electric power stations carry normal currents which often exceed 10 kA. During shortcircuits these currents may jump to 100 kA before the fault is isolated. Two bus bars carrying 100 kA each and placed 1 m apart according to formula (3.72) will be subject to shortcircuit forces of magnitude

$$f = 4\pi \cdot 10^{-7} \frac{(10^5)^2}{2\pi \cdot 1} = 2000 \text{ N/m.} \tag{3.73}$$

Forces of such magnitude may be of destructive nature unless the supporting insulators possess sufficient strength.

Example 3.20 Determine qualitatively the nature of the forces acting between two parallel toroidal conductors carrying currents in the same direction.

SOLUTION: The magnetic field B_1 around conductor 1 is shown in Fig. 3.25 (cf. Fig. 3.16). We now place conductor 2 (shown by dashed lines) in the field of the

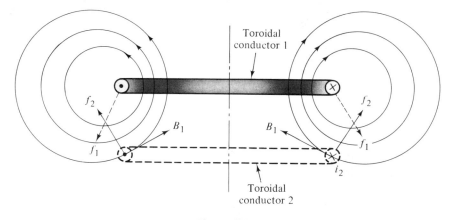

Figure 3.25

---

† Note that we compute the value of B at conductor 2 as caused by the current in conductor 1 *only*. Compare comment on page 94.

first one. As the force f_2 acting on this second conductor must be perpendicular to both the field B_1 and the current i_2, it will have a direction as indicated in the figure.

Force f_1 acting on conductor 1 must be symmetrical as indicated by dashed lines.

> **Conclusion.** The forces tend to attract the conductors axially and to expand them radially. A cylindrical coil, if assumed elastic, will thus become shorter as its current increases.

3.21 CONCEPT OF MUTUAL INDUCTANCE

The coil and conductor in Example 3.15 represent two *coupled magnetic circuits*. A change in the current in circuit 1 (the one containing the long conductor) induces a voltage (or emf) in circuit 2 (the coil). Magnetic coupling forms the basis for some of the most important devices in electric power engineering, for example, transformers and induction motors.

In analysis work it is convenient to account for this coupling by means of a parameter referred to as *mutual inductance*. Let us introduce this parameter by considering the above example. By combining Eqs. (3.53) and (3.54) we obtain the following expression for the coil voltage:

$$v_2 = \frac{d}{dt}\left[\mu_0 \frac{Li_1}{2\pi} N \ln\left(\frac{r_2}{r_1}\right)\right] \quad \text{V.} \tag{3.74}$$

We have here introduced subscripts 1 and 2, and the equation then tells us the magnitude of the voltage induced in circuit 2 (the coil) as a result of current change in circuit 1 (the straight conductor).

We define now the mutual inductance between circuits 1 and 2 as

$$M_{12} \equiv \mu_0 \frac{L}{2\pi} N \ln\left(\frac{r_2}{r_1}\right), \tag{3.75}$$

and can then write Eq. (3.74) in the following simpler way[†]:

$$v_2 = M_{12} \frac{di_1}{dt} \quad \text{V.} \tag{3.77}$$

[†] Note that we can write the equation in this form *only if r_2 and r_1 are constants, i.e., if the two circuits are fixed in relation to each other.*

Should the coils be moving relative to each other (which is the case in rotating machines), then the formula will read

$$v_2 = \frac{d}{dt}(M_{12}i_1), \tag{3.76}$$

The physical unit for M_{12} is volt-seconds per ampere and is given the special name *henry*, (H). M_{12} tells us how many volts are induced in circuit 2 for a current change in circuit 1 of one A/s.

In many situations (for example, in transformers), the opposite situation is of equal importance—what voltage would be induced in circuit 1 as a result of current change in circuit 2? We would now be interested in the mutual inductance M_{21} defined by the equation

$$v_1 = M_{21}\frac{di_2}{dt} \quad \text{V.} \tag{3.78}$$

We will not give the proof here but it can be generally shown that the two mutual inductance parameters are of identical magnitude:

$$M_{12} = M_{21}. \tag{3.79}$$

As two coupled coils are thus characterized by *one* mutual inductance, we will prefer the simpler symbol M.

3.22   CONCEPT OF SELF-INDUCTANCE

Faraday's law tells us that a voltage is induced in a loop or coil if the magnetic flux linked to it undergoes changes. In the previous examples we calculated these voltages, assuming the flux was *due to currents external to the loop or coil in question*. We introduced the mutual-inductance parameter to account for the induced voltage.

We must understand, however, that some of the flux linked to a coil can also be attributed to its *own* current. In fact, if the coil has no coupling to other circuits, *all* its linked flux must be due to its own current. The voltage induced in a coil, therefore, in part (or possibly in its entirety) is due to changes in its own current. We call this phenomenon *self-induction.*

For example, consider a toroidal coil of the type shown in Fig. 3.16 which depicts the magnetic field due to the coil current. This field and thus the corresponding flux is proportional to current. Specifically, if i is the coil current, then we can write for the magnetic flux, Φ_1, due to *one coil turn*:

$$\Phi_1 = ki \quad \text{Wb.} \tag{3.80}$$

If the coil is relatively thin, then each of the N turns contributes an equal share to the *total* coil flux, Φ, and we thus have

$$\Phi = N\Phi_1 = kNi \quad \text{Wb.} \tag{3.81}$$

According to Faraday we have for the induced voltage

$$v = N\frac{d\Phi}{dt} = N\frac{d}{dt}(kNi) = kN^2\frac{di}{dt} \quad \text{V.} \tag{3.82}$$

We define now the self-inductance parameter L of the coil

$$L \equiv kN^2 \quad \text{H.} \tag{3.83}$$

By this definition the self-inductance equals $L = \Phi_c/i$ (see comment on page 88). In other words, it is a measure of the "coupled flux", Φ_c, caused by unit current. (Note the similarity to the definition of capacitance— $C \equiv Q/v$, that is, the charge per unit voltage.)

Now we can write for the self-induced voltage

$$v = L\frac{di}{dt} \quad \text{V.} \tag{3.84}$$

The constant k in formula (3.83) depends upon coil geometry. Actually, even for the very simple geometry of a toroidal coil, the inductance computations become quite tedious and we shall not dwell on this matter. Note that the self-inductance increases as the square of number of turns.

3.23 ELECTROMAGNETIC ENERGY STORAGE

If a mass m (Fig. 3.26a) is being accelerated by a force f, its velocity s can be computed from the following equation which states that the force f must equal the sum of the inertia force plus the friction force f_{fr}:

$$f = m\frac{ds}{dt} + f_{fr} \quad \text{N.} \tag{3.85}$$

Upon multiplication by sdt and integrating we obtain

$$\int fs\,dt = \int ms\,ds + \int f_{fr}s\,dt \quad \text{J.} \tag{3.86}$$

The kinetic energy of the mass equals

$$w_{\text{kin}} = m\int s\,ds = \tfrac{1}{2}ms^2 \quad \text{J.} \tag{3.87}$$

Equation (3.86) thus states that the energy supplied by the force f equals the sum of the kinetic energy imparted to the mass plus the energy dissipated in heat through friction.

If a coil of inductance L and resistance R is being excited by the emf e (Fig. 3.26b), the resulting current can be obtained from the voltage equilibrium equation

$$e = L\frac{di}{dt} + Ri \quad \text{V.} \tag{3.88}$$

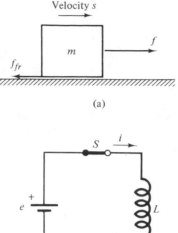

(a)

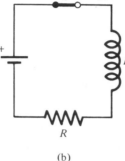

(b)

Figure 3.26

Upon multiplication by $i\,dt$ and integration we have

$$\int ei\,dt = \int Li\,di + \int Ri^2\,dt \quad \text{J.} \tag{3.89}$$

The first term represents the energy supplied by the e-source. The third term is identified as the energy dissipated in the resistor. The second term thus must represent the *magnetic energy*, w_{mag}, stored in the magnetic field:

$$w_{\text{mag}} = \int Li\,di = \tfrac{1}{2}Li^2 \quad \text{J.} \tag{3.90}$$

The analogy between magnetic and kinetic energies should be obvious. Sometimes the inductance is referred to as "magnetic inertia."

Example 3.21   The field coil of a synchronous generator (Chap. 4) has a self-inductance of $L = 12$ H. How much energy is stored in this coil if the field current is $i = 27$ A?

SOLUTION: Formula (3.90) yields:

$$w_{\text{mag}} = \tfrac{1}{2} \cdot 12 \cdot 27^2 = 4374 \text{ J.}$$

The possibility of storing vast amounts of energy in magnetic coils is an interesting one. The main obstacle is the ohmic heat loss in the winding. In Chapter 9 we shall discuss the possibilities of bulk storage of electric energy in *super-conductive* coils.

3.24 MAGNETIC ENERGY STORAGE IN MUTUALLY COUPLED CIRCUITS

Mutually coupled circuits are of great importance in electric power technology. Let us briefly study the magnetic energy situation in such circuits.

Consider the two coils shown in Fig. 3.27(a). The fluxes Φ_1 and Φ_2 resulting from the coil currents i_1 and i_2 are obviously *additive*, and therefore the induced voltages caused by the change of these fluxes must also be additive. If each coil is fed from sources with emfs e_1 and e_2, respectively, we obtain the following equations for voltage balance:

$$e_1 = R_1 i_1 + L_1 \frac{di_1}{dt} + M \frac{di_2}{dt} \qquad \text{V},$$

$$e_2 = R_2 i_2 + L_2 \frac{di_2}{dt} + M \frac{di_1}{dt} \qquad \text{V}. \tag{3.91}$$

By multiplying the equations by i_1 and i_2, respectively, we obtain

$$e_1 i_1 = R_1 i_1^2 + L_1 i_1 \frac{di_1}{dt} + M i_1 \frac{di_2}{dt} \qquad \text{W},$$

$$e_2 i_2 = R_2 i_2^2 + L_2 i_2 \frac{di_2}{dt} + M i_2 \frac{di_1}{dt} \qquad \text{W}. \tag{3.92}$$

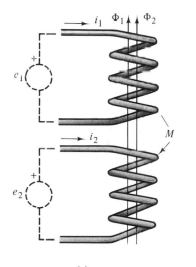

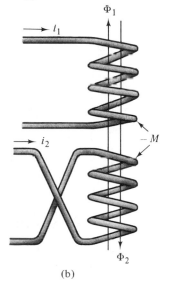

(a)   (b)

Figure 3.27

The expressions on the left of the equals signs are identified as the powers delivered by the two sources, respectively. The quadratic terms on the right of the equals signs represent ohmic losses in the windings. The sum of the remaining four terms must therefore represent the rate of change of the magnetic energy in the two-coil system, that is,

$$\frac{d}{dt}(w_{\mathrm{mag}}) = L_1 i_1 \frac{di_1}{dt} + L_2 i_2 \frac{di_2}{dt} + M\left(i_1 \frac{di_2}{dt} + i_2 \frac{di_1}{dt}\right) \quad \mathrm{W}. \quad (3.93)$$

By integration we obtain:

$$w_{\mathrm{mag}} = \tfrac{1}{2}L_1 i_1^2 + \tfrac{1}{2}L_2 i_2^2 + M i_1 i_2 \quad \mathrm{J}. \quad (3.94)$$

Consider now the coil system shown in Fig. 3.27(b). The fluxes Φ_1 and Φ_2 caused by currents i_1 and i_2 are now bucking each other. The corresponding mutually induced voltages must therefore enter the equations with negative signs. *Clearly the end result would be the same if all equations were to be unchanged and the mutual inductance M were to be assigned a negative sign.*

Example 3.22   Two coupled coils have the following inductance parameters:

$$L_1 = 10\,\mathrm{H},$$
$$L_2 = 2\,\mathrm{H},$$
$$M = \pm 3\,\mathrm{H}\ \text{(sign depending upon polarity)}.$$

The coil currents are:

$$i_1 = 20\,\mathrm{A}, \qquad i_2 = 10\,\mathrm{A}.$$

Find the magnetic energy stored in the system!

SOLUTION

Polarity as shown in Fig. 3.27(a):

$$w_{\mathrm{mag}} = \tfrac{1}{2}\cdot 10 \cdot 20^2 + \tfrac{1}{2}\cdot 2 \cdot 10^2 + 3 \cdot 20 \cdot 10 = 2700\,\mathrm{J}.$$

Polarity as shown in Fig. 3.27(b):

$$w_{\mathrm{mag}} = \tfrac{1}{2}\cdot 10 \cdot 20^2 + \tfrac{1}{2}\cdot 2 \cdot 10^2 - 3 \cdot 20 \cdot 10 = 1500\,\mathrm{J}.$$

3.25 THE MAGNETIC MOMENT

The torque expression (3.71) can be written

$$T = B(Ni)A \sin \beta \quad \mathrm{N \cdot m}, \quad (3.95)$$

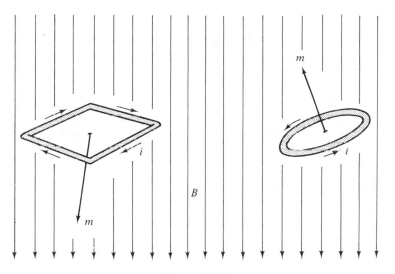

Figure 3.28

where A is the coil area and Ni is the total current circulating around the contour of A. We introduce now the *magnetic moment* defined by

$$m \equiv NiA \qquad \text{At} \cdot \text{m}^2, \tag{3.96}$$

and can then write Eq. (3.95) more simply as

$$T = Bm \sin \beta \qquad \text{N} \cdot \text{m}. \tag{3.97}$$

We can think of the magnetic moment as a vector (see Fig. 3.28) which is orthogonal to the coil surface and pointing into that direction into which a right-hand screw would move when turned clockwise with the current.

The result obtained in Example 3.18 can now be interpreted as follows:

Under the influence of the magnetic field the coil will assume an equilibrium position where its magnetic moment vector is lined up with the magnetic field vector.

We proved this fact for a rectangular coil constrained to rotate around a certain axis. However, it is easy to prove (and we suggest the reader do so) that the rule applies to coils of *any* shape and which are completely free to move in any direction. For example, the circular coil shown in Fig. 3.28 will have to turn almost 180° before it comes into a position of equilibrium.

The concept of magnetic moment vector will prove very useful in interpreting ferromagnetic phenomena.

3.26 FERROMAGNETISM

The magnetic characteristics that we have discussed so far apply to phenomena which in the strictest sense occur in vacuum. However, all formulas developed, for all practical purposes, have validity in air. Should we put a piece of some other "nonmagnetic" material (for example, copper, plastic, wood, etc.) in the magnetic field in Fig. 3.16, the field would not change measurably.†

However, three elements, iron, cobalt, and nickel (which are adjacent in the periodic table of elements) have a most unique magnetic behavior. We refer to these materials (including some alloys containing small quantities of Al, Cu, and Ti) as "ferromagnetic" (derived from the Latin word *ferrum* for iron). They are important, in fact crucial, in the design of most electrical power apparatus, because they permit us to create magnetic fields of very much greater intensity than are obtainable in air or vacuum.

3.26.1 Magnetic "Conduction"

We demonstrate the basic feature of ferromagnetism by means of the arrangement shown in Fig. 3.29. Around a long cable consisting of N strands, each carrying the current i, we have placed a toroidal core made of ferromagnetic

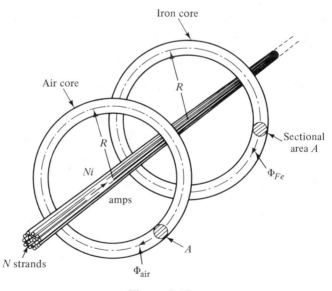

Figure 3.29

---

† If the field is a *variable* one, caused for example by an ac current, then this statement is not true. Currents would now be induced in the copper due to Faraday's law, and these currents would greatly change the field picture.

material. It has the dimensions given in the figure. As we measure the magnetic flux Φ_{Fe} inside the core and compare it with the flux Φ_{air} inside a toroidal "core" of air of exactly the same dimensions, we find that the iron core magnetic flux density and flux both are larger by a factor μ, that is,

$$\frac{B_{Fe}}{B_{air}} = \frac{\Phi_{Fe}}{\Phi_{air}} = \mu. \tag{3.98}$$

We refer to μ as the *relative permeability* of the core material in question; μ will assume values as large as 10^6. Although μ varies widely among various ferrous materials, it is always much greater than unity. As will be explained in Section 3.26.4, μ also varies strongly with flux density—it is *not* a constant for a given material.

What this experiment thus shows is that iron is a very good *magnetic conductor*. The current i, which is the *causative agent* in the sense that it "produces" the field intensity B, will in effect produce μ times more magnetic flux in an iron core than in an air core of equal size.

3.26.2 Ohm's Law for a Magnetic Circuit

The magnetic flux density in the center of the air core depicted in Fig. 3.29 equals, according to formula (3.48),

$$B_{air} = \mu_0 \frac{Ni}{2\pi R} \quad \text{T.} \tag{3.99}$$

The flux density in the iron core is μ times larger:

$$B_{Fe} = \mu\mu_0 \frac{Ni}{2\pi R} \quad \text{T,} \tag{3.100}$$

and for the iron core flux we thus get

$$\Phi_{Fe} = A \cdot B_{Fe} = \mu\mu_0 \frac{ANi}{2\pi R} \quad \text{Wb.} \tag{3.101}$$

By experiment we now can confirm a most interesting characteristic of this flux. If we keep the iron core position fixed but move the cable to any position inside the core window, *the flux will not change.* (The reader should contemplate what happens with the flux in the air core under the same conditions.)

The flux will not change even if we bend the cable into any conceivable shape as long as the total current Ni, remains unchanged and as long as it passes through the core window. In fact, the magnetic flux will remain unchanged should we replace the N-strand cable with an N-turn coil as long as the product Ni stays constant.

These experiments prove conclusively that the *core flux is determined solely by the product Ni* and *is unaffected by the relative positions of coil and core.*

The toroidal iron core in Fig. 3.29 can be viewed as a *magnetic circuit.* The magnetic flux Φ_{Fe} in this circuit is obviously "moved" by the product Ni, which we should therefore think of as a *magnetomotive force* (mmf). We now write Eq. (3.101) as follows:

$$Ni = \Phi_{Fe} \frac{l}{\mu\mu_0 A} \qquad \text{ampere} \cdot \text{turns,} \qquad (3.102)$$

where $l = 2\pi R$ is the length of the magnetic core path.

We introduce the *magnetic resistance* or *reluctance* defined by

$$\mathcal{R} \equiv \frac{l}{\mu\mu_0 A}. \qquad (3.103)$$

Equation (3.102) can now be written

$$Ni = \Phi_{Fe} \cdot \mathcal{R} \qquad \text{A} \cdot \text{t} \qquad (3.104)$$

which is "Ohm's law for the magnetic circuit." The unit for $\mathcal{R}$ is ampere turns per weber, $\text{A} \cdot \text{t/Wb}$. We shall find this law useful in discussing the characteristics of the magnetic circuits in transformers and rotating machines. Note the analogy between the magnetic reluctance $\mathcal{R}$ and the electric resistance R (Eq. 3.30).

Example 3.23 Consider the toroidal core shown in Fig. 3.30. Find the current needed to produce a flux density of $B_{Fe} = 1.2$ T in the following two cases:

1) no airgap,

2) with a 2-mm air gap.

Let $N = 100$ and $\mu = 4000$ in the iron. The iron cross-sectional area is $A = 4$ cm^2.

SOLUTION: The required flux is

$$\Phi_{Fe} \doteq A \cdot B_{Fe} = 0.0004 \cdot 1.2 = 48 \cdot 10^{-5} \text{ Wb.}$$

1) For the reluctance we have

$$\mathcal{R} = \frac{322 \cdot 10^{-3}}{4000 \cdot 4\pi \cdot 10^{-7} \cdot 4 \cdot 10^{-4}} = 1.601 \cdot 10^5 \text{ A} \cdot \text{t/Wb.}$$

Equation (3.104) thus gives

$$100i = 48 \cdot 10^{-5} \cdot 1.601 \cdot 10^5;$$

$$\therefore \; i = 0.769 \text{ A.}$$

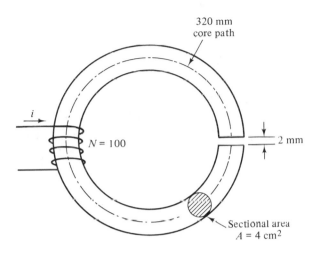

320 mm
core path

N = 100

i

2 mm

Sectional area
A = 4 cm²

Figure 3.30

2) We must now separate the reluctances for the iron path and the air gap. Formula (3.103) yields

$$\mathcal{R}_{Fe} = \frac{320 \cdot 10^{-3}}{4000 \cdot 4\pi \cdot 10^{-7} \cdot 4 \cdot 10^{-4}} = 1.592 \cdot 10^5,$$

$$\mathcal{R}_{air} = \frac{2 \cdot 10^{-3}}{4\pi \cdot 10^{-7} \cdot 4 \cdot 10^{-4}} = 39.79 \cdot 10^5.$$

The total reluctance is the sum of these two:

$$\mathcal{R}_{tot} = \mathcal{R}_{Fe} + \mathcal{R}_{air} = 41.38 \cdot 10^5.$$

From Eq. (3.104) we thus get

$$100\,i = 48 \cdot 10^{-5} \cdot 41.38 \cdot 10^5;$$

$$\therefore\ i = 19.86\ \text{A}.$$

Note that although the air gap is of only 2-mm length, its reluctance is the dominant one. This has great practical significance. For example, in a rotating electric machine, the magnetic iron path must be interrupted by an air gap (see Fig. 3.31). Even a very small air gap increases drastically the need for "magnetizing" current.

....

Example 3.24 The types of magnetic circuits depicted in Fig. 3.31 are extremely important in the electric energy conversion field. Consider first the

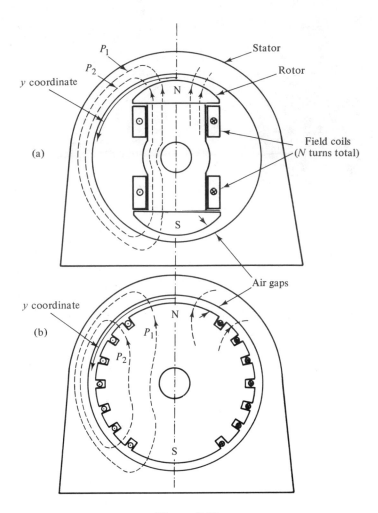

Figure 3.31

"salient rotor" machine shown in Fig. 3.31(a). The dc† current in the rotor *field coils* provides the mmf of magnitude Ni that drives the magnetic flux along the paths indicated. Both paths, P_1 and P_2, enclose the same current, Ni. But because of the tapered poleface, path P_2 contains a wider air gap than path P_1. As a consequence P_2 is characterized by less magnetic flux density than P_1.

---

† The term "dc current" is linguistically a redundant statement. However, ac and dc, originally meant as abbreviations, are now commonly used as adjectives in engineering vocabulary.

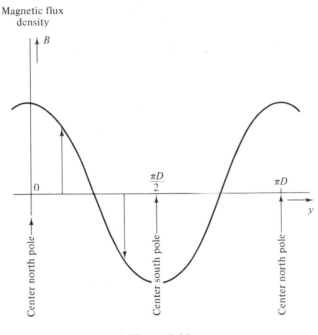

Magnetic flux density

Center north pole

Center south pole

Center north pole

Figure 3.32

If we plot the flux density versus the tangential coordinate y, then we obtain the graph shown in Fig. 3.32. The flux density seems to vary almost sinusoidally. The advantage of this will be explained in the next chapter.

Consider then the "round-rotor" design shown in Fig. 3.31(b). The dc field winding is now placed in rotor slots. The two flux paths, P_1 and P_2, are now traversing equally wide air gaps. However, path P_1 encloses more current than P_2 and thus we can expect greater flux densities along the former path. Again we obtain a sinusoidal flux distribution of the type shown in Fig. 3.32.

3.26.3 The Magnetic Field Intensity

Consider for a moment Ohm's law for an electric circuit. Equations (3.29) and (3.30) can be written:

$$\frac{v}{l} = \rho \frac{i}{A} \quad \text{V/m.} \tag{3.105}$$

The left-hand side of this equation, having the dimension volts per meter, represents the per meter voltage drop, that is, *the electric field intensity E*, as measured along the current path. The ratio i/A equals the current density, measured in A/m^2. Equation (3.105) thus tells us that

$$E = \frac{\text{emf}}{\text{meter}} = \rho \cdot \text{current density.} \tag{3.106}$$

If we now take a look at the situation in the magnetic core we have, according to Eq. (3.102),

$$\frac{Ni}{l} = \frac{1}{\mu\mu_0} \cdot B_{Fe}, \tag{3.107}$$

or

$$H \equiv \frac{\text{mmf}}{\text{meter}} = \frac{1}{\mu\mu_0} \cdot \text{flux density.} \tag{3.108}$$

Here we have introduced, in analogy with E in equation (3.106), *the magnetic field intensity H*, defined as

$$H \equiv \frac{Ni}{l} \quad \text{A} \cdot \text{t/m.} \tag{3.109}$$

The electric field intensity E represents a per-meter emf measured along the current path of the electric circuit. *It is independent of the electric characteristics of the conductor.*

Likewise, the magnetic field intensity H represents a per-meter mmf along the flux path of the magnetic circuit. *It is also independent of the characteristics of the magnetic conductor.* The use of this new quantity will be developed in the next section.

3.26.4 Magnetization Curves For Ferrous Materials

We have exposed some far-reaching similarities between electric and magnetic circuits. Now we shall focus our attention on some important dissimilarities.

As we increase the emf in an electric resistive circuit, the current will likewise increase. *Over a very large range of current densities* the increase in current is directly proportional to the increase in emf. This means (compare Eq. 3.30) that the resistivity parameter ρ is a *constant*. The proportionality ceases only when the current density reaches such high values that the conductor heats up excessively.

As we increase the mmf in the magnetic circuit the flux density will increase *but not in direct proportion to mmf*. Differently expressed, the μ-parameter *is not a constant.*

The actual relationship between flux density and mmf for an iron sample is given in a so-called magnetization curve, a typical example of which is shown in Fig. 3.33. In such curves, the flux density B is always plotted versus H. The

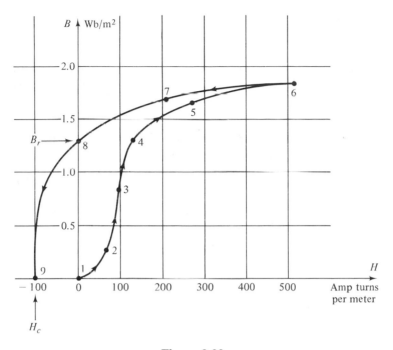

Figure 3.33

advantage of using H rather than mmf is that the data is now presented on a per-unit basis, that is, the data does not refer to a specific core size. Let us explore the main features of ferromagnetic behavior by discussing the curves in Fig. 3.33. We will assume that we begin with a totally demagnetized core.

As we slowly increase the coil current from zero we will be moving on a nonlinear curve (points 1 → 2) corresponding to increasing μ-values maybe in the μ-range 50–200. A further increase in coil current takes us to points 3 → 4 characterized by a very rapid growth of the flux, where μ may take on values of 10^4 or 10^5.

A continued increase in mmf through points 5 → 6 will now see a continued decrease in flux growth, that is, a decrease in μ. This decrease in μ will continue and should we apply mmf's corresponding to 50,000 A · t/m (far beyond the scale of our graph) μ will actually approach unity. This means that the magnetic conductivity is reduced to that of vacuum or air. We say that the iron has reached *magnetic saturation.*

As we turn around at point 6 and as we decrease the mmf, the flux densities will decrease also—but we will now follow a different curve. Should we decrease the mmf to zero (point 8) we note that a certain flux density, B_r, called the *residual density*, remains.† The core sample has now taken on the characteristics of a permanent magnet. The core will be demagnetized only upon application of a *negative* mmf, called *coercive intensity*, $-H_c$.

We should point out that this magnetic behavior applies to iron at normal temperatures. Above 770° C (Curie temperature) iron loses all ferromagnetism and behaves essentially as air.

Example 3.25   In Example 3.23 we studied the magnetic circuit in Fig. 3.30 under the assumption of *constant* permeability. Now we will work the problem under the assumption that the *BH* curve in Fig 3.33 applies.

Specifically we want to find the current needed to produce a flux density of $B = 1.5$ T. As before, the air gap is 2 mm.

SOLUTION: From the *BH* curve we find that a flux density of 1.5 T corresponds to

$$H_{Fe} = \frac{Ni}{l} = 195 \text{ A} \cdot \text{t/m}.$$

The mmf needed for the iron part of the path is thus

$$(Ni)_{Fe} = 195 \cdot 0.320 = 62 \text{ A} \cdot \text{t}.$$

The flux density in the air gap equals that in the iron (if we neglect fringe effects) and from Eq. (3.108) we get (for $\mu = 1$)

$$H_{air} = \frac{B_{air}}{\mu_0} = \frac{1.5}{4\pi \cdot 10^{-7}} = 1.19 \cdot 10^6 \text{ A} \cdot \text{t/m}.$$

The mmf needed for the air portion of the magnetic path is therefore

$$(Ni)_{air} = 1.19 \cdot 10^6 \cdot 0.002 = 2380 \text{ A} \cdot \text{t}.$$

Thus the total mmf is

$$(Ni)_{tot} = 62 + 2380 = 2442 \text{ A} \cdot \text{t}.$$

The desired flux density could be obtained, for example, with a 1000-turn coil carrying 2.442 amperes or a 100-turn coil carrying 24.4 amperes.

The example teaches us the following important facts. Although the air portion amounts to less than one percent of the total magnetic path it requires 97.5% of the total mmf to sustain the required flux density. The iron core thus, in effect, permits us to concentrate or focus practically the total magnetizing force of the coil on the air gap.

---

† If an electric circuit were to behave in a similar manner, we would be able to measure a circulating current after removal of the emf.

3.26.5   A Physical Explanation of Ferromagnetism

Needed at this time is a reasonably satisfactory explanation of the ferromagnetic phenomena described above. To present an exhaustive explanation of ferromagnetism would require a thorough knowledge of modern theory of microphysics—like quantum theory. A meaningful explanation and one that certainly suffices for most engineering purposes can be presented by the help of the earlier introduced concept of magnetic moment.

A. Magnetization—a Result of Line-up of Atomic Magnetic Moments.   In the modern model of the atomic structure of matter, the electrons perform orbital motions around the nucleus of the atom. The orbiting negative electron corresponds to a current in the electron orbit. This then would give the electron a *magnetic orbital moment* m_{orb} as depicted in Fig. 3.34(a).

In addition, each electron is associated with a *magnetic spin moment* m_{spin}. This can be understood if the electron is pictured as a negatively charged spinning sphere (Fig. 3.34b). According to quantum theory, both of these moments can have only *discrete* or *quantized values.*

Consider now an iron core inside an electric coil (Fig. 3.35).

In its unmagnetized state, the above-mentioned elementary atomic magnetic moments have a random orientation. If a coil current is applied, a magnetic flux is created in the core and this flux subjects the elementary electron moments to torques that tend to line them up with the field. Why this line-up takes place in iron but not in copper is a result of the difference in atomic binding or restraining forces.

As the electron moments are being lined up their mmf's are added to that of the coil and the result is an mmf amplification. When the coil mmf is reaching a certain level (corresponding to point 3 in Fig. 3.33), a drastic mmf

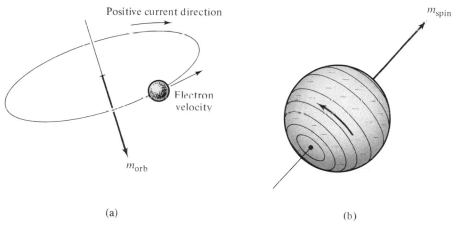

(a)   (b)

Figure 3.34

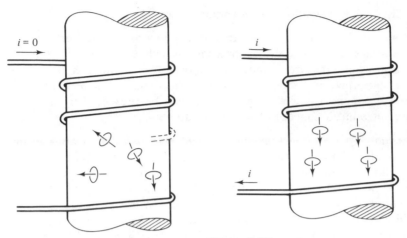

Figure 3.35

amplification takes place. This suggests that the line-up process takes place collectively by the simultaneous "snapping" into position of a large number of elementary moments.

When all or most of the magnetic moments have been lined up, no additional mmf amplification can be achieved—we have now reached the saturation level as shown in Fig. 3.33. Should the coil current be reduced to zero, some of the magnetic moments will lose their orientation. Some will be "frozen" in their lined-up position, and this accounts for the residual magnetism of the core.

B. "Bound" Currents. Visualize for a moment each atomic magnetic moment as a small coil. As these coils are lined up under the influence of the magnetizing field, it becomes obvious (Fig. 3.36a) that the currents in neigh-

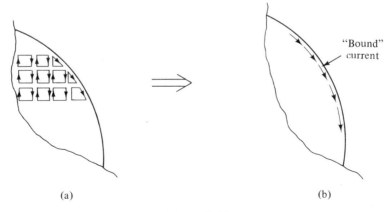

(a)   (b)

Figure 3.36

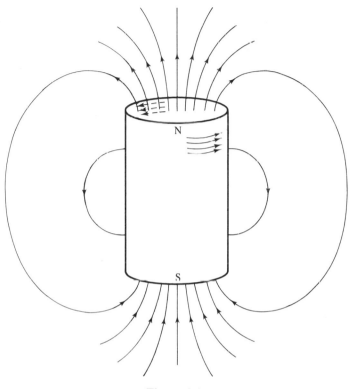

Figure 3.37

boring coil sides neutralize each other. It is assumed that the magnetization (or "magnetic moment density") is uniform throughout the core material. Only those coil currents located at the core surface (Fig.3.36a) remain unneutralized.

The nonneutralized currents thus add up to a surface sheet current referred to as the "bound" current (Fig. 3.36b). These "frozen" currents in the core material itself cannot, of course, be measured. However, their field and force effects are as real as those that would be created by corresponding currents in an equivalent coil.

For example, consider the cylindrical iron sample shown in Fig. 3.37. If magnetized the core sample will possess bound surface currents that constitute a cylindrical shell. The magnetic flux picture in and around the core will thus be similar to that associated with a cylindrical coil shown in Fig. 3.18.

C. Force and Torque Effects of Magnetized Materials.   The forces and torques acting upon magnetized cores have great practical usefulness, because they are of considerable magnitude. The nature of these forces and torques can easily be ascertained by use of the concepts of "bound" surface currents. We give a couple of important examples.

Example 3.26 Consider the lift magnet shown in Fig. 3.38. As the magnet approaches the unmagnetized load object, the magnetic field from the magnet will magnetize the object (Fig. 3.38b). The bound currents in the load object and the actual currents in the lift magnet coil are shown in Fig. 3.38(c). As we determined in Example 3.20, this currrent geometry results in an attraction between the magnet and the load.

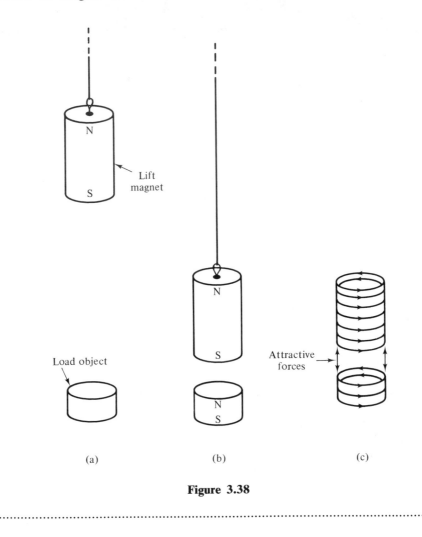

(a)   (b)   (c)

Figure 3.38

Example 3.27 Of great importance in certain types of electric machines is creation of what is called "reluctance torque." Consider the arrangement depicted in Fig. 3.39. The dc current in the stator field coils sends a flux along

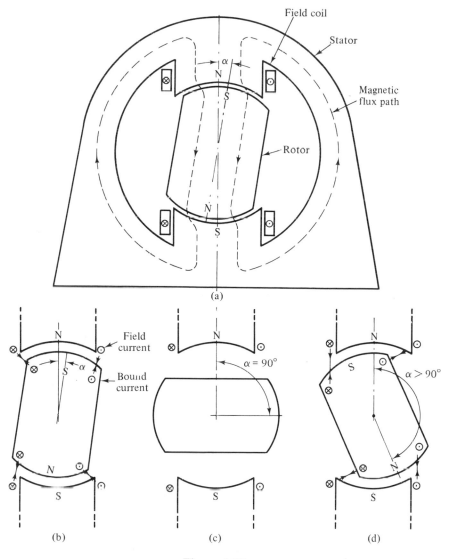

Figure 3.39

the indicated magnetic paths. The flux will magnetize the rotor, but its polarity (N or S) will depend upon its angular position.

Three different rotor positions are shown. In each position we have indicated polarity by N or S and the corresponding bound current directions. In (b) the attractive forces between field currents and bound currents (compare Example 3.20) have been shown. As the rotor angle α is increased, these forces exert a CCW torque upon the rotor (you may think of the forces as rubber bands).

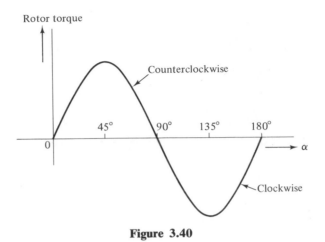

Figure 3.40

The torque reaches a maximum for $\alpha = 45°$ and then decreases to zero as the rotor takes on a horizontal position characterized by zero magnetization.

As the α-angle exceeds 90° the polarity of the rotor reverses and so does the torque. Figure 3.40 shows how the reluctance torque varies as a function of rotor position. Note that the torque completes a *full* cycle for one *half* turn of the rotor. In Chapters 4 and 8, we shall discuss further the practical significance of this type of torque.

3.27 SUMMARY

We have described in this one chapter those electric and magnetic phenomena upon which electric power engineering technology is based. Following a brief comparison with the concepts of mass, gravity, gravitational potential, and energy, we then introduced the concepts of "static" electricity, electric field, potential, and energy. Electrostatic force actions and energy storage possibilities were discussed.

The latter part of the chapter deals with electromagnetic phenomena. Electromagnetic force actions and the induction law occupy central positions in all areas of electric power engineering. A number of examples, all pertaining to practically important apparatus, have been treated. The chapter ends with a brief summary of the most important aspects of ferromagnetism.

The chapter contains a wide spectrum of important material, and it is not to be expected that all can be digested in one reading. The remainder of the book will rest heavily upon this basic chapter. The reader will have many opportunities to return for more penetrating study of many of the covered topics.

EXERCISES

3.1 A plate capacitor is charged from a battery of voltage v_0 and then disconnected from the battery. It is assumed that no charge will subsequently leak across the dielectric. The stored electric energy is

$$w_e = \tfrac{1}{2}Cv_0^2 \quad \text{J.} \tag{3.110}$$

The plates are now moved apart so that the plate distance d is tripled. According to formula (3.25) this will decrease the capacitance to the new value $C/3$.

a) What will be the new capacitor voltage?

b) What will be the new stored electric energy?

c) You will find, if you work part (b) correctly, that the stored energy has *increased*. Where did this additional energy come from?

3.2 One of the two identical capacitors in Fig. 3.41 is charged to V volts. Upon closing the switch S the charge will divide equally between the capacitors. In the new steady state each capacitor will thus have a voltage of $V/2$ volts.
The stored energy *before* closing of S is

$$w_e' = \tfrac{1}{2}CV^2 \quad \text{J.}$$

The stored energy *after* closing will be

$$w_e'' = 2 \cdot \tfrac{1}{2}C\left(\frac{V}{2}\right)^2 = \tfrac{1}{4}CV^2 \quad \text{J.}$$

Half the initial energy has "disappeared." What is your theory as to where it went?

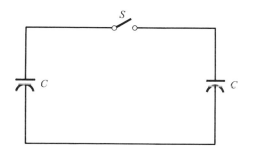

Figure 3.41

3.3 Consider the previous problem again. Put a resistor R into the circuit before you close S.

a) Find the current i versus time upon closing of S! (Compare Example 3.13 in text.)

b) Find and plot *both* capacitor voltages versus time.

    c) Compute the ohmic heat energy dissipated in R during the charge redistribution process!

    d) In view of your findings in part (c), find an explanation to Exercise 3.2!

3.4 Ten equal capacitors are connected in parallel and charged from a 1000-volt battery. The capacitors are disconnected from the battery and then reconnected in series. What will be the voltage across this capacitor pile? It is assumed that no charge leakage is taking place. (This is how high-magnitude impulse test voltages are obtained in high-voltage laboratories)

3.5 Find the capacitance of the layered capacitor in Fig. 3.9. There are a total of 20 conductor foils, each of dimension 20×2000 cm, and 20 insulator sheets, each having a thickness of 0.1 mm. The foils are rolled into a cylindrical bundle. The insulator material is characterized by $\epsilon = 5$.

3.6 A capacitor C is being charged from a battery having an emf E. To limit the inrush current a resistor R is placed in series with the capacitor.

    a) Prove that the charging current following the closing of the switch is of the form

$$i = \frac{E}{R} e^{-(t/RC)} \quad \text{A.} \tag{3.111}$$

    b) When the capacitor is fully charged its stored energy will be

$$w_e = \tfrac{1}{2}CE^2 \quad \text{J.}$$

Prove that exactly the same amount of energy is dissipated in the resistor during the charging process. (This means that the charging efficiency is only 50%.)

3.7 Three equal positive charges of 1 microcoulomb each are placed at the three corners of an equilateral triangle. The distance between the charges is 1 m.

    a) Find the magnitude and direction of the forces acting on each charge.

    b) Compute the magnitude and direction of the electric field at the midbase of the triangle.

    c) Prove that the electric field midpoint between the three charges is zero.

[*Hint*: You know the voltage v and field intensity E emanating from *one* charge. The effects of *several* charges are obtained by addition. Note, however, that E is a *vectorial* quantity. The potential v is a *scalar*.]

3.8 Eight equal charges of 1 microcoulomb each are placed so as to form the corners of a 1-m^3 cube.

    a) Find the electric potential in the center of the cube.

    b) Find the electric field strength in the cube center.

    c) Compute the energy required to move a 1-microcoulomb charge from "infinity" to the cube center.

3.9 An ac current in a conductor can be expressed in the form

$$i = 100 \cdot \sin (314\, t) \qquad \text{A.} \qquad (3.112)$$

a) What is the frequency?
b) The "free electron cloud" in the conductor will evidently oscillate back and forth in the conductor covering a distance d. Find this distance under the following assumptions: (1) conductor cross-sectional area = $10\,mm^2$, (2) one free electron per atom, (3) 10^{20} atoms per mm^3.

3.10 A copper wire of 1-km length and 1-mm^2 cross-sectional area is connected across the terminals of a 12-V automobile battery. Neglect internal battery resistance and find the following:

a) current!
b) ohmic power dissipation in wire!
c) ohmic dissipation per meter wire!

Do you think the wire will melt?

3.11 Work all parts of the previous problem assuming that the wire length is shortened to 1 m (so-called short circuit).

Do you think the wire will melt?

3.12 An HVDC (high voltage dc) transmission system transmits energy over a distance of 900 km. Each of the two conductors has a resistance of 0.014 Ω/km. The load voltage (v_l in Fig. 3.13) equals 600 kV. The load power p_l is 800 MW. Compute:

a) current in each conductor,
b) ohmic power loss in the transmission line,
c) generator voltage, v_g,
d) generator power, p_g,
e) transmission efficiency.

3.13 Consider the coil in the text Example 3.15. If the current in the straight conductor is a sawtooth-type ac current as shown in Fig. 3.42, find the induced coil emf.

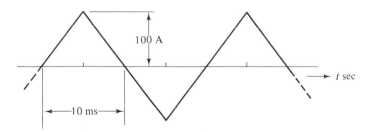

Figure 3.42

3.14 The two coils in Fig. 3.27 are interconnected to form a *single* coil. Two cases are possible depending upon polarity chosen.

Prove that the self-inductance of the resulting coil will be computed from

$$L = L_1 + L_2 \pm 2M \qquad \text{H.} \tag{3.113}$$

Be careful to indicate which cases the $+$ and $-$ signs refer to!

3.15 Twenty axial conductors, each of length 1 m, are attached to the surface of a cylindrical rotor of diameter 1 m. Each conductor carries a current of 100 A magnitude. A magnetic field of density 1 T penetrates the cylindrical rotor surface *radially*. The rotor will be subject to a torque.

Find torque magnitude in $N \cdot m$!

3.16 The N spokes in a wheel (Fig. 3.43) each carry the current i A. The wheel is located in a magnetic field B perpendicular to the wheel plane. Prove that the wheel will be subject to a torque of magnitude

$$T = \tfrac{1}{2}NBi(R^2 - r^2) \qquad N \cdot m. \tag{3.114}$$

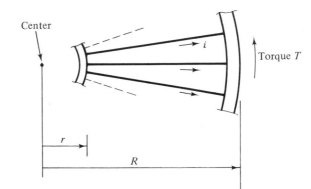

Figure 3.43

3.17 Figure 3.44 depicts an ammeter. The permanent magnet M sends a magnetic flux across the air gaps. We can assume that the flux density in the air gap is radial, uniform, and of magnitude B T. A small rectangular, N-turn coil is free to rotate in the air gaps against the restraining torque of a spring. The spring torque is proportional to the angle of rotation and equals T N $\cdot$ m per degree.

Prove that the coil will rotate through an angle α which is proportional to current i! Find α for the following numerical data:

$$B = 0.3 \text{ T.} \qquad N = 500 \text{ turns,}$$
$$i = 0.1 \text{ mA,} \qquad T = 10^{-5} \text{ N} \cdot \text{m/degree.}$$

3 cm of each coil side is located in the magnetic flux. The coil radius, $R = 2$ cm.

3.18 Find the self-inductance of the 100-turn coil placed on the toroidal core discussed in Example 3.23 in the text. Treat the two cases with and without air gap. [*Hint:* Find the total linked flux caused by a 1 A coil current.]

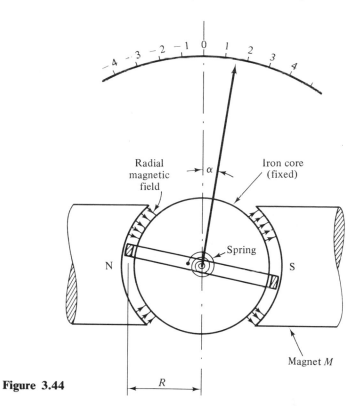

Figure 3.44

REFERENCES

Angrist, S. W. *Direct Energy Conversion.* Boston, Massachusetts: Allyn and Bacon, 1971.

Feynman, R. P., et al. *Feynman Lectures on Physics,* Vol. II. Reading, Massachusetts: Addison-Wesley, 1964.

Hoyt, William H., Jr. *Engineering Electro-Magnetics.* New York: McGraw-Hill, 1974.

Johnk, C. T. A. *Engineering Electromagnetic Fields and Waves.* New York: Wiley, 1975.

Kettani, M. A. *Direct Energy Conversion.* Reading, Massachusetts: Addison-Wesley, 1971.

White, D. C., and H. H. Woodson. *Electromechanical Energy Conversion.* New York: Wiley, 1959.

SYNCHRONOUS ELECTRIC ENERGY CONVERTERS

chapter 4

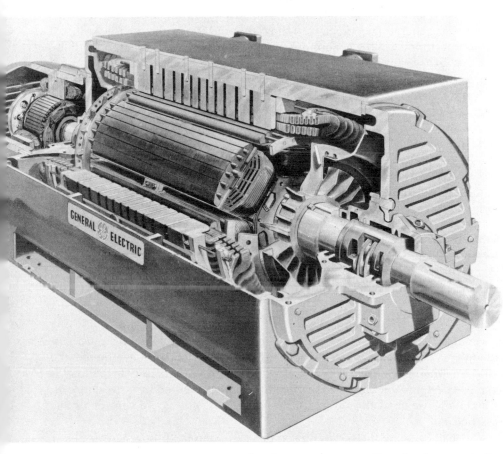

Synchronous generator. Note the damper winding on the rotor. The exciter machine is shown at the left end of the shaft. (Courtesy General Electric)

The previous chapter described the basic physical laws upon which electric energy technology is based. The remainder of the book will be devoted to the various components and systems, the design and operation of which is the concern of the electric power engineer.

It would be impossible and entirely beyond the scope of this treatise to attempt an inclusion of all types of machines that electric power technology has produced. Likewise, it would be unrealistic at this introductory level to develop mathematical models that would describe all phenomena in power apparatus. For example, we will not attempt to describe the behavior of a three-phase synchronous generator subject to unbalanced fault currents. We shall feel content if we are able to convey an understanding of the functioning of this important but rather complex machine under *normal balanced* operation.

In presenting the material we find it natural to follow the flow of the electric energy in our power systems. Thus, starting with the electric generators in the power generating centers, we follow the journey of the energy as it first is being transformed in the generator step-up transformers and thereupon flows out in the coarse meshes of the high-voltage power transmission system. As it reaches the final destination it is again being transformed in the step-down transformers and fed into the fine meshes of the distribution system whence it is finally portioned out to the multitude of users.

Following this organization philosophy, we shall devote this chapter to *electric generators.* Chapter 5 will tell the story of the electric *power transformers.* In Chapter 6, we will turn our attention to the operating features of *power networks.* This will include the *loadflow* problem, the problem of *parallel operation of generators,* and the important aspects of *system control.*

Chapters 7 and 8 will be devoted to the most important electric power consumption devices—the *electric motors.* We will be exposing the important operating characteristics of various motor types, dc and ac. In the final Chapter 9 we shall try to determine the probable future developments in the electric power field.

4.1 DIRECT CURRENT VERSUS ALTERNATING CURRENT

As we unfold the story about electric generating technology, we must immediately settle the question of "ac versus dc." Consider for a moment the simple electric transmission system in Fig. 3.13. Let us assume that the load is of the simplest possible kind and in effect can be represented by a resistance R. For simplicity we shall also disregard the line losses, that is, we neglect all line resistance.

We shall compare the power flow in this system in the following two cases:

CASE 1 *Direct Current.* The "generator" (for example, an automobile battery) can now be represented by an emf e of constant magnitude. The line current will obviously have the value

$$i = \frac{e}{R} \quad \text{A.}$$

Because e is constant, the current also will be constant.
   The power flow on the line will be

$$p = vi = ei = \frac{e^2}{R} \quad \text{W.} \tag{4.1}$$

and it too will be *constant*.

CASE 2   *Single-Phase Alternating Current.* The "generator" may consist of the rotating coil in Fig. 3.22. The emf is now sinusoidal, that is, of the form

$$e = e_{max} \sin \omega t \quad \text{V.}$$

The current will therefore also be sinusoidal:

$$i = \frac{e}{R} = \frac{e_{max}}{R} \sin \omega t \quad \text{A.}$$

For the line power we obtain

$$p = vi = ei = \frac{e_{max}^2}{R} \sin^2 \omega t = \frac{e_{max}^2}{2R} (1 - \cos 2\omega t) \quad \text{W.} \tag{4.2}$$

   We note that the line power is *pulsating*, at the double radian frequency 2ω, between zero and the maximum value

$$p_{max} = \frac{e_{max}^2}{R} \quad \text{W,} \tag{4.3}$$

   The average power equals

$$p_{ave} = \frac{p_{max}}{2} = \frac{e_{max}^2}{2R} \quad \text{W.} \tag{4.4}$$

   A comparison of the power expressions (4.1) and (4.2) immediately reveals the superiority of dc over single-phase ac. The dc power flow is *smooth*. The pulsating single-phase ac power would cause completely unacceptable vibration problems, in both the generator and the load—certainly at power levels in the megawatt range.
   Fortunately, the power pulsations, as we shall see in Sect. 4.6, can be completely eliminated by use of *multi-phase* ac, particularly the important *three-phase* system with which we will be exclusively concerned in this book.

Three-phase ac thus offers the smooth features of dc power plus the following additional ac advantages:

1) Easy generation,
2) Easy transformation to and from high-voltage levels (which permits low-loss transmission),
3) Cheap and simple motor design.

As a result, practically 100% of all electric bulk energy in the world is of the three-phase ac variety. In those relatively few instances where for very special reasons (HVDC transmission,† dc motors, auto batteries, etc.) dc is preferred over ac, one invariably obtains the dc energy by rectification of ac energy.

4.2 SINGLE-PHASE AC POWER FORMULAS

As ac power shall be of dominating concern throughout the remainder of this book, we find it fitting to review some of its important characteristics. We shall present this material for single-phase—but all of it will later find direct applicability to the all-important, three-phase case.

4.2.1 Real and Reactive Powers

Consider again the transmission system shown in Fig. 3.13. Remove now the restriction made in Section 4.1 that the load be resistive. For a general type of load, the current and voltage will *not* be *in phase,* but *out of phase* by an angle φ. We thus have

$$\left.\begin{array}{ll} v = v_{\max} \sin \omega t & \text{V,} \\ i = i_{\max} \sin (\omega t - \varphi) & \text{A.} \end{array}\right\} \tag{4.5}$$

A *positive* φ-value means that the current *lags* the voltage. This case (as shown in Fig. 4.1a) is obtained when the load is *inductive*. A *negative* φ-value implies that the current would *lead* the voltage, which would be the case if the load were *capacitive*.

---

† High-voltage dc (HVDC) transmission is preferred in those cases where ac transmission would be either very impractical or physically impossible. For example, in transmitting electric energy over large bodies of water, one is forced to use submarine cables. A cable is characterized by very large shunt capacitance (because of the proximity between the conductor and the outer shield), and its shunt impedance $\dfrac{1}{\omega C}$ will thus be small. Unacceptably large capacitive leakage currents will thus result. A dc cable has zero capacitive leakage.

HVDC is also preferred over ac when energy is being transmitted over extremely long distances (>600 km) where ac can cause stability problems.

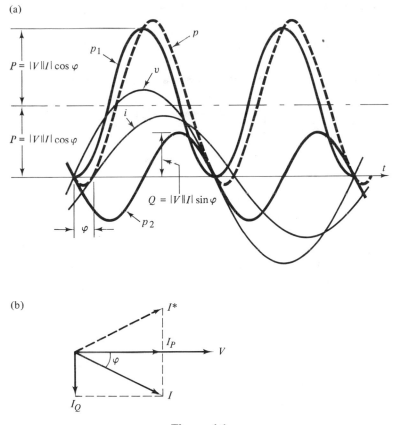

Figure 4.1

The reader no doubt is familiar with *phasor* representation of sinusoidal variables. (Appendix A summarizes the most important features of phasor analysis.) If we represent the voltage and current with the phasors V and I, respectively, then we obtain the *phasor diagram* of Fig. 4.1(b). Note that our phase-angle φ is defined by the relationship

$$\varphi = \underline{/V} - \underline{/I}. \tag{4.6}$$

For the transmitted power we now get

$$p = vi = v_{max}i_{max} \sin \omega t \sin (\omega t - \varphi) \quad \text{W}. \tag{4.7}$$

If we now make use of the trigonometric identity

$$\sin \alpha \sin \beta = \tfrac{1}{2}[\cos (\alpha - \beta) - \cos (\alpha + \beta)], \tag{4.8}$$

we can write the power formula in the following form:

$$p = \frac{v_{max}\, i_{max}}{2}[\cos \varphi - \cos (2\omega t - \varphi)] \qquad \text{W.} \qquad (4.9)$$

We introduce now the *effective, or rms,* values of voltage and current;

$$\left. \begin{aligned} |V| &\equiv \frac{1}{\sqrt{2}}\, v_{max} \qquad \text{V,} \\[2mm] |I| &\equiv \frac{1}{\sqrt{2}}\, i_{max} \qquad \text{A,} \end{aligned} \right\} \qquad (4.10)$$

and can then write Eq. (4.9) as follows:

$$p = |V|\,|I| \cos \varphi - |V|\,|I| \cos (2\omega t - \varphi) \qquad \text{W.} \qquad (4.11)$$

The line power evidently pulsates (Fig. 4.1a) around an *average power* ($|V|\,|I| \cos \varphi$) at double radian frequency 2ω. During certain periods the power actually is *negative*.

Equation (4.11) can be transformed into the alternate form

$$p = \underbrace{|V|\,|I| \cos \varphi\, (1 - \cos 2\omega t)}_{p_1} - \underbrace{|V|\,|I| \sin \varphi \sin 2\omega t}_{p_2}. \qquad (4.12)$$

The power has been decomposed into two components p_1 and p_2. The first component pulsates around the same average value as before but *never goes negative*. The second component has a *zero average value*.

At this juncture we introduce the following two quantities:

$$\left. \begin{aligned} P &\equiv |V|\,|I| \cos \varphi \qquad \text{(\textit{real} power)} \\ Q &\equiv |V|\,|I| \sin \varphi \qquad \text{(\textit{reactive} power)} \end{aligned} \right\} \qquad (4.13)$$

and can then write Eq. (4.12) more compactly:

$$p = P(1 - \cos 2\omega t) - Q \sin 2\omega t \qquad \text{W.} \qquad (4.14)$$

These concepts are of such fundamental importance that we find it appropriate to say a few words about their meaning:

1) The real power P is *defined* as the average value of p and therefore means the *useful* power transmitted.
2) The reactive power Q is by *definition* equal to the *peak* value of that power component that travels back and forth on the line. It results in zero average, and is therefore capable of no useful work.
3) One can associate P and Q with two different current components, I_P and I_Q, shown in Fig. 4.1(b). The "inphase" component I_P evidently has the

magnitude

$$|I_P| = |I| \cos \varphi \qquad \text{A.} \tag{4.15}$$

The "out-of-phase" component I_Q is of magnitude

$$|I_Q| = |I| \sin \varphi \qquad \text{A.} \tag{4.16}$$

According to Eq. (4.13) we obviously have

$$P = |V| |I_P|,$$
$$Q = |V| |I_Q|. \tag{4.17}$$

Both P and Q have dimension watts. To emphasize the fact that the latter represents a nonactive or reactive power, it is measured in *volt-amperes reactive* (VAr). Larger and more practical units are kilovar and megavar, related to the basic unit as follows:

$$1 \text{ MVAr} = 10^3 \text{ kVAr} = 10^6 \text{ VAr}.$$

4.2.2   Effects of Various Load Types

In analyzing the real and reactive power situation for a specific type of load the electrical engineer makes use of concepts like *impedance, phasor,* and *complex algebra.* These topics belong in electric circuits courses. They have been summarized in Appendix A for those readers who wish a compact review.

Table 4.1 summarizes the real and reactive power for the most commonly encountered types of impedance load. (Many loads are not of impedance type (compare Section 6.3.1).)

Note, in particular, that an inductive load draws reactive power from the source. In power lingo, an inductor is said to "consume" reactive power. A capacitive load, on the other hand, draws negative Q. A capacitor is said to "generate" reactive power.

Example 4.1   Let us exemplify how the table is obtained by working out the details for the series RL circuit (4).

SOLUTION: The impedance for this circuit equals

$$Z = R + j\omega L \qquad \Omega. \tag{4.18}$$

Thus

$$|Z| = \sqrt{R^2 + (\omega L)^2} \qquad \Omega, \tag{4.19}$$

and

$$\varphi = \angle V - \angle I = \angle Z = \tan^{-1}\left(\frac{\omega L}{R}\right). \tag{4.20}$$

Table 4.1

Load type	Phasor relation	Phase angle	Power absorbed by load	
			P	Q
1		$\varphi = 0$	$\dfrac{|V|^2}{R}$	0
2		$\varphi = +90°$	0	$+\dfrac{|V|^2}{\omega L} = \omega L |I|^2$
3		$\varphi = -90°$	0	$-\omega C|V|^2 = -\dfrac{1}{\omega C}|I|^2$
4		$\varphi = \tan^{-1}\dfrac{\omega L}{R}$	$\dfrac{R|V|^2}{R^2 \div (\omega L)^2}$ $= R|I|^2$	$\dfrac{\omega L|V|^2}{R^2 + (\omega L)^2}$ $= \omega L|I|^2$
5		$\varphi = \tan^{-1}\dfrac{R}{\omega L}$	$\dfrac{|V|^2}{R}$	$\dfrac{|V|^2}{\omega L}$
6		$\varphi = -\tan^{-1}\dfrac{1}{\omega CR}$	$\dfrac{R|V|^2}{R^2 + \dfrac{1}{\omega^2 C^2}}$ $= R|I|^2$	$-\dfrac{|V|^2}{\omega C(R^2 + \dfrac{1}{\omega^2 C^2})}$ $= -\dfrac{1}{\omega C}|I|^2$
7		$\varphi = -\tan^{-1}\omega CR$	$\dfrac{|V|^2}{R}$	$-\omega C|V|^2$

From this φ-value we then get:

$$\left.\begin{aligned} \sin \varphi &= \frac{\omega L}{\sqrt{R^2 + (\omega L)^2}}, \\ \cos \varphi &= \frac{R}{\sqrt{R^2 + (\omega L)^2}}. \end{aligned}\right\} \qquad (4.21)$$

If the rms of V is $|V|$ then the rms of I equals

$$|I| = \frac{|V|}{\sqrt{R^2 + (\omega L)^2}} \qquad \text{A.} \qquad (4.22)$$

Substitution of these expressions for $|I|$, $\sin \varphi$, and $\cos \varphi$ into (4.13) yields the tabulated values for P and Q.

....

Example 4.2 In this example we demonstrate the physical relationship between reactive power and the stored magnetic energy in the coil of the previous example.

SOLUTION: The instantaneous current in the RL circuit is of the form

$$i = \sqrt{2} \frac{|V|}{\sqrt{R^2 + (\omega L)^2}} \sin(\omega t - \varphi) \qquad \text{A.} \qquad (4.23)$$

According to formula (3.90) we can thus write for the magnetic energy:

$$w_{\text{mag}} = \frac{1}{2} L i^2 = \frac{1}{2} L \left[\sqrt{2} \frac{|V|}{\sqrt{R^2 + (\omega L)^2}} \sin(\omega t - \varphi) \right]^2 \qquad (4.24)$$

The rate of change of w_{mag} is

$$\begin{aligned} \frac{dw_{\text{mag}}}{dt} &= 2\omega L \frac{|V|^2}{R^2 + (\omega L)^2} \sin(\omega t - \varphi) \cos(\omega t - \varphi) \\ &= \omega L \frac{|V|^2}{R^2 + (\omega L)^2} \sin 2(\omega t - \varphi) \qquad (4.25) \\ &= Q \sin 2(\omega t - \varphi) \qquad \text{W.} \end{aligned}$$

The last step follows from Table 4.1. We conclude: The magnetic field energy *rate* varies harmonically with a frequency of 2ω. It has zero average value, *and a peak value that equals Q.*

This, of course, identifies the *reactive power component* in Eq. (4.12) as that power component that *periodically, and twice per voltage cycle, stores and discharges the coil with its magnetic field energy.*

An analysis of the capacitive circuits reveals a similar relationship between the reactive power and the stored electric field energy in the capacitor.

....

4.2.3   Concept of Complex Power

The voltage and current phasors in Fig. 4.1(b) can be written in the form

$$\left.\begin{array}{ll} V = |V|\, e^{j\underline{/V}} & \text{V,} \\ I = |I|\, e^{j\underline{/I}} & \text{A.} \end{array}\right\} \tag{4.26}$$

We introduce now the *conjugate* current defined by

$$I^* \equiv |I|\, e^{-j\underline{/I}}. \tag{4.27}$$

(The conjugate current phasor I^* is shown dashed in Fig. 4.1b.) Then we form the product

$$S = VI^*. \tag{4.28}$$

This product called *complex power* has a very useful property, which we may expose by the following analysis.

Upon substitution of I^* into (4.28) we obtain

$$S = VI^* = |V|\, e^{j\underline{/V}}\, |I|\, e^{-j\underline{/I}} = |V|\,|I|\, e^{j(\underline{/V}-\underline{/I})}. \tag{4.29}$$

In view of Eq. (4.6) we thus get

$$S = |V|\,|I|\, e^{j\varphi} = |V|\,|I|(\cos\varphi + j\sin\varphi) = P + jQ. \tag{4.30}$$

The last step follows directly from the definition of P and Q.
In words:

The real and reactive power can be obtained as the real and imaginary parts of S.

The magnitude $|S|$ of the complex power is referred to as *apparent power*. It can be expressed in any one of several ways:

$$|S| = |VI^*| = |V|\,|I| = \sqrt{P^2 + Q^2}. \tag{4.31}$$

The unit of $|S|$ is obviously volt-amperes (VA), but often we prefer the larger units kVA or MVA. The practical significance of apparent power is *as a rating unit* for generators and transformers (see Sect. 5.5.3).

Example 4.3   Find the real and reactive powers consumed by the circuit depicted in Fig. 4.2. The voltage V has the rms value 100 V.

SOLUTION: We first find the impedance Z_p of the parallel branch:

$$Z_p = \frac{(6 - j3)(5 + j8)}{11 + j5} = 5.199 + j0.637\ \Omega.$$

The total impedance Z_{tot} of the circuit is thus

$$Z_{\text{tot}} = 10 + Z_p = 15.199 + j0.637\ \Omega.$$

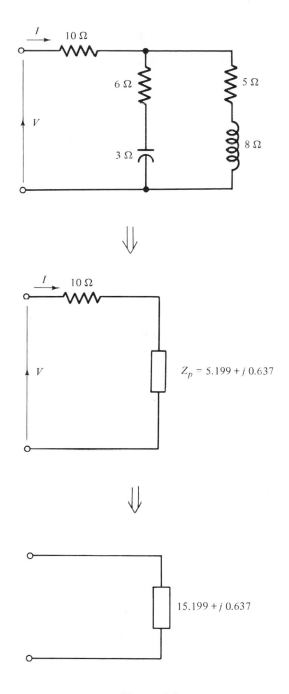

Figure 4.2

For the current, and its conjugate, we then have

$$I = \frac{V}{Z_{tot}} = \frac{100}{15.199 + j0.637} = 6.574 \;\underline{/-2.400°} \quad \text{A;}$$

$$\therefore I^* = 6.574 \;\underline{/+2.400°}.$$

We next compute the complex power:

$$S = VI^* = 100 \cdot 6.574e^{j2.400°} = 656.8 + j27.5.$$

Thus the circuit consumes

$$P_{tot} = 656.8 \text{ W,}$$

$$Q_{tot} = 27.5 \text{ VAr.}$$

Example 4.4   Often in the study of power flows in power transmission lines (Chapter 6) it is useful to find expressions for those powers in terms of the line terminal voltages. Find a formula for the power flowing on the line shown in Fig. 4.3 having the purely reactive impedance $Z = jX$! (The reactive part of a power line impedance usually completely dominates over the resistance.)

SOLUTION: If we turn our attention toward the *sending end* powers P_1 and Q_1 we can, according to formula (4.28), obtain them from the relation

$$S_1 = P_1 + jQ_1 = V_1 I^* \quad \text{VA.} \tag{4.32}$$

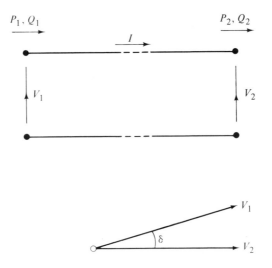

Figure 4.3

The line current I is

$$I = \frac{V_1 - V_2}{Z} \qquad \text{A,}$$

and upon substitution† into Eq. (4.32) we have

$$P_1 + jQ_1 = V_1 \frac{V_1^* - V_2^*}{Z^*} = \frac{|V_1|^2 - |V_1||V_2| e^{j\delta}}{-jX}$$

$$= \frac{|V_1||V_2|}{X} \sin \delta + j \frac{|V_1|^2 - |V_1||V_2| \cos \delta}{X} \qquad (4.33)$$

where

$$\delta \equiv \angle V_1 - \angle V_2.$$

(See Fig. 4.3.)

By separating the real and imaginary parts we thus get

$$P_1 = \frac{|V_1||V_2|}{X} \sin \delta \qquad \text{W,}$$

$$Q_1 = \frac{|V_1|^2 - |V_1||V_2| \cos \delta}{X} \qquad \text{VAr.} \qquad (4.34)$$

These formulas are very important in understanding the limits for power-transmission capability of power lines (Chap. 6).

4.3 THE SINGLE-PHASE AC GENERATOR

Although, as we earlier concluded, the single-phase ac generator is of very limited‡ practical significance, it will serve as a natural takeoff point in our presentation of the three phase generator.

We analyzed the induction of an ac emf in a coil rotating in a uniform magnetic field (Chap. 3). This simple system exemplifies in fact the prototype of

---

† We make use of these easily verifiable identities:

$$(u + b)^* - a^* + b^*$$

$$\left(\frac{a}{b}\right)^* = \frac{a^*}{b^*}$$

$$aa^* = |a|^2$$

‡ Small, single-phase units, where the vibration problem is of controllable dimensions, find limited use in cases when single-phase ac power is desired in relatively small quantities.

all ac generators. A basic difference is, however, that in an ac generator the coils are generally stationary and the magnetic field is rotating.

4.3.1 AC Generator Design

Figure 4.4 depicts the two main components of a *synchronous* ac generator, the *rotor* and the *stator*.

The rotor consists of an even number (four in this case) of *poles* of alternating polarity. On each pole is placed a *field coil,* a detail of which is shown in Fig. 4.4. The field coils are connected together to form a *field winding* (shown schematically in Fig. 4.5). An *exciter*† feeds dc current into the field

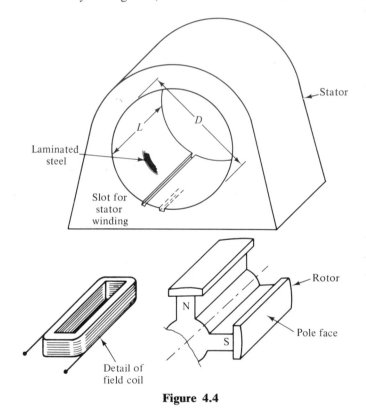

Figure 4.4

---

† The exciter may be a regular dc generator (Chap. 7) driven by the same prime mover that drives the synchronous generator. In this case the dc current is fed into the rotor field winding via brushes and sliprings.

In a "brushless" exciter the dc current is obtained from a separate ac winding placed on a separate rotor, directly connected to the main rotor. The ac voltage is rectified in a rectifier circuit placed on the rotor.

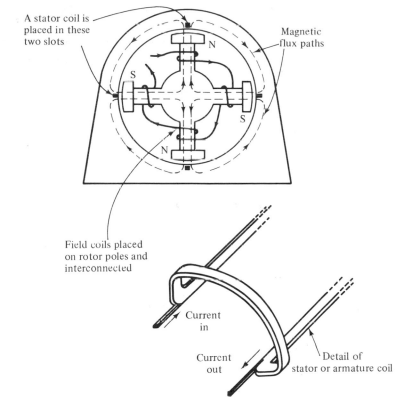

A stator coil is placed in these two slots

Magnetic flux paths

Field coils placed on rotor poles and interconnected

Current in

Current out

Detail of stator or armature coil

Figure 4.5

winding and the resulting mmf creates the magnetic flux in the paths indicated in Fig. 4.5. Note that the flux paths twice cross the *air gaps*, the airspace between rotor and stator.

The stator or *armature winding*, in which the emf's are generated, are placed in equidistant *slots* in the stator surface (only one slot is shown in Fig. 4.4). The stator winding consists of *coils* placed so that the coil sides are one pole division (90° in Fig. 4.4) apart. Figure 4.5 shows a detail of the stator coils.

As the rotor spins, the flux sweeps by the armature winding. The stator iron thus will experience a changing flux. If the stator iron were solid, induced currents would flow in it resulting in core losses and raised temperatures. To prevent this from happening the stator core is made of *laminated* iron sheets individually insulated from each other (compare Sect. 4.9).

4.3.2 Frequency Formula

The dotted flux path shown in Fig. 4.5 tells us that, for the rotor position indicated in the figure, the total magnetic flux linked to one of the armature coils is zero. If the rotor turns 45°, the linked coil flux will reach a maximum. An added rotation of 45° reduces the flux to zero and after 45° additional degrees the coil flux will reach a maximum of opposite polarity, etc. We have a periodic flux change and can thus, according to Faraday, expect a periodic emf induced in the armature coil.

We conclude from this simple reasoning that a full emf cycle will be achieved when the rotor of the 4-pole generator has turned 180 *mechanical* degrees. A full emf cycle represents 360 *electrical* degrees in the voltage wave (compare Figs. 3.22 and 4.1) and by extending this finding to a *p*-pole generator (*p* must always be an *even* integer), we conclude the following important relation between mechanical rotor angles α_{mech} and electrical angles α_{el}:

$$\alpha_{el} = \frac{p}{2} \alpha_{mech}. \tag{4.35}$$

Because $p/2$ emf cycles will be generated for one complete rotor turn and because the rotor completes $n/60$ turns per second, we obtain the following relation between electrical frequency f Hz and mechanical rotor speed n rpm.

$$f = \frac{p}{2} \cdot \frac{n}{60} = \frac{pn}{120} \text{ Hz.} \tag{4.36}$$

Example 4.5   How fast must a 6-pole generator run if operated in a 60 cps network?

SOLUTION: Formula (4.36) gives directly

$$n = \frac{120f}{p} = \frac{120 \cdot 60}{6} = 1200 \text{ rpm.}$$

4.3.3   Saliency and Nonsaliency

The 4-pole generator depicted in Figs. 4.4 and 4.5 is of *salient rotor* design. This is a typical rotor design in cases where *low-speed prime movers* (for example, hydro turbines) are used.† In the United States most electric generators are driven by *steam turbines*. Expanding steam has high velocity and consequently steam-driven generators are *high-speed*. The two most common designs have 2-pole and 4-pole rotors running at 3600 and 1800 rpm, respectively. (In countries where the frequency is 50 Hz the corresponding speeds are 3000 and 1500 rpm.)

---

† The speed of a hydro turbine is determined by the speed of the falling water. From our discussions in Chapter 2, it is clear that the hydro turbine speed decreases (and the pole number thus increases) with decreased water head.

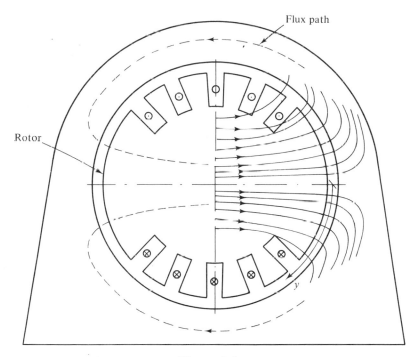

Figure 4.6

The rotor coils in a salient rotor machine would not stand the centrifugal stresses for such high speeds; therefore, steam-turbine driven generators are designed with *nonsalient* rotors as depicted in Fig. 4.6. Here the dc rotor field winding, like the ac stator armature winding, is placed in slots. This design evidently permits a better mechanical support.

4.3.4 The Air Gap Flux in Rotor Coordinates

For a salient-pole rotor (Fig. 3.31a) the pole face is tapered. The width of the air gap thus reaches a minimum at the center of the pole. Consequently, the magnetic flux is maximum at this point, decreasing to zero at the midpoint between the poles and reaching a negative maximum at the center of the adjacent pole (compare Fig. 3.32). By properly increasing the air-gap dimensions as we move away from the pole center we can make the flux density vary *harmonically*† along the periphery. We can write in this case

$$B = B_{\max} \cos \beta y \qquad \text{T}, \qquad (4.37)$$

---

† By *harmonic* function we mean one that can be expressed as a sine or cosine of the independent variable.

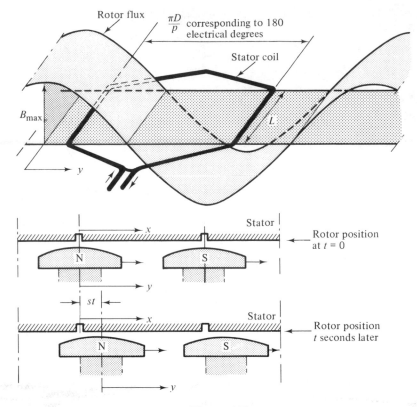

Figure 4.7

where y is a coordinate *fixed with respect to the rotor* (Fig. 4.7). (The coordinate y is "curved" around the air-gap periphery. In all graphs to follow we have "straightened out" the axes for easier drawing.)

   The factor β is determined from the fact that the angle βy must equal $(p/2) \cdot 2\pi$ radians for $y = \pi D$, D being the air-gap diameter shown in Fig. 4.4. Thus

$$\frac{p}{2} \cdot 2\pi = \beta\pi D. \tag{4.38}$$

This equation yields

$$\beta = \frac{p}{D}. \tag{4.39}$$

Equation (4.37) now can be written

$$B = B_{max} \cos\left(\frac{py}{D}\right) \quad \text{T.} \tag{4.40}$$

In the case of a nonsalient rotor the harmonic relationship in Eq. (4.40) is assured by proper shape and distribution of the rotor slots. (Compare comment on p. 109.)

4.3.5 The Traveling Flux Wave

The reason for designing the rotor so as to achieve a harmonic flux distribution along its periphery will be clear as we compute the induced emf in the stator winding. Preparatory to doing this we shall write the flux in terms of a coordinate x, *fixed with respect to the stator* (Fig. 4.7). The reason for doing this is the need for expressing the flux in terms of coordinates fixed with respect to the stator winding.

From Fig. 4.7 we note the relationship

$$x = y + st \qquad \text{m}, \tag{4.41}$$

where s, the tangential speed of the rotor, equals

$$s = 2\pi \cdot \frac{n}{60} \cdot \frac{D}{2} = \frac{n\pi D}{60} \qquad \text{m/s}. \tag{4.42}$$

By substitution into Eq. (4.40) we now have

$$B = B_{\max} \cos\left(\frac{px}{D} - \frac{pn\pi}{60}t\right) \qquad \text{T} \tag{4.43}$$

or, in shorter form,

$$B = B_{\max} \cos(\beta x - \omega t) \qquad \text{T} \tag{4.44}$$

where

$$\omega \equiv \frac{pn\pi}{60} = 2\pi f \qquad \text{rad/s} \tag{4.45}$$

(the last step following from Eq. (4.36)).

Equation (4.44) is the mathematical expression for a traveling wave. If we plot B versus x for two different moments t_1 and t_2, we obtain the plots, shown in Fig. 4.8, of a harmonic wave traveling from left to right with speed s m/s.

4.3.6 The Induced Stator emf

We turn our attention now to an N-turn coil of the type shown in Figs. 4.5 and 4.7. Note that the coil spans 360/p mechanical degrees in order to link the total flux emanating from one pole. The distance in meters between the coil sides is $\pi D/p$, measured along the periphery.

The differential flux penetrating the differential space window of width dx (see Fig. 4.8) is

$$d\Phi = BL\,dx \qquad \text{Wb}, \tag{4.46}$$

where L is the axial rotor length (see Fig. 4.4).

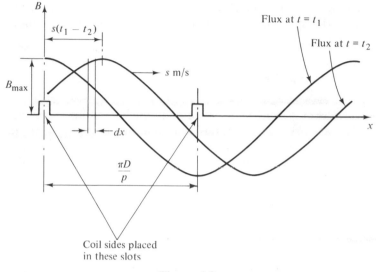

Coil sides placed
in these slots

Figure 4.8

By integration over the total coil span we thus get the total flux, Φ, passing through the coil:

$$\Phi = \int_{x=0}^{\pi D/p} BL\ dx = \int_{x=0}^{\pi D/p} LB_{max} \cos(\beta x - \omega t)\ dx \qquad \text{Wb.} \qquad (4.47)$$

This integral is readily evaluated. Integration yields

$$\Phi = 2\frac{B_{max}LD}{p}\sin \omega t \qquad \text{Wb.} \qquad (4.48)$$

As we proceed to find the total induced emf as caused by this coil flux we make the following observations:

1) The derivative $d\Phi/dt$ will give the emf in *one* turn of the coil placed in the two slots shown in Fig. 4.8.

2) In reality we have N conductors per slot (or N turns per coil).

3) Also, we have similar coils placed in similar slots located under all $(p/2)$ pole-pairs (as shown in Fig. 4.5). Not only are the emfs induced in those coils identical in magnitude but they are also of equal phase.

If e_1 represents the total emf induced in the p equidistant slots, and we assume all conductors to be series connected, as shown in Fig. 4.9, we have (according to Faraday)

$$e_1 = N \cdot \frac{p}{2} \cdot \frac{d\Phi}{dt} = \omega LDNB_{max} \cos \omega t \qquad \text{V;} \qquad (4.49)$$

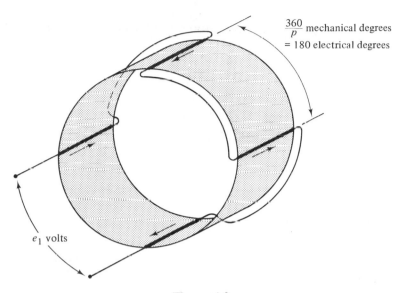

Figure 4.9

or, in view of Eq. (4.45),

$$e_1 = \frac{\pi}{60} pnNLDB_{max} \cos \omega t \quad V. \tag{4.50}$$

The rms value of the generated emf equals

$$|E_1| = \frac{\pi}{60\sqrt{2}} pnNLDB_{max} \quad V. \tag{4.51}$$

We make the following two important observations.

1) *Because the rotor was designed so as to achieve harmonic flux variation with respect to the space coordinates y or x, the coil flux and thus the emf will be harmonic with respect to time t.*†

2) *The emf reaches its peak value at the instant when the pole center is opposite to the slot in which the coil is placed.*

---

† It is important that the emf be harmonic because this means zero content of high-frequency components which would otherwise be present in the emf. Improperly designed rotors cause generation of components of the frequencies 180, 300, 420, ... Hz. These will usually cause trouble in communication networks and also add to the losses. (Compare discussion of harmonics in Appendix B and Exercise 4.15.)

Example 4.6   A 2-pole, 3600-rpm turbogenerator has the dimensions

$$L = 2 \text{ m},$$
$$D = 0.7 \text{ m}.$$

The flux wave has a peak density value of $B_{max} = 1.5$ T. Find the induced emf if $N = 4$.

SOLUTION: Equation (4.51) yields directly

$$|E_1| = \frac{\pi}{60\sqrt{2}} \cdot 2 \cdot 3600 \cdot 4 \cdot 2 \cdot 0.7 \cdot 1.5 = 2239 \text{ V}.$$

....

4.3.7   Distribution Effect

The formula (4.51) gives the rms value of the induced stator voltage assuming that the winding consists of only one coil per pole pair. Of course, in a practical design one would utilize the *total* stator surface by arranging the stator winding in many slots distributed around the periphery, as shown in Fig. 4.10. Assume that the slots are placed α *electrical* degrees apart and that there are a total of q slots per pole. What would be the total emf obtained by connecting all these coils in series?

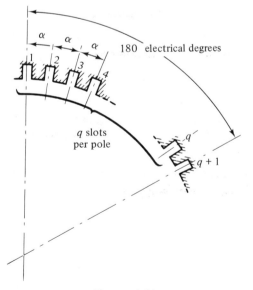

Figure 4.10

Clearly the emf induced in slot 2 will lag the emf induced in slot 1 by α degrees. The emf induced in slot 3 will lag 2α degrees behind in phase, etc. By showing the emfs as *phasors* (Appendix A) designated E_1, E_2, $\cdots$, E_q, respectively, one would thus obtain the phasor diagram in Fig. 4.11. The total stator emf, E, would be the complex sum of phasors E_1, E_2, etc.:

$$E = E_1 + E_2 + \cdots + E_q \quad \text{V.} \tag{4.52}$$

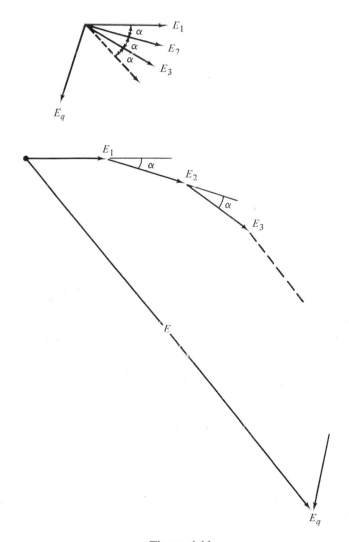

Figure 4.11

But from Fig. 4.11 we have

$$
\left.\begin{aligned}
E_2 &= E_1 e^{-j\alpha}, \\
E_3 &= E_1 e^{-j2\alpha}, \\
&\quad\vdots \\
E_q &= E_1 e^{-j(q-1)\alpha}.
\end{aligned}\right\} \tag{4.53}
$$

By substitution into Eq. (4.52) we obtain

$$
\begin{aligned}
E &= E_1 + E_1 e^{-j\alpha} + \cdots + E_1 e^{-j(q-1)\alpha} \\
&= E_1(1 + e^{-j\alpha} + \cdots + e^{-j(q-1)\alpha}).
\end{aligned} \tag{4.54}
$$

The expression within parenthesis is a geometric series, the sum of which can be written in closed form:

$$
1 + e^{-j\alpha} + \cdots + e^{-j(q-1)\alpha} = \frac{1 - e^{-jq\alpha}}{1 - e^{-j\alpha}}. \tag{4.55}
$$

We thus have for phasor E

$$
E = E_1 \frac{1 - e^{-jq\alpha}}{1 - e^{-j\alpha}}. \tag{4.56}
$$

and for the rms value $|E|$

$$
\begin{aligned}
|E| &= |E_1| \left| \frac{1 - e^{-jq\alpha}}{1 - e^{-j\alpha}} \right| = |E_1| \frac{|1 - \cos q\alpha + j \sin q\alpha|}{|1 - \cos \alpha + j \sin \alpha|} \\
&= |E_1| \frac{\sqrt{(1 - \cos q\alpha)^2 + \sin^2 q\alpha}}{\sqrt{(1 - \cos \alpha)^2 + \sin^2 \alpha}} = |E_1| \frac{\sin\left(\dfrac{q\alpha}{2}\right)}{\sin\left(\dfrac{\alpha}{2}\right)}.
\end{aligned} \tag{4.57}
$$

Example 4.7 Consider the turbogenerator discussed in Example 4.6. The stator winding is placed in 18 equidistant slots. Compute the total stator emf assuming all coils are series connected.

SOLUTION: This is a 2-pole generator and we thus have

$$
q = \frac{18}{2} = 9 \text{ slots/pole.}
$$

Because electrical and mechanical degrees are identical for $p = 2$, we have

$$
\alpha = \frac{360°}{18} = 20°.
$$

We thus have

$$\frac{\sin \dfrac{q\alpha}{2}}{\sin \dfrac{\alpha}{2}} = \frac{\sin 90°}{\sin 10°} = 5.759.$$

The emf per coil is (see Example 4.6)

$$|E_1| = 2239 \text{ V}.$$

From Eq. (4.57) we thus have for total stator emf

$$|E| = 2239 \cdot 5.759 = 12,894 \text{ V}.$$

4.4 THE THREE-PHASE GENERATOR

We have pointed out the one great drawback of single-phase ac power—its pulsating character. The power pulsations can be eliminated by designing the stator windings of the synchronous machine to form a three-phase arrangement.

Before we explain this system consider the hydraulic analog depicted in Fig. 4.12. Each single piston results in the same pulsating shaft torque as would

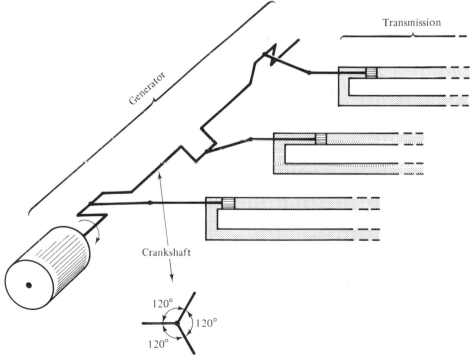

Figure 4.12

be obtained in a single-phase motor (compare also Ch. 8). It is well known, however, how a smoothing of the torque can be achieved by several parallel operating pistons—the system in Fig. 4.12 employs three.

By taking a hint from this analog, we arrive at the conclusion that an m-phase electric generator should provide m single-phase, parallel-working emfs of equal rms value but displaced $360/m$ degrees in relative time phase.

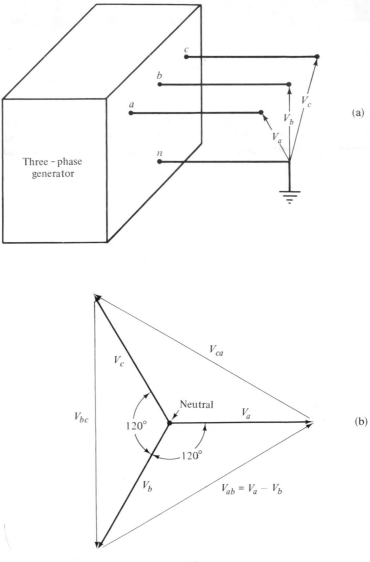

Figure 4.13

We settle for three-phase, that is, $m = 3$.† A three-phase electric generator (Fig. 4.13a) supplies, at the *phase terminals a, b,* and *c,* three *phase voltages* V_a, V_b, V_c, all of equal rms value, $|V|$, but of different phase angles.‡ V_b lags V_a by 120° and V_c lags V_b by an equal angle; we talk about tne generator having the *phase sequence abca*.... The phase voltages are all measured relative to a fourth terminal, the *neutral n*, which is usually grounded. The three voltages are shown in phasor-diagram form in Fig. 4.13(b): they are said to comprise a "symmetrical three-phase set."

If V_a is chosen as the reference voltage, we can express the three-phase voltages thus:

$$\left. \begin{array}{l} V_a = |V|, \\ V_b = |V|\, e^{-j120°}, \\ V_c = |V|\, e^{-j240°}. \end{array} \right\} \tag{4.58}$$

4.4.1 The Three-Phase Winding Design

We proceed now to show how such a set of voltages can be generated.

Consider the two-pole synchronous generator discussed in Example 4.7. We had earlier connected the total number of conductors placed in its 18 slots in series and obtained a single winding capable of supplying a single-phase emf of 12,894 V rms.

Let us now divide the 18 slots into six *phase belts* containing three slots each, as shown in Fig. 4.14(a). The three coils belonging to the phase belts designated a are interconnected to form a winding $(a - a')$ referred to as "phase a" (Fig. 4.14b). We similarly group the coils in the remaining phase belts into windings $(b - b')$ and $(c - c')$ to form "phases b and c," respectively. The three separate windings thus obtained are joined together in a *neutral node* (Fig. 4.14c) and the single winding thus finally arrived at is referred to as a "three-phase" winding.

Consider now the emf E_a generated in phase a. We determined in Example 4.6 that the emf generated in each coil equals 2239 V. The total

---

† The three-phase system has certain cost advantages over other multi-phase choices.

‡ We prefer the letter symbol E for emf and V for terminal voltage. Sometimes the two are equal, sometimes not. For example, should the generator in Example 4.7 be open-circuited (that is, unloaded) then the terminal voltage would be equal to the emf:

$$V = E.$$

If the generator is loaded, a certain stator current I would exist, and this current would cause a voltage drop IZ across the winding impedance, Z. We would now have

$$V = E - IZ;$$
$$\therefore V \neq E.$$

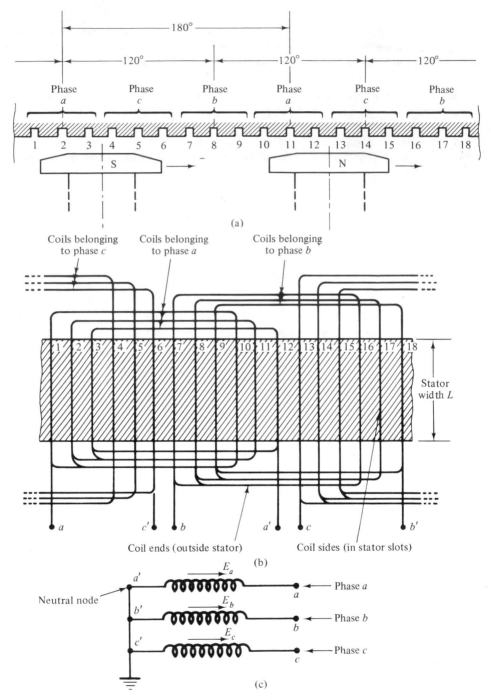

phase emf is obtained by vectorial addition of the three individual coil emfs. The electrical phase angle between the latter equals 20°.

By arbitrarily designating the emf of the center coil 2 (Fig. 4.15a) as reference phasor we obtain the phasor diagram shown in Fig. 4.15(b). The total emf, E_a, of the phase-a winding is obtained from formula (4.57). Note that q in

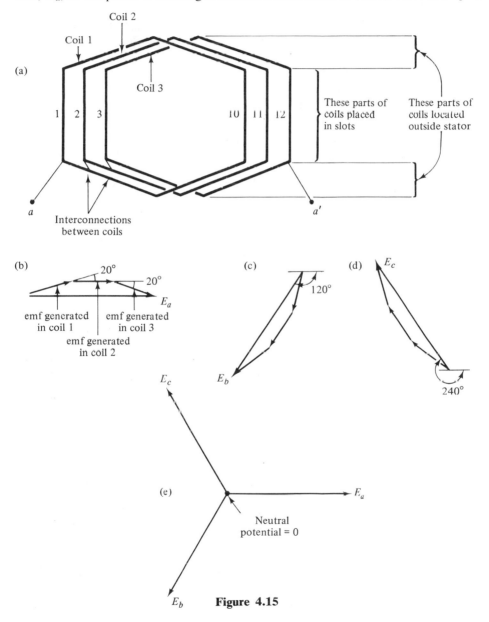

Figure 4.15

this formula, which in the single-phase case represented the number of slots per pole, now must mean number of slots per pole *and phase*.

We thus have

$$q = \frac{18}{2 \cdot 3} = 3.$$

Equation (4.57) gives

$$|E_a| = |E_1| \frac{\sin\left(\dfrac{3 \cdot 20}{2}\right)}{\sin\left(\dfrac{20}{2}\right)} = 2239 \cdot \frac{0.5000}{0.1736} = 6447 \text{ V}. \qquad (4.59)$$

Note also that the phase angle of E_a equals that of the reference emf.

We next turn our attention to the "phase-b" winding ($b - b'$) located in phase belt b. The center slot (8) of phase b is exactly 120° (6 × 20°) behind center slot (2) of phase a. Thus the phasor diagram of phase b must look exactly like the one of phase a but rotated clockwise through the angle 120° (Fig. 4-15c).

Finally, using a similar reasoning, we obtain the emf generated in phase-c winding by a 240° clockwise rotation of E_a (Fig. 4.15d).

In summary, we have obtained the following symmetrical three-phase emf set:

$$\left. \begin{array}{l} E_a = 6447, \\ E_b = 6447e^{-j120°}, \\ E_c = 6447e^{-j240°}. \end{array} \right\} \qquad (4.60)$$

The three phase windings $a - a'$, $b - b'$, and $c - c'$ are finally interconnected as shown in Fig. 4.14(c), and the neutral node thus formed is grounded. The winding we now have obtained is referred to as a *symmetrical three-phase winding*. Should this generator be operated open-circuited (unloaded) the voltages that can be measured at its three phase terminals a, b, and c are those shown in the phasor diagram in Fig. 4.15(e).

It is easy to verify that for the symmetrical three-phase emf set we have the important relation

$$E_a + E_b + E_c = 0. \qquad (4.61)$$

When the individual phase voltages have the relative time phase relationships shown in Fig. 4.15(e), one talks about the generator having the *phase sequence abca* ⋯ . Should the rotor turn in the opposite direction, the induced emfs would have the reversed phase sequence *acba* ⋯ .

4.4.2   Phase and Line Voltages

As was pointed out earlier, should we load the generator in a *balanced* manner (see Sect. 4.5) the voltages at the terminals *a*, *b*, and *c* will be different from the emf's but *they will still constitute a symmetrical three-phase set.*

The voltages V_a, V_b, and V_c (Fig. 4.13b) measured between the phase terminals and ground (or neutral) are referred to as *phase* or *phase-to-ground* voltages. The voltages that can be measured *between* terminals *a*, *b*, and *c* are referred to as *line* or *phase-to-phase* voltages. (Whenever one refers to the voltage level of a three-phase system one invariably implies the *line* voltage.) The three line voltages V_{ab}, V_{bc}, V_{ca} are obtained from the phase voltages thus:

$$\left.\begin{array}{l} V_{ab} = V_a - V_b, \\ V_{bc} = V_b - V_c, \\ V_{ca} = V_c - V_a. \end{array}\right\} \tag{4.62}$$

These voltage phasors are shown in Fig. 4.13(b).

By use of Eq. (4.58) we get

$$V_{ab} = V_a - V_b = V_a - V_a e^{-j120°} = \sqrt{3}\,V_a e^{j30°} = \sqrt{3}\,|V|\,e^{j30°} \tag{4.63}$$

Similarly we find

$$\left.\begin{array}{l} V_{bc} = \sqrt{3}\,|V|\,e^{-j90°}, \\ V_{ca} = \sqrt{3}\,|V|\,e^{-j210°}. \end{array}\right\} \tag{4.64}$$

Note that the rms value of the line voltages equals $\sqrt{3}$ times the rms value of the phase voltage. This can also easily be derived *graphically* from the phasor diagram in Fig. 4.13. The reader should confirm this.

Note also that the line voltages form a symmetrical three-phase set.

Example 4.8   Show how by disconnection of the three-phase winding developed above and by proper reconnection as a single-phase winding we obtain anew the single-phase generator of Example 4.7!

SOLUTION: We start with the three-phase winding in Fig. 4.16, having its neutral grounded. (The winding is graphically displayed in a "Y" so as to give a direct symbolic display of the phase relationship of the three emfs.)

In three easy-to-follow steps we first "dissolve" the neutral, then reconnect the three-phase windings so as to finally obtain a single-phase winding with terminals marked $a' - c'$.

The emf $E_{a'c'}$ which we can measure across these new terminals is clearly

$$E_{a'c'} = E_a - E_b - E_c \qquad \text{V.} \tag{4.65}$$

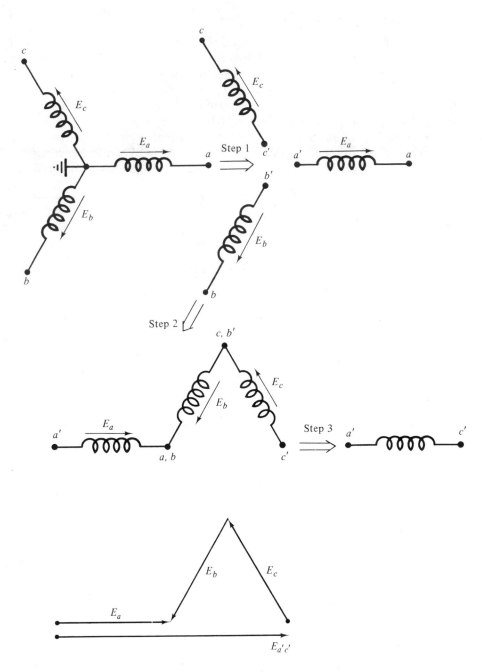

Figure 4.16

But from Eq. (4.61) we have

$$E_b + E_c = -E_a,$$

and by substitution into Eq. (4.65) we get

$$E_{a'c'} = 2E_a \quad \text{V}. \tag{4.66}$$

(An equally simple deduction can be made *graphically* in Fig. 4.16.)

The rms value of $E_{a'c'}$ is written

$$|E_{a'c'}| \doteq 2 |E_a| = 2 \cdot 6447 = 12{,}894 \text{ V}$$

which agrees with the earlier result in Example 4.7.

4.5   BALANCED THREE-PHASE LOADING

Assume that the three-phase generator in the previous section is called upon to supply a load. Loading can be achieved in any of several ways:

1) A single load element, for example an electric heater, is connected either between one phase terminal and ground, or between two phase terminals.

2) Three identical load elements are connected between the phase terminals and ground to form a "Y-connected" three-phase load. (It will be demonstrated in Chapter 8 that a symmetrical three-phase motor is equivalent with a Y-connected load.)

3) Three identical load elements are connected between the phase terminals to form a "Δ-connected" three-phase load.

4) The generator is *synchronized* onto an existing operating network and then made to share the common load, for example, like that of a city.

In the first case the generator would be loaded *unsymmetrically* or *unbalanced*. This would defeat the real purpose of three-phase power and will not be discussed further here.† The three remaining loading methods all result in a *balanced* or *symmetrical* load. We treat them separately.

4.5.1   Balanced Loading Between Phase Terminals and Ground (Y-Connected Load)

This loading case is shown in Fig. 4.17(a).

The three *phase currents* are related to the phase voltages through the

---

† Single-phase loading is very common in domestic distribution systems. However, the power company distributes the customers in a certain area equally between the three phases, so as to achieve balanced *overall* loading (see also Chapter 6).

following equations:

$$\left.\begin{array}{l} I_a = \dfrac{V_a}{Z} \quad \text{A,} \\[3mm] I_b = \dfrac{V_b}{Z} \quad \text{A,} \\[3mm] I_c = \dfrac{V_c}{Z} \quad \text{A.} \end{array}\right\} \qquad (4.67)$$

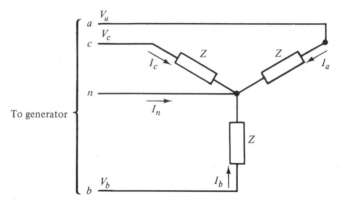

(a)

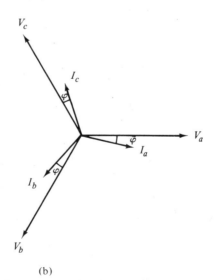

(b)

Figure 4.17

If the impedance Z is written as

$$Z = |Z| e^{j\varphi} \tag{4.68}$$

and if we make use of Eq. (4.58) the phase currents can be written

$$\left. \begin{array}{ll} I_a = \dfrac{|V|}{|Z|} e^{-j\varphi} & \text{A,} \\[2ex] I_b = \dfrac{|V|}{|Z|} e^{-j(120° + \varphi)} & \text{A,} \\[2ex] I_c = \dfrac{|V|}{|Z|} e^{-j(240° + \varphi)} & \text{A.} \end{array} \right\} \tag{4.69}$$

We have shown these current phasors in Fig. 4.17(b), together with the phase voltages.

Conclusion. The currents constitute a symmetrical three-phase set.

By vectorial addition we readily confirm that

$$I_a + I_b + I_c = 0. \tag{4.70}$$

This means that the current I_n in the neutral return lead is zero; in other words, *a return conductor is superfluous.*

Example 4.9 A 60 Hz, 220 V, three-phase generator is loaded by three equal impedances

$$Z = 1.21 + j0.31 \ \Omega$$

connected between each phase and ground. Find the phase currents!

SOLUTION: A three-phase generator is always rated in *line* voltage. For the *phase* voltages we thus have

$$V_a = \frac{220}{\sqrt{3}} - 127.0,$$

$$V_b = 127.0e^{-j120°},$$

$$V_c = 127.0e^{-j240°}.$$

As the impedance is

$$Z = 1.249e^{j14.4°},$$

we obtain from Eq. (4.69)

$$I_a = 101.7e^{-j14.4°},$$

$$I_b = 101.7e^{-j134.4°},$$

$$I_c = 101.7e^{-j254.4°}.$$

4.5.2   Balanced Loading Between Phase Terminals (Δ-Connected Load)

The three equal impedances, Z, are now connected across the three line voltages as depicted in Fig. 4.18(a).

For the three "Δ-currents" we have

$$\left.\begin{array}{l} I_{ab} = \dfrac{V_{ab}}{Z} \quad \text{A,} \\[3mm] I_{bc} = \dfrac{V_{bc}}{Z} \quad \text{A,} \\[3mm] I_{ca} = \dfrac{V_{ca}}{Z} \quad \text{A.} \end{array}\right\} \tag{4.71}$$

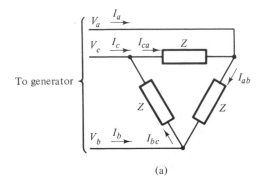

(a)

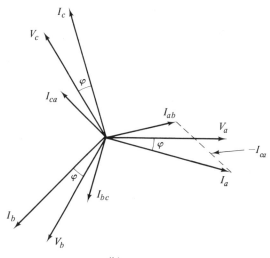

(b)

Figure 4.18

By using the earlier derived expression for the line voltages (Eqs. 4.63 and 4.64) we solve for the Δ-currents:

$$I_{ab} = \frac{\sqrt{3}\,V_a e^{j30°}}{Z} = \sqrt{3}\,\frac{|V|}{|Z|}\,e^{j(30° - \varphi)},$$

$$I_{bc} = \sqrt{3}\,\frac{|V|}{|Z|}\,e^{j(-90° - \varphi)},$$

$$I_{ca} = \sqrt{3}\,\frac{|V|}{|Z|}\,e^{j(-210° - \varphi)}.$$

(4.72)

With a knowledge of the Δ-currents we can, finally, compute the *phase currents*:

$$I_a = I_{ab} - I_{ca} = \sqrt{3}\,\frac{|V|}{|Z|}\,(e^{j(30° - \varphi)} - e^{j(-210° - \varphi)})$$

$$= 3\,\frac{|V|}{|Z|}\,e^{-j\varphi} \quad \text{A},$$

(4.73)

and similarly,

$$I_b = 3\,\frac{|V|}{|Z|}\,e^{-j(120° + \varphi)} \quad \text{A},$$

$$I_c = 3\,\frac{|V|}{|Z|}\,e^{-j(240° + \varphi)} \quad \text{A}.$$

(4.74)

We have shown the "delta" and phase currents in relation to the voltage phasors in Fig. 4.18(b).

Conclusions

1) The phase currents have the same phase relation to their respective phase voltages as in the Y-connected case, but their rms values are three times larger (assuming, of course, the same impedance load).

2) The rms value of the phase currents equals $\sqrt{3}$ times the rms value of the delta currents.

Example 4.10   Reconnect the three impedances in Example 4.9 into a Δ-load. Find all currents!

SOLUTION: The phase currents will have an rms value of $3 \cdot 101.7 = 305.1$ A. The rms value of the Δ-currents will be $305.1/\sqrt{3} = 176.2$ A.

4.5.3   The Generator Operating as Part of a Power Grid

This is the most important loading case. Most generators are operating in this mode. We shall treat this case separately in Sect. 4.8. Here it will suffice to say that even in this case the phase currents will constitute a symmetrical three-phase set.

4.6   THREE-PHASE POWER FORMULAS

In every case of balanced three-phase loading, we have concluded that the phase currents constitute a symmetrical three-phase set. The *phase* voltages and *phase* currents can thus *always* be represented by the type of phasor diagram shown in Fig. 4.17.

   If $|V|$ and $|I|$ represent the rms values of phase voltages and currents, respectively, we can thus write for these variables in the three phases:

$$
\left.
\begin{aligned}
\text{Phase } a: \quad & v_a = \sqrt{2}|V| \sin \omega t && \text{V,} \\
& i_a = \sqrt{2}|I| \sin (\omega t - \varphi) && \text{A;} \\
\text{Phase } b: \quad & v_b = \sqrt{2}|V| \sin (\omega t - 120°) && \text{V,} \\
& i_b = \sqrt{2}|I| \sin (\omega t - 120° - \varphi) && \text{A;} \\
\text{Phase } c: \quad & v_c = \sqrt{2}|V| \sin (\omega t - 240°) && \text{V,} \\
& i_c = \sqrt{2}|I| \sin (\omega t - 240° - \varphi) && \text{A.}
\end{aligned}
\right\}
\tag{4.75}
$$

The *instantaneous* three-phase power $p_{3\Phi}$ is the sum of the phase powers

$$
p_{3\Phi} = v_a i_a + v_b i_b + v_c i_c \quad \text{W.} \tag{4.76}
$$

   Perform now the following three-step analysis!

Step 1. Substitute the expressions in Eq. (4.75) into Eq. (4.76).

Step 2. Use the trigonometric identity in Eq. (4.8).

Step 3. Use the additional trigonometric identity.

$$
\cos \alpha + \cos (\alpha - 120°) + \cos (\alpha - 240°) = 0. \tag{4.77}
$$

   This analysis will reveal the very important result

$$
p_{3\Phi} = 3|V||I| \cos \varphi = 3P \quad \text{W.} \tag{4.78}
$$

In words:

In a balanced three-phase system the sum of the three individually pulsating phase powers adds up to a constant, nonpulsating total power of magnitude three times the real power in each phase.

Example 4.11   The terminal voltage of a three-phase generator measured phase-to-phase equals 13.2 kV. It is symmetrically loaded and delivers an rms current of 1.230 kA per phase at a phase angle of $\varphi = 18.3°$ lagging (meaning current lagging voltage as in Fig. 4.17). Compute the power delivered by the machine.

SOLUTION: We first compute the rms value, $|V|$, of the phase voltage.

$$|V| = \frac{13.2}{\sqrt{3}} = 7.621 \text{ kV/phase.}$$

Formula (4.13) then yields

$$P = 7.621 \cdot 1.230 \cdot \cos 18.3° = 8.900 \text{ MW/phase,}$$

$$Q = 7.621 \cdot 1.230 \cdot \sin 18.3° = 2.943 \text{ MVAr/phase.}$$

The powers in the phases a, b, and c will thus pulsate according to formula (4.14):

$$\left. \begin{array}{l} p_a = 8.900(1 - \cos 2\omega t) - 2.943 \sin 2\omega t, \\ p_b = 8.900(1 - \cos 2(\omega t - 120)) - 2.943 \sin 2(\omega t - 120°), \\ p_c = 8.900(1 - \cos 2(\omega t - 240°)) - 2.943 \sin 2(\omega t - 240°). \end{array} \right\} \quad (4.79)$$

The total three-phase power will be

$$p_{3\Phi} = 3 \cdot 8.900 = 26.700 \text{ MW.}$$

These various powers are plotted in Fig. 4.19.

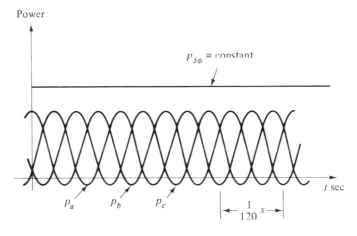

Figure 4.19

Important Note: The fact that the three-phase power is constant tempts us to believe that the *reactive* power in a three-phase system is zero (as in a dc circuit).

However, the reactive power is very much present *in each phase* as evidenced by formulas (4.79). The reactive power *per phase* is 2.943 MVAr. In power engineering lingo one would say in this case that the reactive power produced by the generator is 8.829 $(3 \cdot 2.943)$ *MVAr three-phase*. (This is natural to do because the *real* three-phase power was found to be three times the per-phase value.)

Example 4.12 Find the power delivered to the loads in Examples 4.9 and 4.10!

SOLUTION: In Example 4.9 we had:

$$|V| = 127.0\,\text{V},$$
$$|I| = 101.7\,\text{A},$$
$$\cos\varphi = 0.969;$$
$$\therefore \quad p_{3\Phi} = 3 \cdot 127.0 \cdot 101.7 \cdot 0.969 = 37.5\,\text{kW}.$$

In the Δ-connected case the current tripled. Thus the power will also triple, that is,

$$p_{3\Phi} = 3 \cdot 37.5 = 112.5\,\text{kW}.$$

4.7 TORQUE MECHANISM IN A THREE-PHASE GENERATOR

A loaded three-phase synchronous generator via its three phase terminals delivers a constant electric power $p_{3\Phi}$. This power originates in the prime mover that drives the generator. Let us now consider the important mechanism whereby this mechanical power is transformed into electric power. The mechanical power p_{mech} delivered by the prime mover must equal the electric power $p_{3\Phi}$, plus the losses p_{loss} sustained in the process of transformation. These losses are, in general, relatively small and we can thus write

$$p_{\text{mech}} = p_{3\Phi} + p_{\text{loss}} \approx p_{3\Phi} \quad \text{W.} \tag{4.80}$$

The mechanical power, according to formula (2.24), is associated with a mechanical torque, delivered by the prime mover:

$$T_{\text{mech}} = \frac{p_{\text{mech}}}{\omega_{\text{mech}}} \quad \text{N} \cdot \text{m,} \tag{4.81}$$

where ω_{mech} is the angular synchronous velocity of the rotor measured in rad/s. Both the velocity and mechanical power are constant and, consequently, so is the torque. The direction of the torque is such as to maintain the speed, that is, T_{mech} *tends to accelerate the rotor in the ω-direction.*

The fact that the rotor *does not accelerate* is related to the existence of a counteracting torque T_{em} of equal magnitude but opposite direction. This *electromechanical torque* is created by the interaction between the rotor-bound flux and the stator-bound current. The presence of T_{em} acting on the rotor necessitates a *reaction torque* of equal magnitude but opposite direction acting on the stator. *This reaction torque evidently tends to tilt the stator in the direction of rotation.*

We have summarized the torque and power picture in Fig. 4.20. There is, of course, also a reaction torque acting on the static portion of the prime mover. This torque equals in magnitude T_{mech} but is of opposite direction, that is, it tends to tilt the prime mover housing in the opposite direction of ω_{mech}.

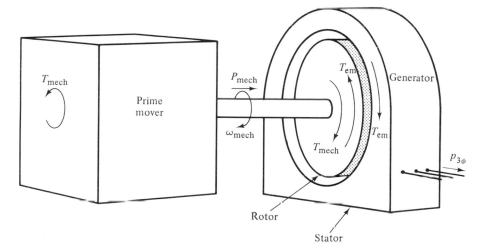

Figure 4.20

Example 4.13 A four-pole synchronous generator delivers an electric power of 1000 MW. Determine the magnitude of T_{mech} and T_{em}.

SOLUTION: The speed of a four-pole generator (assuming 60 Hz) is 1800 rpm. Thus

$$\omega_{\text{mech}} = \frac{1800}{60} \cdot 2\pi = 188.5 \text{ rad/s}.$$

Equation (4.81) then yields

$$T_{em} \approx T_{mech} = \frac{1000 \cdot 10^6}{188.5} = 5.31 \cdot 10^6 \, \text{N} \cdot \text{m},$$

or 541 tonmeters.

4.7.1 The Stator "Current Wave"

The electromechanical torque T_{em} acting across the air gap of the generator evidently is the key factor in the energy transformation process. How is it formed? What characteristics does it have? *All* rotating electric energy converters, motors and generators alike, differ in essential ways only as their electromechanical torque features differ. In our study of these devices we will, therefore, be most interested in these torques. Let us presently seek to explain the torque features of a symmetrically loaded three-phase synchronous machine.

The torque, as already mentioned, arises from the interaction between the rotor flux and the stator current. We know the former—let us learn more about the latter. Preparatory to doing so consider the analog in Fig. 4.21. A template is cut with a sinusoidal contour. It is moved with constant velocity across a piano keyboard where each key will be performing a vertical motion, harmonic with respect to time. Each key has a different time phase, and it is immediately clear that the total result is a key motion which in the combined x and t space represents a traveling wave.

As we learned in previous sections, the currents in each of the three phase belts of the stator winding pulsate sinusoidally but in different time phase. Let

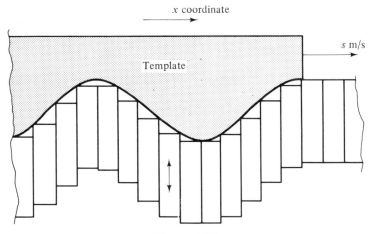

Figure 4.21

us prove how they combine into a "current wave" which travels, with constant amplitude, around the periphery of the stator having a speed identical to that of the rotor.

In Fig. 4.22, we have redrawn the phase belts as they were shown in Fig. 4.14. Note that each of the six phase belts spans 60° of the stator surface. We have also shown the stator currents at two different instants of time. In Fig. 4.22(a) we have plotted the currents at the moment when I_a reaches its peak value (see the inset phasor diagram). At this same instant, the currents I_b and I_c are negative and at 50 percent of their respective peak values.

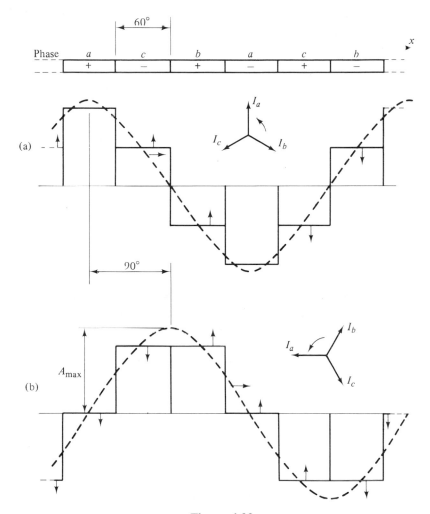

Figure 4.22

Fig. 4.22(b) depicts the stator current one-quarter cycle (or 90°) later. I_a is now zero, I_b positive and at 87 percent of its peak value, and I_c negative and at 87 percent of peak.

Of course, the stator currents exist only in the stator slots. However, if we view the stator surface in a macrosense by "smearing out" the currents over the total surface then the stator current can be considered a *surface* or *sheet* current. Fig. 4.22 shows this surface current, consisting of the six phase segments *each of which performs a pulsation in analogy with the individual piano keys in Fig. 4.21.* The arrows shown with each "current key" indicate the direction of movement.

As we view the *composite* stator current picture we obviously see a surface current wave (shown dashed) traveling to the right. Note that the wave has traveled a distance of one and one-half phase belts, which corresponds to 90° in the synchronous machine, in the time interval elapsed between the two current pictures. We remember that the two current snapshots were taken 90 electrical degrees apart and we thus conclude *that the current wave travels with the same speed as the rotor flux wave.*

Note also that the crest of the current wave is positioned at the center of a phase belt at the instant when the current in that phase is peaking. The *total* current wave (like the piano keyboard wave) is *not* sinusoidal because of the finite width of the "current keys." The dashed wave in Fig. 4.22 represents only the fundamental (but important) component. Like the fundamental ocean-wave that is accompanied by a large number of ripple components of various wavelengths and frequencies, our fundamental current wave is likewise immersed in a multitude of harmonics. (See also Appendix B.)

4.7.2 Torque and Power Formulas

Figure 4.23 summarizes the wave system existing in a loaded synchronous machine. (All high-frequency harmonics have been left out.) We have first the magnetic flux wave *bound to the rotor* and traveling with constant velocity and constant amplitude B_{max}. It is described by formula (4.44) and shown in Fig. 4.23(a).

As the flux wave sweeps by the stator conductors it induces an emf wave which follows the B wave. The crest of the emf wave coincides with that of the B wave (Fig. 4.23b).

Finally, there is the current wave *bound to the stator* (Fig. 4.23c). We shall assume here, for generality, that the current in phase a lags the emf in phase a by the angle γ. This means, of course, that the current wave trails the emf (or flux) wave by γ radians. Because the current wave has the same speed as the flux wave and equal number of maxima, we can, in analogy with Eq. (4.44), express it in the form

$$A = A_{max} \cos\left(\frac{px}{D} - \omega t - \gamma\right) \qquad \text{A/m.} \qquad (4.82)$$

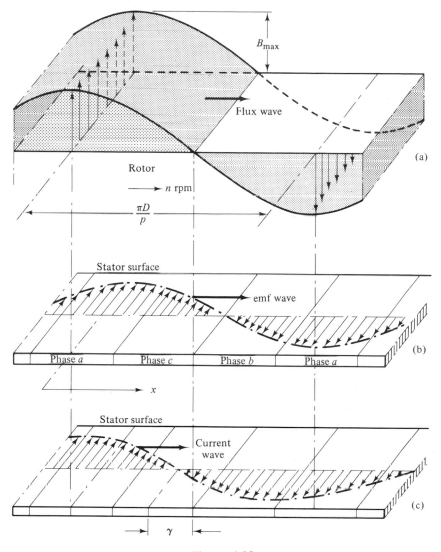

Figure 4.23

(Note that the physical dimension of A is amperes per (tangential) meter, that is, it is a current surface *density*.)

If we focus our attention on a thin element or strip of the stator surface (Fig. 4.24) of axial length L and tangential width dx, it is obvious that the geometry of flux and strip current is such as to favor a *tangential force, df*.

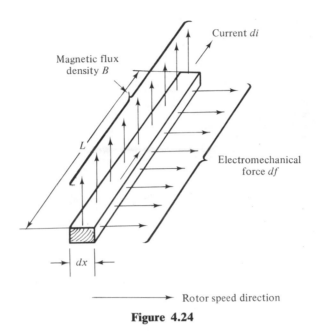

Figure 4.24

According to formula (3.68) the force on the surface strip will be

$$df = LB \, di = LBA \, dx$$

$$= LB_{max} A_{max} \cos\left(\frac{px}{D} - \omega t\right) \cos\left(\frac{px}{D} - \omega t - \gamma\right) dx \qquad \text{N.} \qquad (4.83)$$

Integration from $x = 0$ to $x = \pi D$ yields the total force on stator:

$$f = LB_{max}A_{max} \int_{x=0}^{\pi D} \cos\left(\frac{px}{D} - \omega t\right) \cos\left(\frac{px}{D} - \omega t - \gamma\right) dx$$

Upon performing this integration we obtain

$$f = \tfrac{1}{2}\pi LDB_{max}A_{max} \cos\gamma \qquad \text{N.} \qquad (4.84)$$

By multiplication by $D/2$ we obtain the electromechanical *torque*:

$$T_{em} = \tfrac{1}{4}\pi LD^2 B_{max}A_{max} \cos\gamma \qquad \text{N} \cdot \text{m.} \qquad (4.85)$$

Further multiplication by $\omega_{mech} = (n\pi)/30$ yields the corresponding electrical power:

$$p_{3\Phi} = \tfrac{1}{120}\pi^2 nLD^2 B_{max}A_{max} \cos\gamma \qquad \text{W.} \qquad (4.86)$$

By introducing the rotor volume

$$V_{rot} = \frac{\pi}{4}LD^2 \qquad \text{m}^3,$$

we can write the torque and power formulas more simply as

$$T_{em} = V_{rot}B_{max}A_{max} \cos \gamma \qquad \text{N} \cdot \text{m}, \qquad (4.87)$$

and

$$p_{3\Phi} = \frac{\pi}{30} nV_{rot}B_{max}A_{max} \cos \gamma \qquad \text{W}. \qquad (4.88)$$

4.7.3 Some Practical Observations

We can draw some important conclusions from the formulas just derived.

- The electromechanical force, f and thus the torque, T_{em}, acting on the stator have a directionality which agrees with that depicted in Fig. 4.20.

- As the angle γ is constant both the torque and power must be constant as well. This, or course, reinforces our earlier findings regarding three-phase powers.

- The power and torque increase in direct proportion to volume or machine size. (This is quite obvious.)

- The power also increases in direct proportion to speed. (This means for example that, everything else being equal, one can obtain ten times more "kW per kilogram" from a 3600 rpm steam turbine than from a slow-running 360 rpm hydro generator. (This is true for *all* types of *motors* as well.)

- Power and torque increase in direct proportion to magnetic flux densities used. Magnetic saturation sets a limit here. If we wish to "squeeze" more flux out of the rotor then we must use a disproportinate rotor field current and this causes heating problems in the rotor winding, as well as extra mechanical stresses.

- Power and torque increase in direct proportion to current density in the stator surface. Ohmic losses and resulting winding-temperature elevation set practical limits. In the last two decades we have seen dramatic increases in current densities tolerated: new and better insulating materials withstand higher temperatures, but most importantly, we have improved the methods of *forced cooling*. Typically, we lead high-velocity water or hydrogen through hollow stator conductors thus removing the ohmic heat and allowing higher current densities. By *supercooling* of the windings, the resistance and thus the ohmic losses disappear altogether. One should then be able to use limitless current densities. This technology is still in the experimental stage (compare Chapter 9).

4.7.4 Power Dependence on Trailing Angle γ

An interesting aspect of the formulas (4.87) and (4.88) is the dependence of the generated power on the angle γ. As this angle is permitted to vary

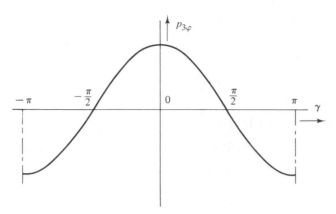

Figure 4.25

throughout the range

$$-\pi < \gamma < +\pi,$$

the power (or torque) will vary as shown by the graph in Fig. 4.25. In the range

$$-\frac{\pi}{2} < \gamma < +\frac{\pi}{2},$$

the force on the stator has the direction shown in Fig. 4.24, the reaction force on the rotor being of opposite direction. This agrees with the torque polarities shown in Fig. 4.20, and the machine is now operating as a *generator*. When γ exceeds $\pi/2$ (or is less than $-\pi/2$) the force and torque shift polarity and the machine is now *motoring*. For $\gamma = \pi/2$ or $-\pi/2$ the torque and power equal zero *although the stator windings may carry full currents*. The machine is now operating as a *synchronous condenser* (compare Example 4.14).

The way the synchronous machine is being operated, as a generator or motor with large or small power output, depends clearly upon the actual value of the angle γ. The relative position between the flux and current waves thus becomes *the* most important factor in controlling the power flow from or to the machine. How this in fact can be accomplished will be our next topic.

4.8   THE SYNCHRONOUS MACHINE AS PART OF A POWER GRID

Essentially 100 percent of all synchronous machines operate in parallel as part of large, often continent-spanning, *power grids*. Such grids operate 24 hours a day. Their common frequency is kept very constant (60.00 ± 0.05 Hz in the United States) and any individual generator, even a 1000-MW unit, is usually very small compared with the combined power capacity of the total system.

Before an individual generator can be connected to such a grid (Fig. 4.26), it must first be *synchronized* onto it. Synchronization means that our machine

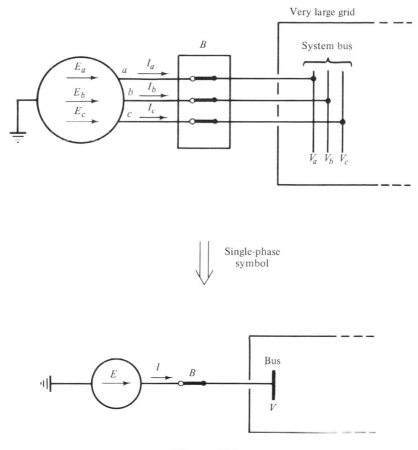

Figure 4.26

must be put into such an operating condition that we can "smoothly lock onto" the running system by closing the generator breaker *B*. Following the closing of this breaker, the generator terminals *a*, *b*, *c* are directly connected to the system terminals. The voltages of the latter are V_a, V_b, and V_c, measured between the respective phasor and ground.

4.8.1 A Mechanical Analog

Consider the huge mechanical system shown in Fig. 4.27. It consists of many parallel working drives that deliver power via gear trains to a multitude of mechanical motors. The system is running on a 24-hour basis and at a constant speed. We wish to connect one additional drive onto the system by meshing its

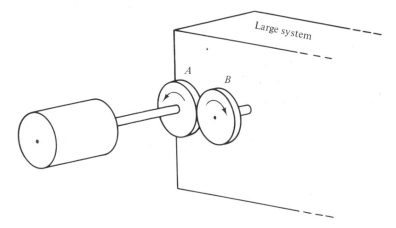

Figure 4.27

gear *A* with the system gear *B*. Clearly, three conditions have to be satisfied before we can do so:

1) The gears must have matching speeds.
2) The gears must have matching teeth.
3) The gears must have equal phase (they cannot be meshed if running tooth-to-tooth).

With a little imagination, we can correlate these conditions with those that must be satisfied to permit synchronization of the electric machine in Fig. 4.26. They are:

1) The machine frequency must match the grid frequency and the phase sequence of the machine must match that of the network.
2) The generator emf *E* must be regulated (by adjusting the rotor field current) so as to match the system bus voltage *V*.
3) The machine emf must have the same phase as the system voltage.

When these conditions are all satisfied the emf and voltage phasors (Fig. 4.28) are running "neck-to-neck" and zero voltage exists across the breaker terminals: the breaker can be closed and synchronization is smoothly accomplished.

4.8.2 Synchronous Machine Controls

Just how the machine behaves versus the network, after synchronization, depends entirely upon how we *control* it. Consider, for comparison, the mechanical analog in Fig. 4.27. If the torque of the drive is controlled *so that it*

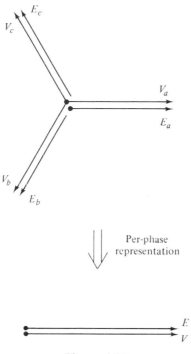

Per-phase representation

Figure 4.28

tends to accelerate the system, the drive will supply power to the system; it will act as a *generator.*

If the drive torque is controlled so that it tends to "load" the system, that is, *tends to decelerate it,* then our "drive" becomes a *motor,* taking power from the system. If we view the gears *A* and *B* via a *stroboscope,* we would see a picture similar to that shown in Fig. 4.29. In the *zero torque* case the gears, *which are assumed elastic,* would retain their shape. In the *generator case* (shown in the figure) the drive gear *A* will run a certain angle *ahead* of system gear *B.* In the motor case the opposite holds true. The magnitude of the angle will increase with the torque in both cases.

In the above mechanical analog we had only one means of control—the torque. In the case of a synchronous machine we have *two* possible control inputs or *control forces:* (1) the prime mover torque, and (2) the field current.

Let us look at the effects of these two controls separately. In doing so we must remember *that the system voltage V will retain essentially a constant magnitude and phase.* This voltage is determined by the combined strength of the many parallel working generators of the network. One added generator usually will not change the overall picture much and the result is a nearly

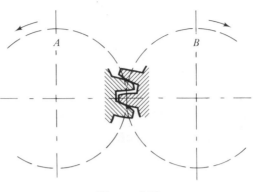

Figure 4.29

constant V. (In power engineer's lingo one says that a system terminal is "infinite" if its voltage is *totally* unaffected by *any* changes in the current injected to or drained from it.)

A. Effects of prime mover torque control.   As we increase the prime mover torque we tend to accelerate the generator away from the network. However, it is locked to the network and, rather than pulling away, the rotor *advances* a certain angle. As the emf wave follows the rotor (Fig. 4.23) the effect will be an *advancement* in the phase of E. This is shown in Fig. 4.30(a). The *power angle* δ is a measure of the real power $p_{3\phi}$ delivered by the machine. It is positive when E leads V.

Were we to apply a *negative* torque by letting our machine pull some mechanical load, thus acting as a *motor*, then E would lag V and the power angle would become negative. The network is now pulling our machine.

It should not be difficult to note the analogy between Figs. 4.29 and 4.30(a).

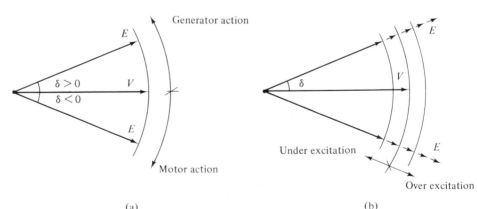

Figure 4.30

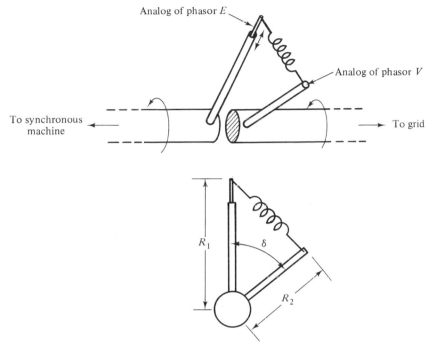

Figure 4.31

B. Effects of field-current control.   As we control the field current no changes can take place in the real power or torque delivered by the prime mover. Thus no change can take place in the real power delivered by the generator, and no *direct* change of the power angle δ can occur. (An *indirect* change will occur which will be explained below.)

The field current affects the magnitude of B_{max} and thus the magnitude of F as shown in Fig. 4.30(b). An increase in field current results in an increase in $|E|$, thus making $|E| > |V|$. We refer to this case as "overexcitation." A decrease in field current will decrease $|E|$, thus making $|E| < |V|$. This is underexcitation.

In summary: A change in *torque* moves the tip of E tangentially along concentric circles. A change in *field current* moves the tip of E radially between these circles. Figure 4.31 depicts a mechanical coupling that fairly well analogs the effect on the power angle of *both* torque and field-current control. As the shaft torque is increased, the angle between the two rods is increasing. An increase in length of the telescoping rod represents an increase in $|E|$, that is, a field current increase. Note that if the rod length increases as the shaft torque is kept constant, the angle will *decrease*. This is the indirect effect upon δ referred to above.

4.8.3 Phasor Diagram

As the machine emf E changes both in phase and magnitude but the grid voltage V stays essentially constant, we obviously obtain a difference voltage:

$$\Delta V = E - V \qquad \text{V/phase.} \tag{4.89}$$

This will give rise to a current I (defined positive *away* from the machine, that is, in generator sense) of value

$$I = \frac{\Delta V}{Z_s} \qquad \text{A/phase,} \tag{4.90}$$

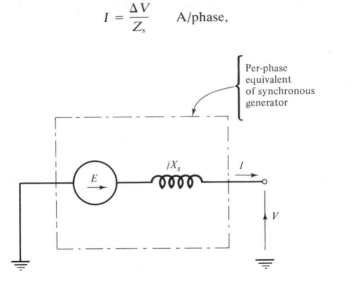

(a)

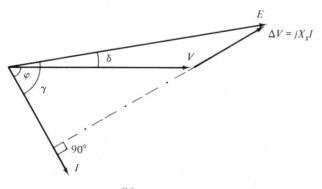

(b)

Figure 4.32

where Z_s represents the "synchronous impedance" of the stator winding.† Z_s has a real and a reactive part and can be written

$$Z_s = R_s + j\omega L_s \qquad \Omega/\text{phase}. \qquad (4.91)$$

R_s is the resistance and L_s is the inductance, both measured per phase. The inductance which is a measure of the coupled flux per amp is relatively large. This is due to the fact that the magnetic path, except for the small air gap, is iron. For a typical machine we therefore have

$$\omega L_s \gg R_s, \qquad (4.92)$$

and with good approximation we can thus set

$$Z_s \approx j\omega L_s \equiv jX_s \qquad \Omega/\text{phase}, \qquad (4.93)$$

X_s being referred to as the *synchronous reactance*. The generator may thus be viewed as an emf E "behind" the synchronous reactance X_s, which corresponds to the "equivalent circuit" shown in Fig. 4.32(a).

As the three phases are completely symmetrical, the voltage differences, ΔV, will constitute a symmetrical three-phase set. Consequently, the same will apply to the currents.

We are now able to draw the phasor diagram of Fig. 4.32(b). Note that, in view of Eq. (4.93) the current will lag ΔV by 90°. Note also that we have identified the angles γ and φ earlier defined in Figs. 4.23 and 4.17, respectively. In view of the three-phase symmetry we need to show one phase only.

Let us use the phasor diagram to work the following example.

Example 4.14   Following synchronization a generator is subject to field current control. Consider these two cases:

CASE A. The field current is raised so that $|E|$ is increased by twenty percent (overexcitation).

CASE B. The field current is lowered so that $|E|$ is decreased by twenty percent (underexcitation).

Give phasor diagrams and compute currents and powers, assuming the following machine data:

$$\text{rated terminal voltage} = 15.0\,\text{kV},$$

$$\text{synchronous reactance} = 11.0\,\Omega/\text{phase}.$$

We also assume zero prime mover torque.

---

† We have neglected the impedance between the generator terminals and the system bus—V therefore equals the terminal voltage of our machine.

SOLUTION: We construct the two phasor diagrams in Fig. 4.33. Note that because the torque is zero, $\delta = 0$ in both cases. Note also that $\gamma = \varphi = \pm 90°$, confirming zero real power both from formulas (4.78) and (4.88).

We first compute the *phase* voltage:

$$|V| = \frac{15.0}{\sqrt{3}} = 8.660 \text{ kV/phase.}$$

For *both* cases we have for the current

$$|I| = \frac{|\Delta V|}{X_s} = \frac{0.2 \cdot 8.660}{11.0} = 0.1575 \text{ kA/phase.}$$

Real powers are zero in both cases but for the *reactive* powers we have (compare Eq. 4.13):

CASE A $\varphi = +90°,$

$Q = 8.660 \cdot 0.1575 \cdot \sin(+90°) = 1.364 \text{ MVAr/phase,}$

(or 4.092 MVAr three-phase).

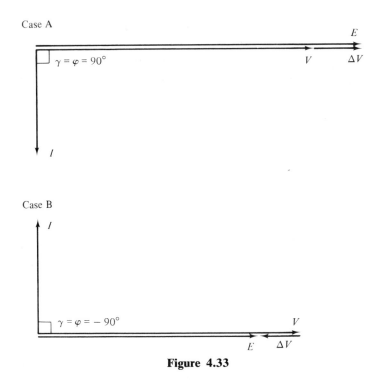

Figure 4.33

CASE B $\varphi = -90°$,

$$Q = 8.660 \cdot 0.1575 \cdot \sin{(-90°)} = -1.364 \text{ MVAr/phase,}$$

(or -4.092 MVAr three-phase).

Summary

The *overexcited* machine *delivers* 4.09 MVAr *to the network. From the network point of view it acts like a capacitive load.*

(Operated in this manner the synchronous machine is referred to as a "synchronous condenser.")

The *underexcited* machine *draws* 4.09 MVAr *from the network. From the network point of view it acts like an inductive load.*

The example thus teaches us that by field-current control we can make the synchronous machine *either generate or consume reactive power.* This feature finds important uses in power systems operation, as will be demonstrated in Chapter 6.

4.8.4 Practical Power Formulas

An advantage of formulas (4.87) and (4.88) is that they explain the formation of torque and power of the machine in terms of *internal physical variables.*

From a practical operational viewpoint it would be better to express the power and torque in terms of *externally measurable* variables. We can readily obtain such formulas from our phasor diagram in Fig. 4.32.

By projecting E upon the reference vector V (and noting that ΔV leads V by the angle $90° - \varphi$) we get

$$|E| \cos \delta = |V| + X_s |I| \sin \varphi \qquad \text{V/phase.} \qquad (4.94)$$

By projecting E orthogonally to V we similarly obtain

$$|E| \sin \delta = X_s |I| \cos \varphi \qquad \text{V/phase.} \qquad (4.95)$$

We solve from these equations:

$$\left. \begin{aligned} |I| \cos \varphi &= \frac{|E|}{X_s} \sin \delta & \text{A/phase,} \\[2mm] |I| \sin \varphi &= \frac{|E| \cos \delta - |V|}{X_s} & \text{A/phase.} \end{aligned} \right\} \qquad (4.96)$$

From Eq. (4.13) we then have our "practical" power formulas:

$$P = \frac{|E||V|}{X_s} \sin \delta \qquad \text{W/phase}, \qquad (4.97)$$

$$Q = \frac{|V||E|\cos \delta - |V|^2}{X_s} \qquad \text{VAr/phase}. \qquad (4.98)$$

Example 4.15 Consider the generator in the previous example. It is over-excited so that $|E|$ is 20% in excess of $|V|$. The prime mover torque is set at a value such that the machine delivers 12 MW (three-phase) to the network. Find δ, Q, and $|I|$!

SOLUTION: As $|E| = 1.20 \cdot |V| = 10.39$ kV/phase, we obtain from Eq. (4.97)

$$P = \frac{p_{3\varphi}}{3} = \frac{12}{3} = 4 = \frac{10.39 \cdot 8.660}{11.0} \sin \delta;$$

$$\therefore \quad \delta = 29.3°.$$

From Eq. (4.98) we then have

$$Q = \frac{8.660 \cdot 10.39 \cos 29.3° - 8.660^2}{11.0} = 0.32 \text{ MVAr/phase},$$

(or 0.96 MVAr three-phase), delivered *to* the network.

For the current we have from formula (4.31)

$$|I| = \frac{\sqrt{4.00^2 + 0.32^2}}{8.660} = 0.463 \text{ kA/phase}.$$

....

4.8.5 Pullout Power

Consider the mechanical analog in Fig. 4.29. As we increase the torque the mechanical power angle between gears A and B will slowly increase. Eventually a point is reached, assuming that the gear material is fully elastic, when we will "skip a tooth." Or, for a still better analog, contemplate what will happen in the transmission shown in Fig. 4.31 when the torque is increased. It can be shown (compare Exercise 4.16) that the transmitted shaft torque is

$$T_{\text{mech}} = kR_1 R_2 \sin \delta \qquad \text{N} \cdot \text{m} \qquad (4.99)$$

where R_1 and R_2 represent the lengths of the two rods with δ being the angle between them. Clearly, as the torque is being increased the angle δ grows. When the angle reaches a value of 90° the transmission will collapse.

A synchronous machine behaves very similarly. If the torque is slowly increased, the power angle δ grows and eventually the rotor will "skip poles." Let us see when this will happen, by considering the power formula (4.97). If the field current is kept constant, then $|E|$ will be fixed. We have earlier pointed out that $|V|$ is essentially a constant. Under these assumptions P will reach a maximum, when δ equals 90°. A negative maximum occurs for δ = −90°. In Fig. 4.34 we have plotted P versus δ.

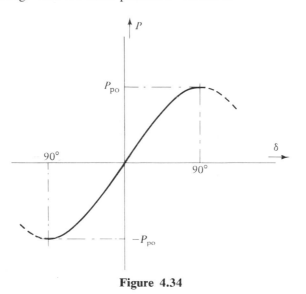

Figure 4.34

If we increase the torque further, an increase in δ will result in a decrease in electric power. We lose synchronism at this point. The maximum power, the "pullout" power, equals

$$P_{po} = \frac{|E||V|}{X_s} \quad \text{MW/phase.} \tag{4.100}$$

"Losing synchronism" implies that the rotor-bound B-wave and the stator-bound A-wave lose the "grip" between them. The rotor will now "slip" or "skip poles" and will continue to do so until the excessive torque is removed. Excessive current will flow in the stator at the moments when the E and V phasors are in opposite phase. Dangerous heating may result.

Example 4.16 Consider the synchronous machine of Example 4.15. Let us assume that it is excited so that $|E| = |V|$. Find real and reactive power and current at the point of pullout! Give a phasor diagram!

SOLUTION: As the torque is increased, the tip of E follows the circle in Fig. 4.35. Pullout occurs when $\delta = 90°$. The current phasor is at this point advancing V by 45°, that is, $\varphi = -45°$.

We have

$$P_{po} = \frac{8.660 \cdot 8.660}{11.0} = 6.818 \text{ MW/phase},$$

or 20.5 MW total three-phase.

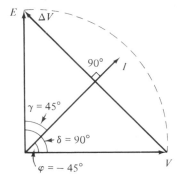

Figure 4.35

The current $|I|$ is obtained most easily from the phasor diagram. We note that $X_s \cdot |I| = |\Delta V| = \sqrt{2}|V|$. Therefore,

$$|I| = \frac{\sqrt{2} \cdot 8.660}{11.0} = 1.11 \text{ kA/phase}.$$

Because $\varphi = -45°$,

$$\sin \varphi = -\cos \varphi = -0.707.$$

Thus, $Q = -P = -20.5 \text{ MVAr}$ (three-phase).

This means that at the point of pullout the generator *absorbs* 20.5 MVAr (three-phase) from the grid. Note that at the point of pullout the angle γ equals 45°. (The reader should explain to his own satisfaction why the pullout power *does not* correspond to the maximum power in Fig. 4.25.)

···

4.9 SUMMARY AND SOME FINAL OBSERVATIONS

We have tried in this chapter to present a fairly complicated machine—the synchronous three-phase generator—in terms as simple as possible. In order not to break up the continuity of our story we have on occasion taken "shortcuts," avoided unnecessary complications, or skipped important practical considerations altogether.

In summary here are our most important findings:

1) The three-phase synchronous generator is the electric power engineer's "workhorse." Worldwide, well in excess of 99% of all bulk electric energy is generated by this type of machine.

2) Although single-phase ac power pulsates at twice the system frequency, balanced three-phase ac power is constant.

3) The link between the mechanical power obtained from the prime mover and the electrical power delivered from the phase terminals of the generator is the electromechanical air-gap torque. This torque arises from an interaction between the rotor flux and the stator current.

Our limited scope has not permitted us to cover some important but not (for our story) essential topics. We summarize here in briefest of terms some points that need to be made in order to complete our basic presentation of the synchronous machine:

■ Every part of the stator (but *not* the rotor) experiences an ac magnetic flux. Faraday's law informs us that emf's are generated around the flux path (Fig. 4.36) and, if solid iron cores were used, ac "eddy currents" would flow perpendicular to the flux. This would cause intolerable heating effects and considerable losses. *Laminated* cores are used to break up the current paths and reduce these eddy current losses. Adding silicon to the core material gives an alloy with high resistivity, thus further reducing these losses. The ac flux also causes socalled *hysteresis losses* in the iron. These losses are due to the reorientation of the magnetic moments that must take place 60 times per second. (One may think of an "internal magnetic friction" that must be overcome to turn the moments around.)

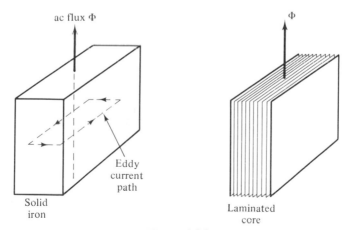

Figure 4.36

Eddy current plus hysteresis losses are collectively referred to as *iron* or *core losses.*

■ We assumed that the impedance of the stator winding could be represented by the reactance X_s. This is not quite correct. To understand this statement, consider the two current wave positions shown in Fig. 4.37. When the current wave is in position A, the wave will give rise to a flux wave (flux A) that will be lined up with the pole ("direct" or d-axis).

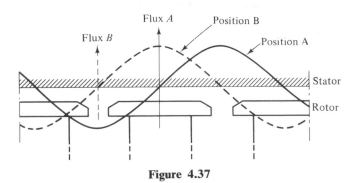

Figure 4.37

When the current wave is in position B, the flux will have its center between the poles ("quadrature" or q-axis). As the flux will encounter a higher magnetic reluctance in q-direction than in d-direction, the stator reactance will vary with the position of the current wave between a maximum value X_d (direct-axis value) and a minimum value X_q (quadrature-axis value).

When one takes this *saliency effect* into account (and does not make use of an *average* value as we have done in the text) the power formulas get slightly modified. The modification results in a second-order correction term.

■ In practice, the stator winding is somewhat differently designed than was described in the text. For example, certain advantages (elimination of harmonics) can be gained by some overlap of the different phase belts. Also the "coil pitch" is not always 180° as we assumed in text.

■ What happens with the higher harmonics present in the stator current wave? An analysis of their combined effect is more complex than that of the fundamental wave presented in the text and is beyond our scope. Generally, it can be said that their effect on power and torque is rather negligible. Not only is their amplitude much smaller than that of the fundamental wave but their effect can be minimized and even nullified by proper design of the winding (see previous point).

■ To simplify our previous analysis we assumed that our generator fed into an *infinite* network bus, the voltage of which did not "budge" as we varied the torque and emf of the generator.

In reality these control actions will be felt on the network, more so the larger the generator unit is. An increase in the torque will not only *tend* to accelerate the network—it *will* accelerate it. The frequency will, in fact, increase. Actually (see Chap. 6) by varying the prime mover torque of individual generators we can, in effect, control the frequency of the power system (so-called load frequency control).

Similarly, control of the generator field current (that is, the emf magnitude) will be felt not only in the terminal voltage but, actually, as a change in the total voltage profile of the system.

■ In practice, synchronous machines are provided with a *damper* (or *amortisseur*) winding on the rotor. This winding can be seen in the photograph which opens this chapter. The winding consists of short-circuited copper bars, resembling the cage windings in induction motors (Chap. 8).

Under normal operation the damper winding carries no currents and therefore has no influence on the torque. However, due to some system disturbance, should the rotor be subject to transient position changes relative to the stator current wave (that is, changes in γ) emf's and currents will be induced in the damper winding. According to Lenz' law these currents will have such directions as to counteract the position changes. This is tantamount to saying that the currents will have a damping effect on the transient rotor swings. (Without this damping the rotor like a pendulum would be very oscillatory.)

EXERCISES

4.1 A single-phase generator delivers at its terminals a voltage of 600 V rms and a current of 30 A rms. The real power delivered from its terminals is $P = 11.5$ kW. Find the reactive power Q! Note that there are *two* solutions.

4.2 Prove item 6 in Table 4.1!

4.3 The 2-pole rotor of the generator discussed in the text Example 4.6 is *replaced* by a 4-pole rotor. This rotor is now run at 1800 rpm. The stator winding *remains unchanged*. Describe the terminal voltages which you would now measure! We assume same B_{max} as before. [*Hint:* Before you dive headlong into formulas look at things physically first.]

4.4 Consider a 3-phase generator the terminal voltage of which measures 8300 V rms line to line. Connect three equal 50-ohm resistors between each phase terminal and ground. We assume that the terminal voltage remains unchanged. (The current would cause a voltage drop across the synchronous reactance, and the terminal voltage would drop. We assume that a *voltage regulator* (Chapter 6) eliminates this drop by increasing the field current.)

a) Use Table 4.1 to compute the real power dissipated in each resistor.

b) Compute total 3-phase power.

c) How many kilowatt-hours of energy are dissipated in the above "load" over a period of eight hours?

4.5 A 2-pole synchronous generator generates 60 cps. Its stator has 36 equidistant slots. Each slot has two conductors. Each coil spans exactly 180 degrees. The magnetic flux is sinusoidally distributed in peripheral direction. The total flux leaving one pole is 2.5 Wb. Compute the rms value of the generated emf if the stator winding is connected as

a) a single-phase winding with all coils in series,

b) a 3-phase winding.

4.6 Formula (3.58) gives the emf induced in a conductor which "cuts" a perpendicular magnetic field B at speed s m/s.

Use this formula to derive the Equation (4.51).

4.7 A synchronous generator of salient pole design is driven by a slow-running hydroturbine at the rated speed $n = 150$ rpm. The generator is generating 60 cps. There are 576 equidistant stator slots with two conductors per slot. The air gap dimensions are $D = 6.30$ m and $L = 1.11$ m. The maximum flux density is $B_{max} = 1.3$ T.

a) Compute the stator emf (rms) if all conductors are connected in such a manner as to produce maximum terminal single-phase stator voltage.

b) Assume the winding to be arranged as a balanced 3-phase winding. Find the emf generated per phase in this winding.

4.8 Three impedances, Z, consist each of a 10-Ω resistor, a 40-mH inductor and a 300-μF capacitor, all connected in series. These impedances are connected in Δ across the phase terminals of a 3-phase generator that delivers a line voltage of 1 kV, 60 Hz. Find: (a) delta currents (rms); (b) phase currents (rms); (c) delivered power, three-phase, real and reactive.   *phase to neutral*   *LINE to Live*

4.9 A 2-pole, 3-phase synchronous generator has the air-gap diameter $D = 1$ m. The stator has 72 equidistant slots, with four conductors per slot. The machine supplies a current of 50 A rms per phase. (This means that we have $4 \cdot 50 = 200$ A/slot.)

"Smear out" the slot currents over the stator surface as was done in Fig. 4.22, and find the actual dimensions of the "current keys" in that figure. Then make a Fourier analysis to obtain the value for A_{max}. Give your answer in A/m.

4.10 A 3-phase synchronous generator operates onto a grid bus of voltage 12 kV (line value). The synchronous reactance is 5 Ω/phase. The magnitude of the generator emf equals the magnitude of the bus voltage. The machine delivers 18 MW to the grid. Find: (a) the power angle δ; (b) phase current, magnitude and phase (relative to V); (c) magnitude and direction of reactive power.

4.11 Consider the generator in the previous example. The prime mover torque is kept constant at a value corresponding to 18 MW real power output. The magnitude of E is now lowered by a decrease in the field current. By how many percent can $|E|$ be decreased before the machine steps out of synchronism?

4.12 If you worked Exercise 4.10 correctly, you found that the machine absorbs reactive power from the network. Now keep the prime mover torque constant and increase $|E|$. As you do so explain by means of the power formulas why

a) the power angle δ decreases,

b) the absorbed reactive power decreases.

c) By how many percent must you increase $|E|$ in order for the reactive absorption to reach zero (you now operate the generator with unity power factor)!

4.13 Consider again the 3-phase generator in Exercise 4.10. It is delivering 10 MW and 5 MVAr (3-phase values) to the 12-kV grid bus. Find: (a) power angle δ; (b) phase angle φ; (c) emf magnitude $|E|$.

4.14 In the previous exercise keep the torque constant corresponding to 10 MW real power. Is it possible by decreasing $|E|$ to reverse the reactive power flow from +5 MVAr to −5 MVAr? If it is possible, what will be the corresponding $|E|$?

4.15 In the 3-phase generator discussed in Section 4.4.1, the rms value of the induced phase emf was 6447 volts, based upon a peak flux value of $B_{max} = 1.5$ T. According to Eq. (4.40) the flux wave can be expressed as

$$B = 1.50 \cos\left(\frac{py}{D}\right) \quad \text{T.}$$

By increasing the rotor field current by 30% one tries to increase the flux peak value by 30% to 1.95 T, thereby also increasing the emf by 30% to 8381 V. Due to magnetic *saturation* the flux density increases only to 1.65 T and, furthermore, it is flattened out as indicated in Fig. 4.38.

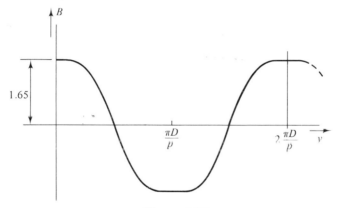

Figure 4.38

A "harmonic analysis" (App. B) of this wave reveals that it contains a considerable third harmonic. In fact we find that the flux wave can be written

$$B = 1.80 \cos\left(\frac{py}{D}\right) - 0.15 \cos\left(3\frac{py}{D}\right) \quad \text{T.} \tag{4.101}$$

a) Prove that the induced emf in each phase now contains a 180-cps component. Find the rms value of this component.

b) Prove that the 180-cps, emf components in the three-phase windings are *in phase.*

c) As a consequence of the finding in part (b), show that *no* 180-cps component appears in the *line voltages.*

d) What will be the rms value of the 60-cps emf component in each phase?

[*Hint:* Find first, by use of formula (4.47), the flux linked by one coil. As you later take the distribution effect into account, note that the α-value in formula (4.57) refers to 60 Hz.]

4.16 Consider the mechanical transmission in Fig. 4.31.

Prove that the transmitted torque equals the value given by formula (4.99). The constant k in the formula is the spring constant of the spring expressed in newtons per meter elongation. Make the assumption that the spring is linear, that is, the spring force f is proportional to its length.

REFERENCES

Fitzgerald, A. E. et al. *Electric Machinery.* New York: McGraw-Hill, 1971.

Hindmarsh, J. *Electrical Machines.* New York: Pergamon, 1965.

Meisel, J. *Principles of Electromechanical Energy Conversion.* New York: McGraw-Hill, 1966.

Nasar, S. A. *Electromagnetic Energy Conversion Devices and Systems.* Englewood Cliffs, N. J.: Prentice-Hall, 1970.

Walsh, E. M. *Energy Conversion.* New York: Ronald, 1967.

Woodson, H. H., and J. R. Melcher. *Electromechanical Dynamics*, Vol. 1. New York: Wiley, 1968.

Young, C. C. "The Synchronous Machine." (IEEE tutorial course on modern concepts of power system dynamics.) IEEE Publication No. 70M62–PWR.

chapter 5

THE POWER TRANSFORMER

*One single-phase unit of a three-phase transformer
bank, rated at 1000 MVA, 765 kV. (Courtesy
General Electric)*

5.1 WHY TRANSFORMERS?

It is impossible to transmit *directly*, even over modest distances, the electric power as it is being generated in the synchronous generators. Unacceptably high losses and voltage drops would result. Let us demonstrate by considering a nuclear-powered generator with a nameplate capacity of 1000 MW, a power rating well within today's engineering capabilities. Let us assume that we wish to transmit this block of power over a distance of only 20 kilometers (about 12 miles).

The rated terminal voltage of a modern generator is of the order of 20–30 kV (line to line). The generator voltage magnitude is limited in practice (compare Chap. 4) by the number of conductors that physically can be placed in the stator slots. We must remember that the conductors must have a minimum cross-sectional area in order to carry the required stator current. We assume that our example generator is rated at 20 kV. The current magnitude $|I|$ is computed from the formula (4.13):

$$|I| = \frac{P}{|V|\cos\varphi} \qquad \text{A/phase.}$$

We have in this case

$$P = \frac{1000}{3} = 333.3 \text{ MW/phase,}$$

$$|V| = \frac{20}{\sqrt{3}} = 11.55 \text{ kV/phase.}$$

Thus,

$$|I| = \frac{333.3}{11.55} = 28.86 \text{ kA/phase.}$$

(We have assumed that the generator delivers the power at unity power factor, that is, $\cos\varphi = 1$. Other values of $\cos\varphi$ would yield still higher current values.)

We will assume that we carry this current in three overhead identical bare copper conductors (depicted in Fig. 5.1) having a radius R of 25 mm (approximately 1 inch). A smaller conductor would probably not accommodate the current (see below). The distance D between the conductors is 5 meters, since a smaller distance would probably not be tolerated in view of insulation safety.

Copper has a resistivity of $1.75 \cdot 10^{-8}$ ohm $\cdot$ meter. According to formula (3.30) the resistance of the 20 km line would therefore be

$$R = 1.75 \cdot 10^{-8} \frac{20 \cdot 10^3}{\pi \cdot 25^2 \cdot 10^{-6}} = 0.178 \ \Omega/\text{phase.}$$

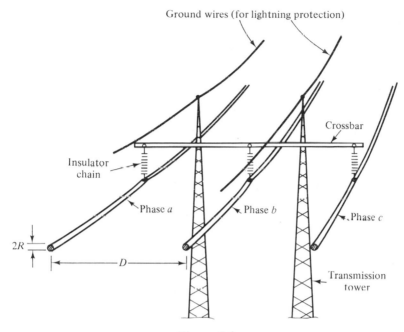

Figure 5.1

We can now compute the ohmic loss power:

$$P_\Omega = R \cdot |I|^2 = 0.178 \cdot 28.86^2 = 148 \text{ MW/phase},$$

or

$$\frac{148 \cdot 10^3}{20 \cdot 10^3} = 7.4 \text{ kW/m conductor}.$$

(A higher loss power would probably melt the conductor. This is the reason why we did not choose a smaller dimension.)

The above loss power represents

$$\frac{148}{333.3} \cdot 100 = 44.4\%$$

of the generator output. In other words, almost half the generated power would be lost in transmission.

Even more disturbing results are obtained if we compute the voltage drop along the 20-km line caused by the above current.

For an overhead transmission line the series reactance is typically much larger than the resistance. Without proof† we give the following formula for

---

† For a discussion of transmission line parameters see Elgerd, 1971.

the series reactance of the line in Fig. 5.1:

$$X = \omega \frac{\mu_0}{2\pi}\left(\tfrac{1}{4} + \ln\left(\frac{\sqrt[3]{2}\,D}{R}\right)\right) \qquad \Omega/\text{m and phase.} \qquad (5.1)$$

The reactance of our 20-km line (computed for 60 Hz) would, therefore, be

$$X = 377 \cdot \frac{4\pi \cdot 10^{-7}}{2\pi}\left(\tfrac{1}{4} + \ln\frac{\sqrt[3]{2} \cdot 5}{25 \cdot 10^{-3}}\right) \cdot 20 \cdot 10^3 = 8.72\ \Omega/\text{phase.}$$

The reactive voltage drop would thus be

$$X \cdot |I| = 8.72 \cdot 28.86 = 251.7\ \text{kV/phase.}$$

This result is, of course, absurd, because the voltage drop cannot exceed the generator voltage, which was 11.55 kV per phase.

Conclusion. It is physically impossible to transmit 1000 MW over this 20-km line if we attempt to do it at a 20-kV voltage level.†

Note how very different the above results would have turned out had our generator voltage been 200 kV instead of 20 kV:

The phase current would then be only 2.89 kA instead of 28.9 kA.

The loss power would be only 1.48 MW per phase instead of 148 MW, that is, only 0.44 percent of generated power.

The reactive voltage drop would be 25.2 kV per phase which is only 21.8 percent of the generator voltage (115.5 kV per phase).

Both losses and voltage drops would be within acceptable and physically realizable limits!

The example tells us vividly that power transmission, even over short distances, is possible only if we can work with voltage levels far exceeding those that can be directly generated in rotating machines. Power transformers make this possible. They transform the generator voltage to levels at which transmission becomes feasible over distances as high as about 1000 km.

5.2 THE SINGLE-PHASE TRANSFORMER—BASIC DESIGN

The power transformer is always designed for a single frequency, for example, 60 Hz. It comes in either single-phase or three-phase units. Sizes range from a few kVA for small-distribution transformers to more than 1000 MVA for

---

† In Chapter 6 we will find that this line can transmit, at most, about 50 MW at 20-kV transmission voltage.

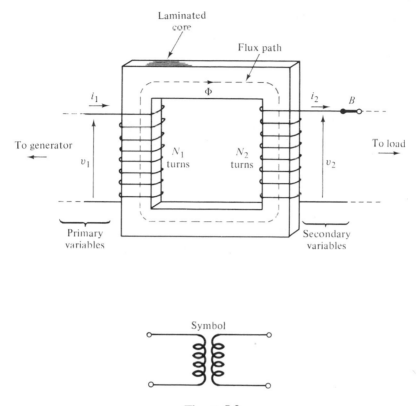

Figure 5.2

large-transmission transformers. Compared to the fairly complex synchronous generator, the power transformer is a relatively simple machine, due mainly to the fact that it is a *static* device.

In simplest terms a power transformer (Fig 5.2) consists of two windings on an iron core. We shall refer to the windings as HV and LV (high and low voltage), respectively. The designations *primaries* and *secondaries* are also common. Often additional windings (*tertiaries*, etc.) are added.

5.3 THE CONCEPT OF "IDEAL" TRANSFORMER (IT)

We shall now proceed to develop a mathematical transformer model which will help us understand the operational features of the device. It is possible to achieve considerable simplifications by making, initially, certain assumptions that will add only insignificant model errors. The model thus derived is called an "ideal" transformer. The actual physical power transformer behavior is very close to that of the ideal transformer (IT).

The IT assumptions are as follows:

1) The transformer windings lack resistance. This means in effect that we neglect both the ohmic power losses and resistive voltage drops in the actual device.

2) The transformer core is made of iron having infinite permeability. This assumption implies two things:

    a) It takes zero mmf to create the magnetic core flux.

    b) All flux is confined to the core, thus reducing all "leakage fluxes" to zero.

3) The transformer core lacks losses. In other words, there exist neither hysteresis nor eddy-current losses.

5.3.1 The Ideal Transformer at No-Load

Consider the two-winding transformer depicted in Fig. 5.2. The N_1-turn primary winding is energized from a single-phase generator having a terminal voltage v_1. The N_2-turn secondary winding feeds a load (not shown). For the present, we consider the transformer operating at "no-load," that is, the load circuit breaker B is open, and the secondary current is zero.

In view of IT assumption 1, we can write, according to Faraday, the following voltage equations for the two windings:

$$v_1 = N_1 \frac{d\Phi}{dt} \qquad \text{V}, \tag{5.2}$$

$$v_2 = N_2 \frac{d\Phi}{dt} \qquad \text{V}, \tag{5.3}$$

where Φ is the core flux identified in Fig. 5.2.

A. Voltage Relation.   Elimination of $d\Phi/dt$ between Eqs. (5.2) and (5.3) yields the first important IT relationship:

$$\frac{v_1}{v_2} = \frac{N_1}{N_2} \equiv a. \tag{5.4}$$

In words: The ratio between primary and secondary voltages equals the winding turns ratio a, referred to as the *transformer ratio*. Note that Eq. (5.4) holds for *arbitrary* voltage waveshapes (except dc, of course). If the voltages are *sinusoidal* then, of course, Eq. (5.4) applies to the voltage phasors V_1 and V_2 as well:

$$\frac{V_1}{V_2} = a. \tag{5.5}$$

B. Magnetic Flux Relation in Sinusoidally Excited Transformer.   Assume that the generator in Fig. 5.2 delivers a *sinusoidal* voltage of $|V_1|$ volts rms. We

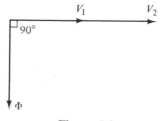

Figure 5.3

can then express the primary voltage as

$$v_1 = \sqrt{2}\,|V_1| \sin \omega t \qquad \text{V} \qquad (5.6)$$

From Eq. (5.2) we get for the flux derivative,

$$\frac{d\Phi}{dt} = \frac{\sqrt{2}\,|V_1|}{N_1} \sin \omega t \qquad \text{Wb/s.}$$

Integration of this equation gives

$$\Phi = -\frac{\sqrt{2}\,|V_1|}{\omega N_1} \cos \omega t = \frac{\sqrt{2}\,|V_1|}{\omega N_1} \sin\left(\omega t - \frac{\pi}{2}\right) \qquad \text{Wb.} \qquad (5.7)$$

Note that the flux lags V_1 (and V_2) by 90°.

If we depict the IT voltages and core-flux as phasors we obtain the phasor diagram shown in Fig. 5.3.

The peak flux Φ_{max} is obtained from Eq. (5.7):

$$\Phi_{max} = \frac{\sqrt{2}\,|V_1|}{\omega N_1} \qquad \text{Wb.} \qquad (5.8)$$

C. The VPT (voltage per turn). In terms of frequency f, effective† core cross-sectional area A, and maximum flux density B_{max}, Eq. (5.8) can be written as

$$\frac{|V_1|}{N_1} = \frac{\omega \Phi_{max}}{\sqrt{2}} = \frac{2\pi f}{\sqrt{2}} AB_{max} = 4.44 f A B_{max} \qquad \text{V/turn.} \qquad (5.9)$$

The ratio $|V_1|/N_1$ is the "voltage per turn" (VPT) ratio, an important transformer design parameter. Note that in view of Eq. (5.5) we have

$$\text{VPT} = \frac{|V_1|}{N_1} = \frac{|V_2|}{N_2} \qquad \text{V/turn.} \qquad (5.10)$$

---

† The *effective* core area refers to the actual iron area (the gross area minus the insulation between the laminations).

Because f is a *system constant* (60 cps in the United States) and B_{max} a *material constant*, formula (5.9) thus tells us that

$$VPT \sim A \qquad V/turn. \qquad (5.11)$$

Example 5.1   Find the VPT of a transformer having a core area $A = 0.4\,m^2$. The core is operated at a peak flux density of $B_{max} = 1.5\,T$. Frequency $f = 60$ Hz.

SOLUTION: Equation (5.9) yields directly

$$VPT = 4.44 \cdot 60 \cdot 0.4 \cdot 1.5 = 159.8 \qquad V/turn.$$

....

D. Dependency of Transformer Size on Frequency. As mentioned, B_{max} is a material constant. We can thus write formula (5.9) as follows:

$$A \sim \frac{VPT}{f}. \qquad (5.12)$$

This formula tells us that transformer size is inversely proportional to frequency. This means, for example, that European transformers are bulkier than United States transformers of equal power rating, because the frequency is only 50 Hz. (The formula (5.12) also informs us that a "dc transformer" ($f = 0$) will be of infinite size.)

Example 5.2   An ideal transformer is characterized by the following data: $N_1 = 100$ turns, $N_2 = 300$ turns.

The LV winding is connected to a single-phase generator delivering a sinusoidal voltage of 3 kV rms and 60 Hz.

Find VPT, core flux, and secondary voltage at no-load operation.

SOLUTION: The turns ratio is

$$a = \frac{100}{300} = 0.3333.$$

As the 3-kV generator voltage is applied to the 100-turn primary we get

$$VPT = \frac{|V_1|}{N_1} = \frac{3000}{100} = 30 \text{ V/t}.$$

The secondary voltage follows from Eq. (5.5):

$$|V_2| = \frac{|V_1|}{a} = \frac{3000}{0.3333} = 9000 \text{ V}.$$

The core flux (peak value) is computed from Eq. (5.8):

$$\Phi_{max} = \frac{\sqrt{2} \cdot 3000}{377 \cdot 100} = 0.113 \text{ Wb.}$$

....

5.3.2   The Ideal Transformer Under Loading Conditions

A. Voltage, Current, and Flux Relationships.   If the circuit breaker *B* in Fig. 5.2 is closed, the secondary transformer voltage v_2 is applied to the "load." The latter may consist, in the simplest case, of a single impedance, or it may, in a more general case, be made up of a number of varied load objects, like for example a whole city.

In any case, the result will be a secondary current drain, i_2, as shown in Fig. 5.2. This current will cause a change in the mmf in the amount $N_2 i_2$. Ohm's law for the magnetic circuit must, however, be satisfied. The only way in which this can be achieved is for a primary current, i_1, to arise. The magnitude of i_1 will be such as to restore the mmf balance of the core.

In mathematical terms this mmf balance reads

$$i_1 N_1 - i_2 N_2 = \mathcal{R} \cdot \Phi \qquad A \cdot t \tag{5.13}$$

where $\mathcal{R}$ is the magnetic reluctance of the core. (Compare Eq. 3.104.)

However, we had assumed that the core material was characterized by infinite magnetic permeability. This is tantamount to saying that $\mathcal{R} = 0$. Because Φ is finite, Eq. (5.13) thus reduces to

$$i_1 N_1 - i_2 N_2 = 0, \tag{5.14}$$

or

$$\frac{i_1}{i_2} = \frac{N_2}{N_1} = \frac{1}{a}. \tag{5.15}$$

Equation (5.15) applies to the instantaneous values for currents. Clearly, if all variables are sinusoidal, the equation must apply also to current phasors, that is,

$$\frac{I_1}{I_2} = \frac{1}{a}. \tag{5.16}$$

We had assumed zero resistance in each winding. Consequently the above currents will cause zero resistive voltage drops. Thus, the voltage equations (5.4) and (5.5) still are valid. *The flux and voltage picture as it existed at no-load thus remains unchanged even under loading conditions.*

B. Power Relationships. The IT was assumed lossless. This means that the input power

$$p_1 = v_1 i_1 \quad \text{W}$$

must equal the output power

$$p_2 = v_2 i_2 \quad \text{W},$$

that is,

$$v_1 i_1 = v_2 i_2 \quad \text{W}. \tag{5.17}$$

(This formula can also be derived by multiplication of Eqs. (5.4) and (5.15).)

In words: The instantaneous power *that enters the primaries of an ideal transformer exits from the secondaries in undiminished magnitude.* The power passing through the device is referred to as *transformed power.*

As this rule applies to instantaneous power values it will, of course, apply equally to real and reactive powers in the case of sinusoidal voltage and current variables, that is,

$$V_1 I_1^* = V_2 I_2^*. \tag{5.18}$$

Example 5.3   An impedance of value

$$Z_2 = 100 + j30 \quad \Omega$$

is connected across the secondary terminals of the transformer in Example 5.2. Find the resulting currents and also compute the transformed power.

SOLUTION: On the assumption that the current drain will not reduce the generator voltage,† the primary and secondary transformer voltages would remain at 3000 V and 9000 V, respectively. Let us choose (like in Fig. 5.3) these voltages as reference phasors. We then have for the secondary current

$$I_2 = \frac{V_2}{Z_2} = \frac{9000}{100 + j30} = 86.20 e^{-j16.70°}.$$

Equation (5.16) gives for the primary current

$$I_1 = \frac{1}{0.3333} I_2 = 258.6 e^{-j16.70°}$$

Currents and voltages are plotted in the phasor diagram in Fig. 5.4.

---

† In a practical situation a *voltage regulator* would maintain the generator voltage constant. If the generator emf were uncontrolled, the current would cause a voltage drop across the synchronous reactance resulting in a lowered terminal voltage.

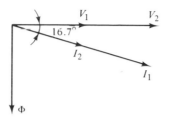

Figure 5.4

The complex power supplied to the load impedance is

$$S_2 = V_2 I_2^* = 9000 \cdot 86.2 e^{j16.70°} = (743.1 + j222.9) \cdot 10^3.$$

Thus, according to Eq. (5.18),

$$P_1 = P_2 = 743.1 \text{ kW},$$
$$Q_1 = Q_2 = 222.9 \text{ kVAr}.$$

C. Impedance Transformation Across the IT. In the previous example the *secondary* voltage-impedance relationship reads

$$\frac{V_2}{I_2} = Z_2. \tag{5.19}$$

By making use of Eqs. (5.5) and (5.16), the above relationship can be re-written thus:

$$\frac{V_2}{I_2} = \frac{V_1/a}{aI_1} = Z_2.$$

In terms of *primary* variables we thus have

$$\frac{V_1}{I_1} = a^2 Z_2 \equiv Z_2'. \tag{5.20}$$

In words: The impedance Z_2 connected across the transformer secondaries has the same effect (that is, it draws equal power) on the primary generator as an equivalent impedance $Z_2' = a^2 Z_2$ connected directly across the generator terminals. Z_2' is called "the secondary impedance referred to the primary."

Example 5.4 If connected directly across the 3-kV generator terminals in Example 5.3, what size impedance would cause exactly the same power drain as that resulting from the actual impedance?

SOLUTION: Formula (5.20) immediately gives the answer:

$$Z_2' = 0.3333^2 \cdot (100 + j30) = 11.11 + j3.333 \ \Omega.$$

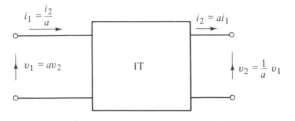

Figure 5.5

D. Equivalent IT Circuit. The basic IT performance Eqs. (5.4) and (5.15) can both be summarized in the "equivalent circuit" depicted in Fig. 5.5. The circuit gives at a glance the relationships between primary and secondary current and voltage variables. As will be shown later this diagram is useful in all kinds of circuit analyses involving transformers.

In the ideal transformer there exists a direct proportionality between primary and secondary currents and voltages. A voltage v_2 applied to the secondary winding will give rise to the same core flux as the voltage av_2 applied to the primary. We refer to av_2 as "the secondary voltage referred to the primary" and give it the special symbol v_2', that is,

$$v_2' \equiv av_2. \tag{5.21}$$

Similarly, the secondary current i_2 gives rise to the same core mmf as the current i_2/a flowing in the primary. We define

$$i_2' \equiv \frac{1}{a} i_2 \tag{5.22}$$

as "the secondary current referred to the primary."

Later examples will demonstrate the practical significance of these "referred" variables.

E. A Mechanical IT Analog. The IT transforms the voltages in ratio N_1/N_2 and the currents in inverse ratio. The transformed power remains invariant.

An analogous situation prevails in a mechanical gear train (Fig. 5.6). Let the *gear ratio* be defined as the ratio of the radii:

$$a \equiv \frac{R_1}{R_2}. \tag{5.23}$$

As the secondary axis turns a degrees for every degree of the primary axis, it becomes obvious that the primary and secondary angular speeds ω_1 and ω_2 are related through the formula

$$\frac{\omega_1}{\omega_2} = \frac{R_2}{R_1} = \frac{1}{a}. \tag{5.24}$$

This equation clearly corresponds to the current equation (5.15).

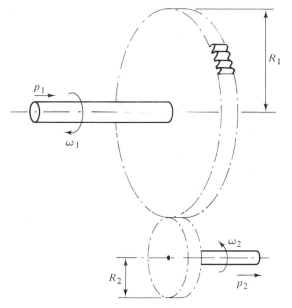

Figure 5.6

As the primary and secondary torques T_1 and T_2 are proportional to the respective radii, we have

$$\frac{T_1}{T_2} = \frac{R_1}{R_2} = a. \tag{5.25}$$

This equation is the analog of the voltage equation (5.4).

The primary mechanical power p_1 equals

$$p_1 = \omega_1 T_1. \tag{5.26}$$

For the secondary power p_2 we have

$$p_2 = \omega_2 T_2 = (a\omega_1)\left(\frac{T_1}{a}\right) = \omega_1 T_1 = p_1 \qquad \text{W.} \tag{5.27}$$

This equation proves that the magnitude of the transformed mechanical power remains invariant (provided the gear train is lossless).

Example 5.5   A flywheel of inertia I kgm^2 is connected to the secondary axis in Fig. 5.6. How is this inertia felt from the primary side?

SOLUTION: Newton's law reads

$$T_2 = I\frac{d\omega_2}{dt} \qquad \text{N} \cdot \text{m.} \tag{5.28}$$

By the use of Eqs. (5.24) and (5.25) we can write Eq. (5.28) as follows:

$$\frac{T_1}{a} = I\frac{d}{dt}(a\omega_1);$$

thus

$$T_1 = (a^2 I)\frac{d\omega_1}{dt}. \qquad (5.29)$$

Clearly the inertia I is "felt" on the primary side as a "referred" inertia I' defined by

$$I' \equiv a^2 I \qquad \text{kgm}^2. \qquad (5.30)$$

Compare the similarity with Eq. (5.20).

5.4   THE PHYSICAL VERSUS IDEAL TRANSFORMER

The IT performance is embodied in the basic equation pair (5.4) and (5.15) and the corollary power equation (5.17). Within very good accuracy limits these equations also describe the behavior of a *physical* power transformer. In fact, the errors made by modeling an actual power transformer by this simple set of equations amount normally, at most, to a few percent–often less than one percent.

However, the simple IT model, in certain instances, is totally inadequate in describing an actual physical transformer. For example, an ideal transformer is characterized by zero power losses. *This means that there is no limit to the power which an IT can transform.* This, of course, does not apply to a real transformer. The actual losses which occur in the physical device, although *relatively* small, set a definite limit to its power capability.

We proceed now to investigate how to adjust the simple IT model in Fig. 5.5 so as to make it more fully representative of the physical device. We do this by removing, one by one, the ideal (but physically unrealistic) assumptions earlier made.

5.4.1   Accounting for Finite Permeability

A. The Magnetizing Current.   We first remove the IT assumption of *infinite permeability*. This assumption led to Eq. (5.15). Should an IT be operated with the secondaries open, that is, with $i_2 = 0$, then Eq. (5.15) tells us that the primary current is also zero. This would imply that the open-circuit impedance, as viewed into the primary, is infinite. In reality, the primary current will be small (compared to the rated winding current) *but not zero*. In fact, according to

Eq. (5.13), this primary *open-circuit*, or *magnetizing current*, will be

$$i_{1m} = \frac{\mathcal{R}\Phi}{N_1} \qquad \text{A.} \qquad (5.31)$$

In the case of *sinusoidal* primary voltage, the magnetizing current, according to Eq. (5.7), will be

$$i_{1m} = \sqrt{2}\frac{|V_1|\,\mathcal{R}}{\omega N_1^2}\sin\left(\omega t - \frac{\pi}{2}\right) \qquad \text{A.} \qquad (5.32)$$

B. The Magnetizing Reactance. We pointed out in Chapter 3 that the reluctance $\mathcal{R}$ strictly speaking is *not* a constant, because of the variable core permeability μ. However, let us for the moment disregard this fact and consider it constant. Equation (5.32) under this assumption informs us that i_{1m} is also sinusoidal and lags the voltage by 90°. We represent it, therefore, by the phasor I_{1m}, having the rms value

$$|I_{1m}| = \frac{|V_1|\,\mathcal{R}}{\omega N_1^2} \qquad \text{A.} \qquad (5.33)$$

Equation (5.32) can thus (compare App. B) be expressed in phasor form as

$$I_{1m} = \frac{V_1\mathcal{R}}{j\omega N_1^2} \equiv \frac{V_1}{jX_m} \qquad \text{A} \qquad (5.34)$$

where

$$X_m \equiv \omega\frac{N_1^2}{\mathcal{R}} \qquad (5.35)$$

is the *magnetizing reactance* as viewed from the primary winding.

Example 5.6   Consider the toroidal iron core discussed in Example 3.23, case (1). The core is wound with primary and secondary coils of 100 turns and 40 turns, respectively. A primary 60 Hz sinusoidal voltage of 16 V rms is applied. What will be the magnetizing current?

SOLUTION: In Example 3.23, we computed the reluctance

$$\mathcal{R} = 1.601 \cdot 10^5 \qquad \text{A} \cdot \text{t/Wb.}$$

From Eq. (5.35) we compute the open circuit reactance:

$$X_m = 377 \cdot \frac{100^2}{1.601 \cdot 10^5} = 23.5 \ \Omega.$$

The magnetizing current will thus be

$$|I_{1m}| = \frac{16}{23.5} = 0.681 \qquad \text{A rms.}$$

....

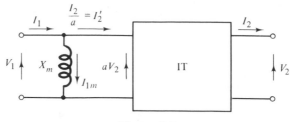

Figure 5.7

C. Adjusting the IT Equivalent Circuit. Equation (5.13) can be written as

$$i_1 = \frac{N_2}{N_1} i_2 + \frac{\mathcal{R}\Phi}{N_1} = \frac{i_2}{a} + i_{1m} \quad \text{A.} \tag{5.36}$$

In the case of sinusoidal time variation, the equation can be written in phasor form:

$$I_1 = \frac{I_2}{a} + I_{1m} = I_2' + I_{1m} = I_2' + \frac{V_1}{jX_m}. \tag{5.37}$$

This equation informs us that the primary current consists of two components:

1) The magnetizing current I_{1m} which is proportional to the voltage V_1.
2) The load component $I_2' = I_2/a$, which is proportional to the secondary loading.

As I_{1m} is proportional to V_1 we can obviously interpret the presence of this current component by a *reactor* connected across the input terminals of the IT circuit in Fig. 5.5. We thus arrive at the adjusted equivalent circuit shown in Fig. 5.7. To put things in proper perspective, the current component I_{1m} is normally of the order of a few percent of rated transformer current. Often it is therefore neglected, which means that Eq. (5.37) reduces to the simpler IT equation (5.16).

5.4.2   Accounting for Core Losses

Were we to operate our transformer at no-load (that is, with $i_2 = 0$) the model just derived would tell us that

$$I_1 = I_{1m} = V_1/jX_m.$$

Furthermore (cf. item 2 in Table 4.1) we note that the power drain from the generator would be

$$P_1 = 0 \quad \text{W,}$$
$$Q_1 = \frac{|V_1|^2}{X_m} \quad \text{VAr.} \tag{5.38}$$

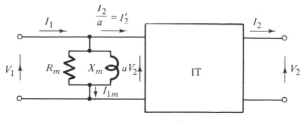

Figure 5.8

In reality, a physical transformer operated in the above mode in addition to reactive power also draws real power from the source—the so-called no-load power or core losses. As in the case of a generator, these losses consist of hysteresis and eddy-current losses. This real power drain can be accounted for by the addition (Fig. 5.8) of a shunt resistance, R_m.

5.4.3   Effects of Core Nonlinearity

The above model corrections were made on the assumption that the relative permeability of the core iron was *constant*. In reality it varies greatly with flux density (Fig. 3.33). This brings about certain nonlinear effects which are of great practical significance in power system operations. We demonstrate these effects by working the following example.

Example 5.7   Study the character of the no-load current i_1 in the primary winding of a small power transformer of the following design:

Core: Sectional area $= 10 \text{ cm}^2$. Total flux path length $= 60 \text{ cm}$. The B–H characteristics of the core have *experimentally* been found to be as shown in Fig. 5-9.

Primary winding: $N_1 = 1000$ turns. A sinusoidal 60-cps voltage of magnitude $|V_1| = 493.$ V rms is applied to the primary.

SOLUTION: We first compute the maximum core flux density using formula (5.9):

$$\frac{493}{1000} = 4.44 \cdot 60 \cdot 10 \cdot 10^{-4} \cdot B_{\text{max}}.$$

Thus

$$B_{\text{max}} = 1.85 \text{ T}.$$

From the core data of Fig. 5.9 we are next able to determine the current waveshape. This must be done *graphically* as follows. From a knowledge of its peak value we are able to plot the *flux density* wave which we know is

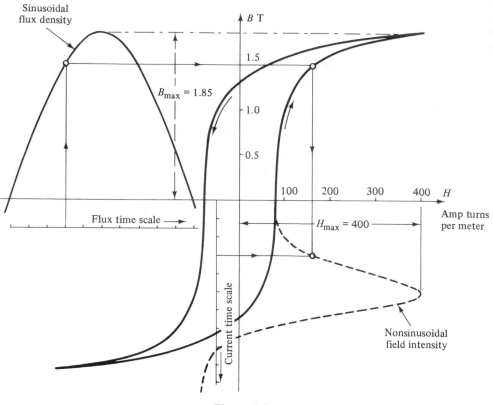

Sinusoidal
flux density

B T

1.5

$B_{max} = 1.85$

1.0

0.5

100   200   300   400 H

Amp turns
per meter

Flux time scale

$H_{max} = 400$

Current time scale

Nonsinusoidal
field intensity

Figure 5.9

essentially *sinusoidal*. A half cycle is shown in Fig. 5.9. Point by point we can next construct the corresponding waveshape of the field intensity H (shown dashed in the figure). The construction of one point is depicted in the figure and should be self-explanatory.

Note that the resulting H-wave is *highly nonsinusoidal.* Because H is proportional to i_1 *the waveshapes of i_1 and H are identical.* The proportional constant is established from Eq. (3.109) which in this case yields

$$H = \frac{1000}{0.60} i_1 = 1667 i_1. \tag{5.39}$$

For example, we note from Fig. 5.9 that

$$H_{max} = 400 \text{ A} \cdot \text{t/m}.$$

Thus

$$i_{1max} = \frac{400}{1667} = 0.24 \quad \text{A.}$$

This realistic example teaches us the following important facts:

1) The no-load component of the transformer current is highly *nonsinusoidal*. A harmonic analysis (Appendix B) of the current reveals that the 180- and 300-Hz components sometimes can be of quite annoying magnitude (through induction they can cause trouble in communication networks).

   We shall also find later (Sect. 5.8.3) that these high-frequency components can cause problems in three-phase transformer connections.

2) By focusing attention on the 60-cps component of i_1 and by accepting some degree of approximation, we can still work with the *linear* equivalent networks. (Differently expressed: In working with the *linear* equivalent circuits in Figs. 5.7, 5.8, and 5.10, one in effect neglects the higher harmonics.)

5.4.4 Modeling of Winding Losses

The model in Fig. 5.8 with sufficient accuracy incorporates the nonideal features of the transformer *core*. Let us now proceed to incorporate into our model the nonideal features of the transformer *coils*. These are of two types.

First we have the winding *resistance* which will cause ohmic heat losses in the coils and also resistive voltage drops. As the resistance is present in both windings it is natural to account for its presence by adding *series* resistances to both leads entering the IT box in Fig. 5.8.

We learned earlier how impedances can be "referred" from one side or the other by use of the formula (5.20). We find it thus practical to lump the effects of both winding resistances into one equivalent series resistor R_s as has been done in Fig. 5.10. It does not matter on what side of IT the resistance is placed. Of course, R_s will have different values (differing by the factor a^2) depending upon where it is placed.

One can also argue whether R_s should be placed left or right (as we have done) of the R_m-X_m shunt. This is of no practical importance because, as we noted above, the series paths carry the overwhelming part of the current.

We have also added in series with R_s the *leakage reactance* X_s. This equivalent reactance will model the effect of the transformer leakage fluxes, which are schematically depicted in Fig. 5.11. The leakage fluxes (note their polarity) are those flux portions which are not mutually linked to both windings. Their magnitudes are very small in relation to the core flux Φ because the leakage paths are dominantly air. This latter fact also results in good proportionality between flux magnitude and its causative current. The equivalent reactance X_s is therefore with good accuracy a constant, independent of current magnitude. (The shunt reactance X_m, in contrast, is not constant because of the iron nonlinearity.)

As will be shown in examples the shunt impedance elements in Fig. 5.10

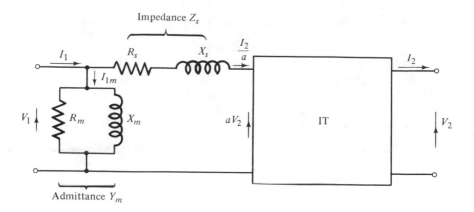

$$Z_s = R_s + jX_s$$
$$Y_m = \frac{1}{R_m} + \frac{1}{jX_m}$$

Figure 5.10

are vastly larger than the series elements, that is,

$$R_s \lll R_m,$$
$$X_s \lll X_m. \tag{5.40}$$

In most power transformers it is also true that

$$X_s \gg R_s. \tag{5.41}$$

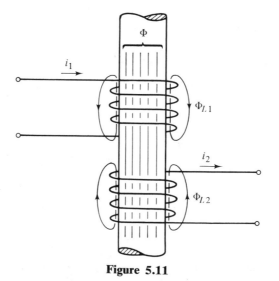

Figure 5.11

Let us show in the two following examples how, in a practical situation, one would construct the above transformer models from test data and what use can be made of them.

Example 5.8   In this example we shall construct a more accurate model of the transformer which we discussed in Examples 5.2 and 5.3. We first give the following more complete set of ratings for the transformer:

$$\text{rated voltage} = 3 \text{ kV primary}$$
$$9 \text{ kV secondary}$$
$$\text{rated frequency} = 60 \text{ Hz}$$
$$\text{rated current} = 333.3 \text{ A primary}†$$
$$111.1 \text{ A secondary}$$

As before we have the transformation ratio:

$$a = \frac{3}{9} = 0.3333.$$

The transformer is subjected to the following two tests designed to render the required circuit data:

Open-Circuit (OC) Test.   In this test the transformer is fed from a 100% voltage source (from either the primary or the secondary) with the other side *open-circuited. Real* and *reactive* power consumptions are measured.

Assumed test results:

$$P_{OC} = 4.31 \text{ kW}, \qquad Q_{OC} = 9.11 \text{ kVAr}.$$

Short-Circuit (SC) Test.   In this test the transformer is fed from a *reduced* voltage source (from either the primary or secondary side) with the other side *short-circuited.* The voltage source is adjusted until 100% current circulates in the coils. *Real* and *reactive* power consumptions are measured.

Assumed test results:

$$P_{SC} = 8.31 \text{ kW}, \qquad Q_{SC} = 50.3 \text{ kVAr}.$$

Model Construction.   Assuming that both the series and shunt elements in Fig. 5.10 are placed on the 3-kV (primary) side, we can now compute those elements as follows. Using Table 4.1 (item 5) we compute directly the *shunt*

---

† This current rating corresponds to a transformer power rating of 1000 kVA (see Sect. 5.5.3 for explanation).

elements from the OC test data:

$$R_m = \frac{3000^2}{4310} = 2088 \ \Omega, \quad X_m = \frac{3000^2}{9110} = 988 \ \Omega.$$

Item 4 in Table 4.1 tells us how to obtain the *series* elements from the SC test data:

$$R_s = \frac{8310}{333.3^2} = 0.0748 \ \Omega, \quad X_s = \frac{50300}{333.3^2} = 0.453 \ \Omega.$$

Note that the above computed data certainly confirm the inequalities (5.40) and (5.41). Note also that in computing R_s and X_s we completely neglected the power drain in the shunt elements. Why could we do so? (What values would the above four impedances have if placed on the opposite side of IT?)

Example 5.9 Use the more accurate model thus arrived at and rework Example 5.3. Also find the transformer efficiency!

SOLUTION: We connect the impedance $Z_2 = 100 + j30$ across the secondaries and obtain the total equivalent circuit shown in Fig. 5.12(a). In one self-explanatory step this circuit is further reduced to the simpler one shown in Fig. 5.12(b).
From this latter circuit we obtain by inspection,

$$I_2' = \frac{I_2}{a} = \frac{V_1}{Z_s + a^2 Z_2}, \tag{5.42}$$

$$V_2' = a V_2 = a^2 Z_2 \cdot \frac{I_2}{a} = V_1 \frac{a^2 Z_2}{Z_s + a^2 Z_2}, \tag{5.43}$$

$$I_1 = \frac{I_2}{a} + I_{1m} = \frac{I_2}{a} + V_1 Y_m. \tag{5.44}$$

We introduce now the following numerical values:

$$V_1 = 3000 \ \underline{/0°},$$
$$a = 0.3333,$$
$$Z_s = 0.0748 + j0.453,$$
$$Y_m = \frac{1}{2088} + \frac{1}{j988} = (0.479 - j1.012) \cdot 10^{-3}.$$

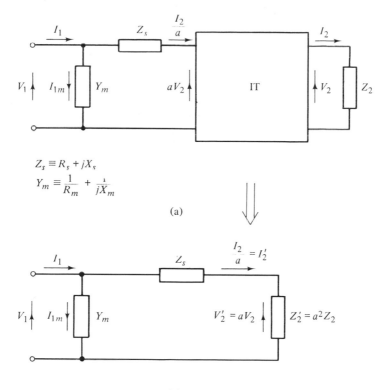

$$Z_s \equiv R_s + jX_s$$
$$Y_m \equiv \frac{1}{R_m} + \frac{1}{jX_m}$$

(a)

(b)

Figure 5.12

The formulas (5.42) to (5.44) then yield

$$I_2 = 84.69 \underline{/-18.70°},$$
$$V_2 = 8841 \underline{/\ 2.00°},$$
$$I_1 = 256.4 \underline{/-19.24°}.$$

Compare these values with the results obtained in Example 5.3. (Note that the errors in either voltage or current magnitudes in no case exceed 2%.)

The transformer efficiency is defined as

$$\eta = \frac{P_2}{P_1} \cdot 100 = \frac{P_2}{P_2 + P_{\text{loss}}} \cdot 100\%.$$

We have

$$P_2 = R_2 |I_2|^2 = 717.2 \cdot 10^3 \text{ W},$$
$$P_{\text{loss}} = P_{Fe} + P_{Cu} = 4310 + 0.0748 \cdot 256.4^2 = 9227 \text{ W}.$$

Thus

$$\eta = \frac{717.2}{726.4} \cdot 100 = 98.73\%.$$

Example 5.10   If the secondaries in Example 5.9 are accidently shorted out, how large will the short-circuit current become?

SOLUTION: A "metallic" short-circuit across the secondaries represents a secondary impedance $Z_2 = 0$.
   From Eq. (5.42) we get

$$I_1 \cong \frac{I_2}{a} = \frac{V_1}{Z_s} = \frac{3000}{0.0748 + j0.453};$$

$$\therefore |I_1| = 6540 \text{ A}.$$

This is 19.6 times rated current! Currents of such magnitudes can be highly destructive—and means must be taken for their speedy interruption. (Were we to use the IT model in the above analysis we would come to the absurd conclusion that the SC current would be of infinite magnitude.)

5.5 SOME PRACTICAL DESIGN CONSIDERATIONS

In this section we shall comment briefly on some important practical aspects of transformer design.

5.5.1 Core and Coil Design

The transformer core is always made *laminated* in order to minimize core losses. Figure 5.13 depicts a typical arrangement showing three separate windings of *concentric coil* design; multiwinding transformers are discussed in Section 5.6. Due to the greater insulation distances needed for increased voltage levels, the high-voltage windings are placed farthest from the core. Core plus coils are immersed in oil, which serves a double purpose; it improves the insulation strength and also serves as a transport medium to the outside for the iron and copper losses generated in the core and coils, respectively. The power is brought in and out of the tank through *bushings* (Fig. 5.14).

5.5.2 Cooling Methods

Required insulation distances and minimum core and coil dimensions determine the overall size of a power transformer. The removal of losses plays a crucial role.

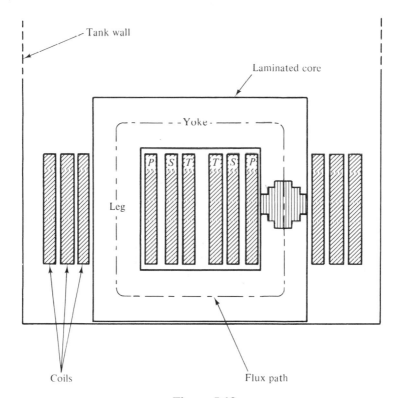

Figure 5.13

The losses consist of core losses and ohmic losses. The former may be minimized by core lamination but they cannot be entirely eliminated.

The ohmic losses in the windings depend on the resistance. To minimize these losses, power transformer windings usually are made of copper. Core and ohmic losses are proportional to the volume of core and windings, respectively.

If L is a linear dimension of core plus windings, the above reasoning thus tells us that the *total* power loss is proportional to the total volume of the active parts, that is,

$$P_{\text{loss}} \sim L^3 \quad \text{W.} \tag{5.45}$$

The amount of heat that can be conducted through a medium is proportional to the surface area of conduction. The amount of heat that can be removed via conduction from the core plus windings thus can be written

$$P_{\text{removed}} \sim L^2 \quad \text{W.} \tag{5.46}$$

We thus have

$$\frac{P_{\text{loss}}}{P_{\text{removed}}} \sim L. \tag{5.47}$$

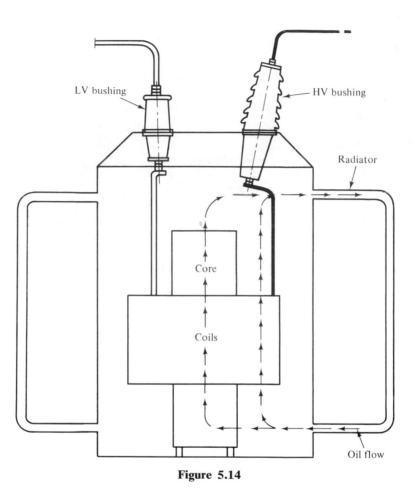

Figure 5.14

This formula shows us the increasing difficulty of heat removal as the transformer size grows. In a small self-cooled transformer (Fig. 5.14), the self-generated flow (convected flow or flow due to natural convection) of heated oil up, via the active parts, and down the cooling tubes of the tank is sufficient to transport the heat to the outside and thus maintain an acceptable temperature balance.

For *forced cooled* larger units, the oil must be pumped and led via heat exchangers for loss power removal.

5.5.3 Transformer Ratings

The transformer core losses depend upon the magnitude of the core flux, which, according to formula (5.8), is proportional to the voltage. The transformer *ohmic*, or *copper losses* depend upon the winding currents. As the *total* losses determine the maximum transformer temperatures, its ratings must include information on *both* voltage and current. The product of current and voltage would evidently be a meaningful rating measure. This explains why it is customary to rate a transformer in terms of its kVA (or MVA) load. As the operating voltage is kept nearly constant the maximum permissible kVA thus in effect sets limits on the *current* load.

For example, the ratings of a particular transformer may read:

$$\text{voltage} = 50/10 \text{ kV}, 60 \text{ Hz},$$
$$\text{power} = 6000 \text{ kVA}.$$

From these ratings we immediately learn that maximum current load is as follows:

$$\frac{6000}{50} = 120 \text{ A} \qquad \text{(on the HV side)},$$

$$\frac{6000}{10} = 600 \text{ A} \qquad \text{(on the LV side)}.$$

Note that kW ratings would be meaningless. A transformer may deliver zero kW (if the load is purely reactive) and still have to withstand maximum permissible core *and* copper losses.

5.6 MULTIWINDING TRANSFORMERS

It is common practice in electric power systems for power to be transformed between more than two voltage levels in a single transformer. Consider, for example, the three-winding transformer shown in Fig. 5.15. If the *primary* 600-turn winding is connected to a 6-kV source, the 200-turn *secondary* will deliver 2 kV and the 100-turn *tertiary* 1 kV to separate loads.

As mmf balance must be maintained, the following equation will determine the current loading of the transformer:

$$I_1 N_1 = I_2 N_2 + I_3 N_3 \qquad \text{A} \cdot \text{t}. \tag{5.48}$$

(We have neglected the magnetizing current.)

Example 5.11 The three windings in the transformer in Fig. 5.15 have the following kVA ratings:

$$\text{primary} = \text{maximum } 300 \text{ kVA},$$
$$\text{secondary} = \text{maximum } 150 \text{ kVA},$$
$$\text{tertiary} = \text{maximum } 200 \text{ kVA}.$$

The secondary is delivering full rated kVA to a purely resistive load. The

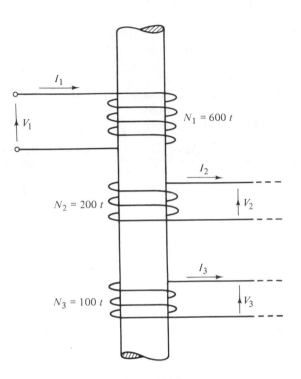

Figure 5.15

tertiary load consists of a variable reactor. As this reactor load is increased, at what point will the kVA ceilings be reached? Use IT modeling in the analysis!

SOLUTION: The three voltages, V_1, V_2, and V_3 will be in phase (neglecting the voltage drops) as indicated in Fig. 5.16.

The secondary is full-loaded and we therefore obtain

$$|I_2| = \frac{150}{2} = 75 \text{ A}.$$

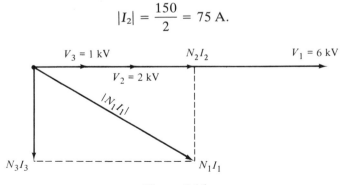

Figure 5.16

As the secondary load is purely resistive, I_2 will be in phase with V_2. As the tertiary load is purely reactive, I_3 will lag V_3 by 90°. In Fig. 5.16 we have plotted $N_2 I_2$ and $N_3 I_3$ as Eq. (5.48) tells us that the mmf's, rather than the currents, determine the loading.

Let us assume that we have adjusted the reactor load until the tertiary is full-loaded. This means that its current, I_3, will have an rms value of

$$|I_3| = \frac{200}{1} = 200 \text{ A}.$$

According to Eq. (5.48) we thus have for I_1

$$600 I_1 = 200 I_2 + 100 I_3.$$

For the rms value $|I_1|$ we therefore have

$$600 |I_1| = |200 I_2 + 100 I_3| = |200 \cdot 75 + 100 \cdot (-j200)|$$

$$= \sqrt{15000^2 + 20000^2} = 25000;$$

$$\therefore |I_1| = \frac{25000}{600} = 41.7 \text{ A}.$$

The rated primary current is

$$\frac{300}{6} = 50 \text{ A}.$$

This means that the primary is not loaded to its capacity. However, we cannot increase the tertiary load any further because it is already "current-loaded" to its full capacity.

5.7 AUTOTRANSFORMERS

If a two-winding transformer is reconnected as an *autotransformer*, its power rating can be considerably increased. To demonstrate this, consider the 30-kVA transformer depicted in Fig. 5.17(a). It has a turns ratio of 250/50 and the voltage rating 500/100 V. The current rating (based upon 30 kVA) is thus 60/300 amperes.

If the windings are reconnected as shown in Fig. 5.17(b), we obtain a new primary with 300 turns. It is therefore possible to energize this new primary from a 600-V source *without change in the VPT (and thus the flux)*. We obtain a new transformer with a voltage rating of 600/500 V. (The alternate voltage rating 600/100 V is obviously also available.)

As shown in Fig. 5.17(b) the new 500 V secondary can be loaded with a maximum of 360 amperes *with neither of the two windings being "current-overloaded."* At this point the primary current will be 300 A, and the transformer has a "through power" of $300 \cdot 600 = 360 \cdot 500 = 180$ kVA.

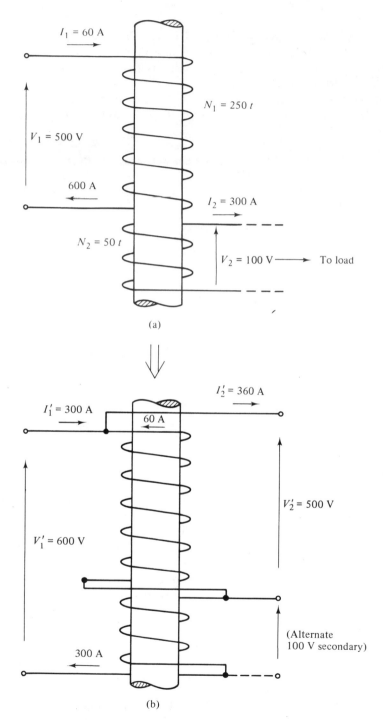

$I_1 = 60$ A

$N_1 = 250\ t$

$V_1 = 500$ V

600 A

$I_2 = 300$ A

$N_2 = 50\ t$

$V_2 = 100$ V $\longrightarrow$ To load

(a)

$I_2' = 360$ A

$I_1' = 300$ A

60 A

$V_2' = 500$ V

$V_1' = 600$ V

(Alternate
100 V secondary)

300 A

(b)

Figure 5.17

Result. The 30-kVA transformer has had its power increased to 180 kVA.

The explanation is the following. In the original 30-kVA transformer the two windings had no metallic connections, and so the 30 kVA had to pass the transformer via the magnetic flux. The new autotransformer has its windings interconnected and thus $180 - 30 = 150$ kVA will pass the transformer without being magnetically transformed.

Had we used the autotransformer in the alternate voltage ratio 600/100 V, the new power rating would have been only 36 kVA, a very modest increase. One can easily prove that the best power-rating increase is obtained when the auto voltage ratio is close to unity. Autotransformers are thus used in power systems as links between voltage levels of nearly equal magnitudes (see Fig. 6.1).

5.8   THREE-PHASE POWER TRANSFORMATION

Transformation of three-phase power can be achieved in several ways. Let us briefly study the most important ones.

5.8.1   Single-phase Units Connected Y–Y (Bank Arrangement)

The simplest way of transforming three-phase power is shown in Fig. 5.18. Three identical single-phase transformer units are connected in parallel—one in each phase. If the HV side is symmetrically loaded, the HV currents I_a'', I_b'', and I_c'' will constitute a symmetrical current set. Because Eq. (5.16) will apply in each transformer unit the LV currents I_a', I_b', and I_c' will likewise possess three-phase symmetry. Consequently, there will be no ground currents through either neutral (due to Eq. 4.70).

Figure 5.19 shows in phasor form the relationship between primary and secondary phase voltages. This phasor diagram explains the designation Y–Y. The figure also shows a symbolic way to display the transformer windings. Each winding is given an angular orientation that corresponds to its voltage phasor.

Example 5.12   The Y–Y connected transformer *bank* in Fig. 5.18 shall serve as a step-up transformer between a 1000-MVA generator of 22-kV rating and a 340-kV grid. Determine the voltage and current rating of each identical unit.

SOLUTION: Each transformer must accept on its LV side a phase voltage of $|V'| = 22/\sqrt{3} = 12.70$ kV and deliver on the HV side the phase voltage

$$|V''| = \frac{340}{\sqrt{3}} = 196.3 \text{ kV}.$$

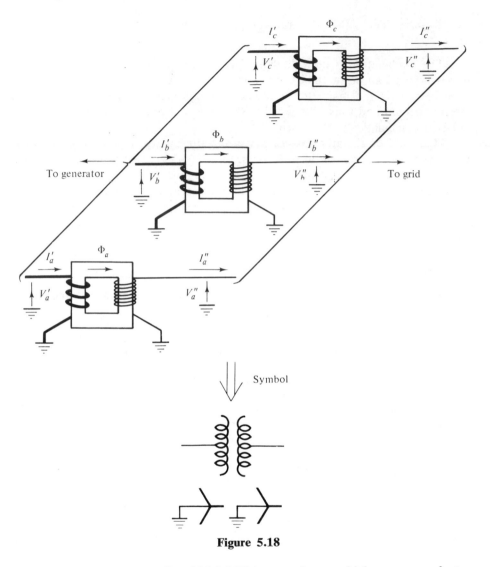

To generator

To grid

Symbol

Figure 5.18

The generator supplies 333.3 MVA per phase, which corresponds to a rated LV current of

$$|I'| = \frac{333.3}{12.70} = 26.24 \text{ kA per phase.}$$

We have for transformer the ratio

$$a = \frac{22/\sqrt{3}}{340/\sqrt{3}} = 0.0647;$$

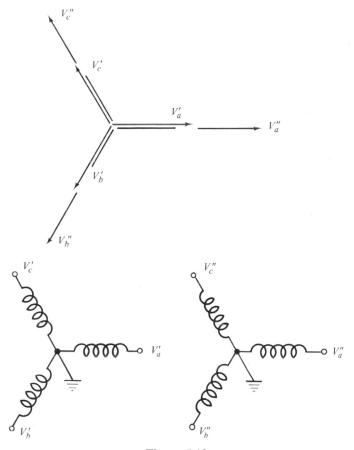

Figure 5.19

the rated HV current will be

$$|I''| = 0.0647 \cdot 26.24 = 1.698 \, \text{kA} \quad \text{per phase.}$$

In summary each unit must have these ratings:

$$\text{power rating} = 333.3 \, \text{MVA,}$$
$$\text{voltage rating} = 12.70/196.3 \, \text{kV,}$$
$$\text{current rating} = 26.24/1.698 \, \text{kA.}$$

5.8.2   Three-phase Core Arrangement

The three core fluxes Φ_a, Φ_b, and Φ_c in the bank arrangement in Fig. 5.18 constitute a symmetrical three-phase set. Consequently, their algebraic sum is

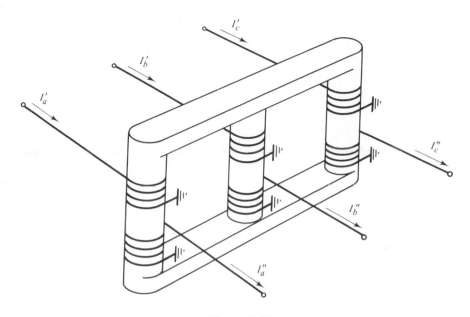

Figure 5.20

identically zero. For this reason, the three separate cores can be replaced by one called a "three-phase core" (Fig. 5.20) *lacking a flux return path.* Advantageously, this means a core containing less iron and of less cost. The disadvantage of the core arrangement is that a short circuit of a winding will disable the whole three-phase unit. A similar mishap in the bank arrangement disables only one-third of the total three-phase transformer.

Electrically, the arrangements in Figs. 5.18 and 5.20 are essentially identical.

5.8.3   The Y–Δ Connection

The Y–Y connection shown in Figs. 5.19 and 5.20 would work beautifully if all currents and voltages in our systems were of 60 cps. However, we have already shown (Fig. 5.9) that the magnetization current in a power transformer contains harmonics—particularly the third harmonic of frequency 180 Hz. This current harmonic can cause serious problems in a Y–Y connected three-phase transformer. Let us examine this.

We have shown in Fig. 5.21 the 60-cps voltage waves in all three phases. Shown also are the 180-cps current components in each phase. *Note that these latter components are all in time phase.*

This latter circumstance would mean that the three 180-cps phase components would add up to a neutral current of a magnitude three times that of each

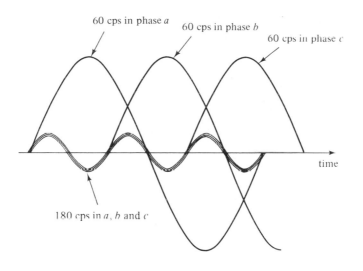

Figure 5.21

phase component. We could take action in several ways:

1) provide a fourth return conductor for this neutral current, or

2) let it exit into ground and "vagabond" back to the generator neutral, or

3) isolate the neutral and thus prevent the formation of the 180-cps component.

Neither of the first two alternatives is very attractive, from cost and communication disturbance point of view. The third alternative seems the simplest. But when you prevent the formation of the third current harmonic you will instead *distort the flux and thus the voltage waveshape*. (This is easily seen in Fig. 5.9. If you *forcefully* remove the 180-Hz component from the current, the flux and thus the voltage wave can on longer remain sinusoidal in time.)

The problem is solved by providing a so-called Δ-connected winding, either in the form of a separate tertiary or, as done in Fig. 5.22, by reconnecting the windings into Δ–Y. The Δ-loop permits the formation of the required 180-cps current component. It will circulate freely in the Δ, without entering the network. The fluxes can therefore take on the sinusoidal waveshape in all three cores, and no distortion of voltage waveshapes takes place.

It is not immediately clear how the 60-cps voltages and currents are related. Let us analyze the situation by considering the following specific example.

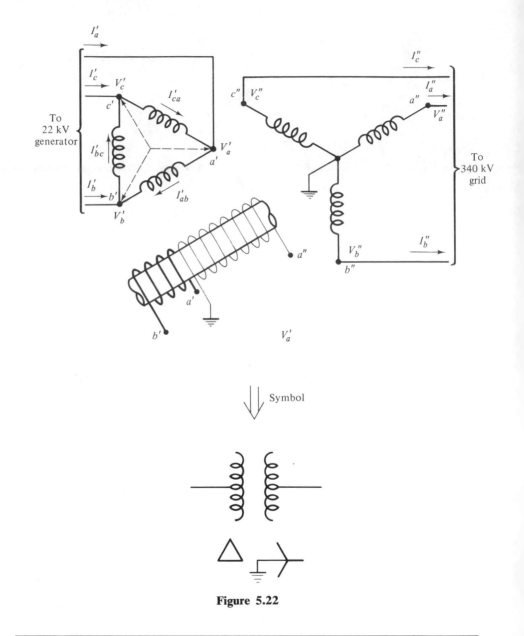

Figure 5.22

Example 5.13 The Δ–Y connected three-phase transformer in Fig. 5.22 serves as a step-up transformer between a 22-kV generator and a 340-kV grid. Determine the voltage ratings of each transformer unit and also compute all the currents and voltages which are identified in Fig. 5.22! It is assumed that the transformer delivers 700 MW and 90 MVAr to the grid.

SOLUTION: The LV side of each transformer unit is connected to the *line* voltage of the generator, that is, 22 kV. The HV side of the unit must output the *phase* voltage of the 340-kV grid, that is, 196.3 kV.

The voltage rating of each transformer unit will thus be 22.0/196.3 kV, that is, each single phase unit will have a transformation ratio of

$$a = \frac{22.0}{196.3} = 0.1121.$$

Voltage analysis. For an analysis of currents and voltages we shall for simplicity assume that each transformer unit is modeled as an IT. The generator neutral is *grounded* and by choosing V'_a as our *reference voltage* we can thus write for the primary phase voltages

$$\left. \begin{aligned} V'_a &= 12.70e^{j0°} \quad \text{kV,} \\ V'_b &= 12.70e^{-j120°} \quad \text{kV,} \\ V'_c &= 12.70e^{-j240°} \quad \text{kV.} \end{aligned} \right\} \quad (5.49)$$

These voltage phasors are shown dashed in Fig. 5.22.

Consider now the transformer unit (the core of which is depicted in Fig. 5.22) which delivers the phase voltage V''_a on its secondary. Because its input voltage equals $V'_a - V'_b$ we obtain from Eq. (5.5) for the secondary phase voltage:

$$V''_a = \frac{1}{a}(V'_a - V'_b) = \frac{1}{0.1121}(12.70 - 12.70e^{-j120°})$$

$$= \frac{1}{0.1121}(12.70 + 6.35 + j10.99); \quad (5.50)$$

$$\therefore \quad V''_a = 196.3e^{j30°}.$$

(This result could be obtained just as simply by *graphically* considering the length and direction of the phasor V''_a $V'_b = V'_a + (-V'_b)$.)

We similarly find the secondary phase voltages in the remaining two phases. We summarize the results:

$$\left. \begin{aligned} V''_a &= 196.3e^{j30°} \quad \text{kV,} \\ V''_b &= 196.3e^{-j90°} \quad \text{kV,} \\ V''_c &= 196.3e^{-j210°} \quad \text{kV.} \end{aligned} \right\} \quad (5.51)$$

Note that the directions of all these voltage phasors agree with the symbolic winding orientation in Fig. 5.22.

The results (5.51) indicate that all secondary phase voltages lead the corresponding primary ones by 30°. (Remember that in the Y–Y case they were all in phase.)

Current analysis.   From a knowledge of secondary power we can first compute the *secondary* phase currents. Formula (4.28) yields, if applied to phase *a*,

$$V_a''(I_a'')^* = \frac{700}{3} + j\frac{90}{3} = 233.3 + j30.0. \tag{5.52}$$

We solve for the conjugate phase current:

$$(I_a'')^* = \frac{233.3 + j30.0}{196.3 \cdot e^{j30°}} = 1.198e^{-j22.67°} \text{ kA.}$$

For the secondary phase current we thus have

$$I_a'' = 1.198e^{j22.67°} \text{ kA.}$$

As the secondary phase currents must possess three-phase symmetry (you can easily confirm this by analysis), we have

$$\left.\begin{array}{l} I_a'' = 1.198e^{j22.67°} \text{ kA,} \\ I_b'' = 1.198e^{-j97.33°} \text{ kA,} \\ I_c'' = 1.198e^{-217.33°} \text{ kA.} \end{array}\right\} \tag{5.53}$$

If we now apply formula (5.16) to the core depicted in Fig. 5.22, we get for the primary Δ-current I_{ab}',

$$I_{ab}' = \frac{1}{a} I_a'' = 10.69e^{j22.67°} \text{ kA.}$$

Similarly for the Δ-current I_{ca}',

$$I_{ca}' = \frac{1}{a} I_c'' = 10.69e^{-j217.33°} \text{ kA.}$$

From Fig. 5.22 we note that the primary phase current I_a' is

$$I_a' = I_{ab}' - I_{ca}'. \tag{5.54}$$

Thus

$$I_a' = 10.69e^{j22.67°} - 10.69e^{-j217.33°} = 18.52e^{-j7.33°} \text{ kA.}$$

The primary phase currents also possess three-phase symmetry:

$$\left.\begin{array}{l} I_a' = 18.52e^{-j7.33°} \text{ kA,} \\ I_b' = 18.52e^{-j127.33°} \text{ kA,} \\ I_c' = 18.52e^{-j247.33°} \text{ kA.} \end{array}\right\} \tag{5.55}$$

We have summarized the results of the above analysis in the phasor diagram in Fig. 5.23. The primary and secondary phase currents and voltages

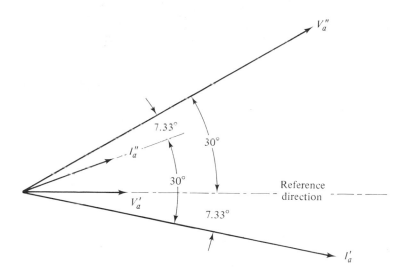

Note: Phasors are *not* shown in proper magnitude scale.

Figure 5.23

are shown. Because complete three-phase symmetry exists we have only included the phasors belonging to phase *a*.

The following important observations are made:

1) The *phase* voltages are transformed in the rms ratio

$$\frac{|V'_a|}{|V''_a|} = 0.0647;$$

that is, *in the ratio existing between primary and secondary line voltages.*

2) The *phase* currents are transformed inversely in the same ratio:

$$\frac{|I'_a|}{|I''_a|} = \frac{1}{0.0647} = 15.46.$$

3) Secondary phase voltages *and* currents are advanced 30° relative to the corresponding primary variables.

4) The magnitude of the primary phase current is $\sqrt{3}$ times the primary *delta current*. (Note that the *delta voltage* is $\sqrt{3}$ times the primary phase voltage.)

5) It should be noted that the primary transformer windings can be interconnected to form a Δ in *two* ways. The two alternatives are indicated in Fig. 5.24. (Alternative 1 was analyzed in Example 5.13). If we perform a similar analysis of the second alternative we shall find that the only

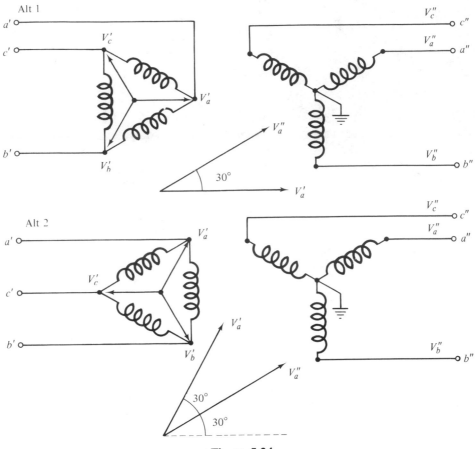

Figure 5.24

difference is a change of phase shift from $+30°$ to $-30°$ between primary and secondary phase variables.

6) The fact that the Δ winding introduces a phase shift of $\pm 30°$ has one very important consequence: *One must be careful in connecting three-phase transformers in parallel.* For example, one cannot operate in parallel Y–Y and Δ–Y connected transformers even if their line-voltage ratios are identical. Why?

5.8.4 Equivalent Circuits for Three-phase Transformers

As we compare the results of the analysis in Examples 5.12 and 5.13 we conclude that both Y–Y and Δ–Y transformers are *identical* in one important respect:

In both transformer types the magnitudes of primary and secondary phase voltages and currents are transformed in equal ratio—this ratio being computed as the ratio between line voltages.

The transformers are *nonidentical* in respect to the ±30° phase shift introduced in the Δ–Y transformer.

In power system studies it is always of great importance to know how three-phase transformers affect the *phase* values of current and voltage. The power systems engineer would thus be helped by an equivalent circuit that shows the relationship between primary and secondary *phase* variables.

In view of the above findings we can conclude that the equivalent circuits for Y–Y and Δ–Y transformers should be identical in all respects except that the Δ–Y equivalent circuit, in addition, must reflect the ±30° phase shift. The two circuits in Fig. 5.25 satisfy these requirements. The equivalent circuit in Fig. 5.25(a) represents a Y–Y connected transformer, the one in Fig 5.25(b) a Δ–Y connected transformer.

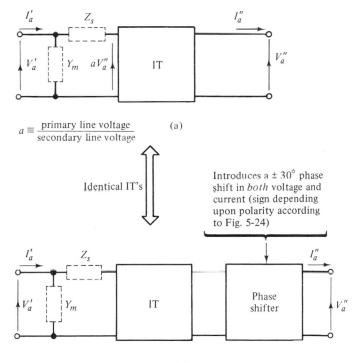

Figure 5.25

The box labeled "phase shifter" shall be thought of as capable of rotating *both* the input current and voltage phasors through the angle $\pm 30°$ (sign depending upon polarity according to Fig. 5.24).

We have also added to the diagram (dashed elements) series and shunt impedances that will account for nonideal features.

Example 5.14   Consider the 1000-MVA three-phase transformer treated in Examples 5.12 and 5.13. What voltage must be maintained on the generator terminals if the secondary voltage level is to be held at 340 kV? We assume the same delivered power as in Example 5.13. From a SC test the transformer series impedance has been found to be

$$Z_s = 0.00130 + j0.0711 \qquad \Omega/\text{phase},$$

as referred to the LV side.

Before we proceed with the solution we make two observations:

1) If the transformer were simply modeled as an IT we could immediately give the answer (22 kV). Clearly we are (as in real life) interested in the voltage drop caused by the series impedance Z_s.

2) Because we are interested in the *magnitude* of the generator voltage we are not concerned whether this is a Δ–Y or Y–Y transformer. We shall thus prefer to work with the simpler circuit (Fig. 5.25a).

SOLUTION: We choose the secondary phase voltage in phase a as our reference phasor:

$$V''_a = \frac{340}{\sqrt{3}} e^{j0°} = 196.3 \text{ kV}.$$

From a knowledge of secondary power we then compute secondary phase current I''_a:

$$V''_a (I''_a)^* = \frac{700}{3} + j\frac{90}{3}.$$

Solution for the current yields

$$I''_a = 1.198 e^{-j7.33°} \text{ kA}.$$

The line-to-line voltage transformation ratio equals

$$a = \frac{22.0}{340.0} = 0.0647.$$

A knowledge of this ratio permits us to compute the primary phase current (assuming Y–Y connection):

$$I'_a = \frac{1}{0.0647} \cdot 1.198 e^{-j7.33°} = 18.52 e^{-j7.33°} \text{ kA}.$$

The equivalent diagram (Fig. 5.25a) gives for the primary phase voltage

$$V_a' = aV_a'' + I_a'Z_s \quad \text{kV}, \tag{5.56}$$

or numerically,

$$V_a' = 0.0647 \cdot 196.3 + 18.52e^{-j7.33°}(0.00130 + j0.0711);$$
$$\therefore \quad V_a' = 12.96e^{j5.77°} \text{ kV/phase}.$$

Conclusion. We must maintain the generator terminal voltage at

$$\sqrt{3} \cdot 12.96 = 22.45 \text{ kV}$$

if we wish the transformer to deliver 340 kV to the grid.

5.9 PER UNIT ANALYSIS

In the work of analysis, electric power engineers prefer to use *per unit* (pu) or *percent* (%) measures of all types of system variables. These measures generally give a better *relative* sense of the variables in question.

5.9.1 Base Power

Consider for example the previous transformer problem. If we arbitrarily designate the rated MVA as the *base* MVA, then other powers can be expressed in percent or per unit of this base power. The transformer had a three-phase rating of 1000 MVA, or 333.3 MVA for single-phase. Assume (as in the previous example) that this transformer delivers 700 MW and 90 MVAr three-phase, or 233.3 MW and 30 MVAr single-phase. According to formula (4.31) this corresponds to

$$\sqrt{700^2 + 90^2} = 705.8 \quad \text{MVA, three-phase,}$$

or

$$\sqrt{233.3^2 + 30^2} = 235.3 \quad \text{MVA, single-phase.}$$

In terms of the chosen basepower this loading will be

$$\frac{235.3}{333.3} = \frac{705.8}{1000} = 0.7058 \text{ pu or } 70.58\%.$$

We may also state that the transformer outputs a complex power of

$$S = \frac{700 + j90}{1000} = 0.700 + j0.090 \text{ pu or } 70.0 + j9.00\%.$$

Note the importance of keeping track of three-phase and single-phase power values!

5.9.2 Base Voltage

The transformer in our example had a voltage rating of 22/340 kV line to line. If these voltages are chosen as *base* values, then, for example, the primary line-to-line voltage 22.45 kV represents 102.0% or 1.020 pu. A secondary line-to-line voltage of 320 kV similarly represents 0.941 pu in terms of *secondary* base values. The result arrived at in Example 5.13 can be thus stated:

If our transformer delivers 70.58% of MVA at a lagging power factor of 0.992 (cos 7.33°) it is necessary to maintain a primary (generator) voltage of 102.0% in order to keep the secondary voltage at an even 100%.

Note that the % measures, in addition to rendering relative value information, also disposed of the need to include the transformer ratio in the statement. Note also, however, the importance of *distinguishing between phase and line, primary and secondary voltages.*

For example, if we wish to express the *primary phase* voltage of 12.96 kV in pu, we must do so in terms of the primary *phase* base voltage $22/\sqrt{3} = 12.70$ kV/phase. We would obtain

$$\frac{12.96}{12.70} = 1.020 \text{ pu or } 102.0\%.$$

5.9.3 Base Current

Rated currents in a transformer (or *any* other power device) are determined by its rated apparent power and voltage. Recall that in Example 5.12 the rated primary current was obtained from the power rating (333.3 MVA) and the primary voltage rating (12.70 kV) thus:

$$|I'| = \frac{333.3}{12.70} = 26.24 \text{ kA}.$$

Similarly, if we wish to define a *base* current, we must do it in terms of already defined base values for power and voltage. If in the cited example the rated power is chosen as our base power $|S_b|$ MVA and the rated voltage as our base voltage $|V_b|$ kV, then the base current $|I_b|$ would be

$$|I_b| = \frac{|S_b|}{|V_b|} \quad \text{kA}. \tag{5.57}$$

In defining base values for power, voltage, and current, one has the option of arbitrarily choosing *any* pair of the three. But following this choice the third base value must always be computed from (5.57).

5.9.4 Base Impedance

Relatively little would be gained if pu and % values could be used as measures in expressing only computed values for voltage, power, and current. Maximum advantage can be gained if they are also *used in the analysis process itself.* To make this possible, it is, of course, necessary to express impedances in pu or %. For this purpose we need an *impedance base.* It is obtained by the following reasoning:

Connect an impedance Z across the base voltage V_b. The current drawn by the impedance has the rms value

$$|I| = \frac{|V_b|}{|Z|} \quad \text{A.}$$

The value for $|Z|$ that will make $|I|$ equal to base current $|I_b|$ will by definition be our *impedance base* $|Z_b|$, that is,

$$|Z_b| \equiv \frac{|V_b|}{|I_b|} \quad \Omega. \tag{5.58}$$

By using Eq. (5.57) we get the alternate formula:

$$|Z_b| = \frac{|V_b|}{|I_b|} = \frac{|V_b|^2}{|V_b| \cdot |I_b|} = \frac{|V_b|^2}{|S_b|} \quad \Omega. \tag{5.59}$$

Any impedance Z_Ω† can then be expressed in pu simply by relating it to $|Z_b|$:

$$Z_{pu} \equiv \frac{Z_\Omega}{|Z_b|} \quad \text{pu,} \tag{5.60}$$

or

$$Z_\% \equiv \frac{Z_\Omega}{|Z_b|} \cdot 100 \quad \%. \tag{5.61}$$

By use of formula (5.59), Eq. (5.60) can also be written as

$$Z_{pu} = Z_\Omega \frac{|S_b|}{|V_b|^2} \quad \text{pu.} \tag{5.62}$$

Example 5.15   In a certain system analysis the following power and voltage bases have been chosen:

$$|S_b| \equiv 333.3 \, \text{MVA/phase} \quad \text{(or 1000 MVA, three-phase)}$$
$$|V_b| \equiv 12.70 \, \text{kV/phase} \quad \text{(or 22.00 kV, three-phase)}$$

a) What will be the base impedance $|Z_b|$ expressed in ohms/phase?

---

† The subscript Ω indicates that the impedance is measured in ohms. The subscripts pu and % likewise indicate that the impedance is measured in pu and percent, respectively.

b) Express the impedance $Z_\Omega = 0.00130 + j0.0711$ in pu and % of the base impedance.

SOLUTION: a) Formula (5.59)† yields

$$|Z_b| = \frac{(12.70)^2}{333.3} = 0.4839 \ \Omega/\text{phase}.$$

b)

$$Z_{pu} = \frac{0.00130 + j0.0711}{0.4839} = 0.00270 + j0.1469 \ \text{pu},$$

or

$$Z_{\%} = 0.270 + j14.69 \ \%.$$

....

Example 5.16   Solve Example 5.14 by working exclusively with pu values. Select the same base values as those used in Example 5.15.

SOLUTION: We follow closely the solution procedure that was used in Example 5.14, so that the reader can make comparisons at each step.

We choose the secondary phase voltage as our reference phasor. Because its magnitude equals $(340/\sqrt{3})$ kV (that is, the base value) we thus have

$$V_a'' = 1.000 \cdot e^{j0°} = 1.000 \ \text{pu}.$$

The secondary power is

$$S_2 = \frac{700}{1000} + j\frac{90}{1000} = 0.700 + j0.090 \ \text{pu}.$$

For the secondary phase current we thus get

$$V_a'' \cdot (I_a'')^* = 0.700 + j0.090.$$

By solving for I_a'' we have

$$I_a'' = 0.700 - j0.090 \ \text{pu}.$$

---

† Note that one really should use voltamperes and volts in the formula. However, if MVA and kV are used the formula still gives the correct value. Check this!

Note also that one can write

$$|Z_b| = \frac{12.70^2}{333.3} = \frac{(12.70 \cdot \sqrt{3})^2}{333.3 \cdot 3} = \frac{22^2}{1000} = 0.4839. \tag{5.63}$$

Thus, the formula provides the correct answer if the three-phase value is used for power and line-to-line value for voltage.

The primary current expressed in pu is *identical*:

$$I_a' = I_a'' = 0.700 - j0.090 \, \text{pu}.$$

The voltage drop across the impedance Z_s (the pu value of which was computed in Example 5.15) becomes

$$I_a' \cdot Z_s = (0.700 - j0.090)(0.00270 + j0.1469) = 0.0151 + j0.1026 \, \text{pu}.$$

If we add this voltage drop to the secondary voltage, we will obtain the primary voltage:

$$V_a' = 1.000 + 0.0151 + j0.1026 = 1.0203 e^{j5.77°} \, \text{pu}.$$

Expressed in kV line to line this computed result represents

$$1.0203 \cdot 22 = 22.45 \, \text{kV}.$$

This result evidently agrees with that obtained earlier in Example 5.14!

The reader will note the obvious simplicity in the pu method of analysis. *We could completely ignore the transformation ratio.*

5.9.5 Transformation Between Different Base Systems

We shall work a final problem that will demonstrate the use of the pu method when *several base systems* are involved. This is normally the case in analysis work involving real systems.

Example 5.17 A 500-MVA, 22-kV, synchronous generator feeds into the 1000-MVA transformer discussed in previous examples. The generator has a synchronous reactance of 90% *based upon its own ratings*. Initially the secondary transformer breaker is open and the generator emf is set at such a value as to excite the transformer to 100% voltage. In this condition a secondary metallic short circuit occurs between all three phases. Compute the resulting SC current.

SOLUTION: Initially, the generator delivers only a very small magnetizing current to the transformer. The generator current is thus essentially zero. According to the equivalent circuit in Fig. 4.32, the emf E must thus equal the terminal voltage V. The latter equals the transformer primary voltage, and because the transformer voltage is 100% we conclude that $|E|$ must also be 100% or 1 pu.

Following the secondary short-circuit we can thus represent the generator–transformer by the equivalent circuit shown in Fig. 5.26. We wish to determine the magnitude of the SC current, I_{SC}, being pumped into the circuit by the 100% emf.

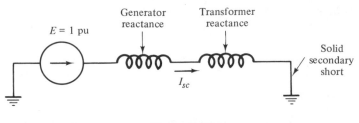

Figure 5.26

We make the following comments about the diagram:

1) We have neglected all resistances. As we have noted in earlier examples the resistive parts of all impedances are very small compared with the reactive parts.

2) The IT portion of the transformer network has been left out. (We have noted that pu currents and voltages are identical on both sides of the transformer.)

3) The SC current in a generator, as measured during the *first cycles* following the fault, is limited not by the synchronous reactance but by the *transient* reactance, which is only a fraction of the synchronous reactance.

   In using the latter our analysis renders the *steady state* SC current which will prevail some seconds later.

Before we can add the generator and the transformer reactances we must express them in pu *of the same base.* The generator reactance equals 0.900 pu on 500-MVA base. The transformer reactance is 0.1469 pu (see Example 5.15) on 1000-MVA base.

We need therefore a formula that will tell us how to shift a pu impedance from one base system (superscript (1)) to another (superscript (2)). Consider Eq. (5.62). An impedance Z_Ω expressed in pu value in terms of the first base system is written as follows:

$$Z_{pu}^{(1)} = \frac{Z_\Omega |S_b^{(1)}|}{|V_b^{(1)}|^2} \quad \text{pu.} \tag{5.64}$$

In terms of the second base system we similarly get

$$Z_{pu}^{(2)} = \frac{Z_\Omega |S_b^{(2)}|}{|V_b^{(2)}|^2} \quad \text{pu.} \tag{5.65}$$

By dividing these two equations we thus obtain the sought transformation formula:

$$Z_{pu}^{(2)} = Z_{pu}^{(1)} \frac{|S_b^{(2)}| |V_b^{(1)}|^2}{|S_b^{(1)}| |V_b^{(2)}|^2} \quad \text{pu.} \tag{5.66}$$

In the present case the two voltage bases are identical and the formula then simplifies to

$$Z_{\text{pu}}^{(2)} = Z_{\text{pu}}^{(1)} \frac{|S_b^{(2)}|}{|S_b^{(1)}|} \qquad \text{pu.} \tag{5.67}$$

Let us recompute the transformer reactance to the generator base:

$$Z_{\text{pu}}^{(2)} = 0.1469 \frac{500}{1000} = 0.0735 \text{ pu.}$$

The sum of the two reactances in Fig. 5.26 is

$$X_{\text{tot}} = 0.900 + 0.0735 = 0.974 \text{ pu} \quad \text{(on 500 MVA base).}$$

The steady state SC current is thus

$$|I_{\text{SC}}| = \frac{|E|}{X_{\text{tot}}} = \frac{1.00}{0.974} = 1.029 \text{ pu} \quad \text{(on 500 MVA base).}$$

Result: The steady state SC current represents only a 2.9% current overload of the generator. Not a dangerous current level!

(Had we instead computed the *instantaneous* or *transient* SC current we would have found its magnitude to be maybe four to five times the rated generator current; in other words, considerably in excess of tolerated current values.)

5.10 SUMMARY

Modern power systems span continents and thousands of megawatts of electric power often must be transmitted over distances measured in hundreds and even thousands of kilometers. Synchronous generators cannot generate power at voltage levels in excess of about 25–30 kV. We have seen that it is impossible to transmit bulk energy at such low voltage levels. We need to transform the power to and from voltage levels in the hundreds of kV. This is the job of the power transformer.

In this chapter we have first learned the basic functioning of the simple single-phase transformer. We have also developed mathematical models that can be used to predict the transformer behavior in system studies. These models were first developed for an "ideal transformer." Corrections were then added to account for realistic nonideal core and winding behavior. Brief discussions of multiwinding and autotransformers have been included.

All power systems are operated in the three-phase mode and it is necessary therefore to use transformers in this same mode of operation. We have discussed the basic Y–Y and Δ–Y connections. Numerous examples have been worked to demonstrate the theory.

We have ended with a discussion of the very important "per-unit" method of analysis. In practical power systems work, this method is overwhelmingly preferred over the regular "volt-amp-ohm" method. The simplicity of the method is demonstrated in several examples.

EXERCISES

5.1 A single-phase transformer is designed to operate at 60 Hz. Its voltage ·ratings are: primary, 500 V; secondary, 200 V. The maximum permissible load is 30 kVA.

a) What will be the magnitudes of primary and secondary currents when the device is full-loaded!

b) Loading is accomplished by an impedance connected across the 200-V terminals. How many ohms will correspond to full-load of the transformer? (Use IT model.)

5.2 Assume that you were to use the transformer in Exercise 5.1 in a 50-Hz power network. If you thoughtlessly connected it to a 500-V source, what values would you measure for (a) core flux? (b) secondary voltage? Answer in percent of design values. Would this nonintended use damage the transformer in your opinion? Explain!

5.3 The 30-kVA transformer in Exercise 5.1 is made subject to a SC test. One winding is short-circuited and the other winding is fed from a 60-Hz voltage source. The voltage is raised until rated current is circulated in the windings. This occurs when the applied voltage equals 5.11% of rated winding voltage. The transformer consumes 290 W during the test.

a) Compute the series impedance $Z_s = R_s + jX_s$ of the transformer referred to primary *and* secondary sides.

b) Compute the core flux during the SC test. (Express its magnitude in percent of normal operating flux.)

c) Why is it permissible to assume that *all* of the 290 W constitute ohmic losses in R_s and no part of it is core loss?

5.4 Assume that the transformer in Exercise 5.3 is operated from a very strong source. If a secondary short occurs, what would be the winding currents? Express your answer in amps and also percent of rated currents. Work the example also by using pu values for all variables.

5.5 The transformer in Exercise 5.3 is fed from a 500-V source. A load impedance $Z_L = 1.03 + j0.72 \ \Omega$ is connected across the secondary.

a) Find the currents in both windings and the secondary voltage by use of the IT model.

b) Same as in part (a) but include now the transformer impedance in your analysis. Take note of the change in your answers.

c) Is the transformer current-overloaded?

5.6 The 30-kVA transformer in Exercise 5.1 is made subject to an OC test. It is fed from a 500-V source with the secondaries open. The transformer consumes 230 W.

Based upon the SC and OC test data compute the efficiency of the transformer when loaded with the Z_L impedance in Exercise 5.5.

5.7 The 500/200-V, 30-kVA transformer in Exercise 5.1 is reconnected as a 700/500-V autotransformer. Compute the new kVA rating of the device.

5.8 The terminals of the 500/200-V windings in the previous transformer can be interconnected in *four* different ways, *two* of which will result in a 700/500-V autotransformer. Assume that you have interconnected the windings in the *wrong* way, but that you are of the belief that you did it the *right* way. In other words, you think that you have a 700/500-V autotransformer when in fact you have something else.

As you now connect the "700-V terminals" of your device to a 700-V source, you expect to obtain 500 V between what you presume to be "500-V terminals." To your surprise you get an entirely different voltage.

a) What voltage do you get?

b) What will happen to your transformer in this kind of treatment?

5.9 Small power transformers used as variable voltage supplies (in labs, for example) are often connected auto as shown in Fig. 5.27.

Compute the ratio between the two winding currents, I' and I'', when:

a) the sliding contact is placed on 50% voltage,

b) when it is placed on 10% voltage. (Note that the maximum voltage obtainable is 100%.)

c) Can you envision any danger of burning the winding in an extreme contact position?

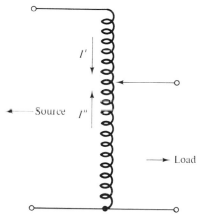

Figure 5.27

5.10 The "hydraulic transformer" shown in Fig. 5.28 transforms mechanical power between a high-force primary piston to a low-force secondary piston. The piston velocities are in inverse ratio to the forces. Thus the primary and secondary piston powers (force times velocity) are equal. This mechanical transformer can often be used as a good analog of an electric one.

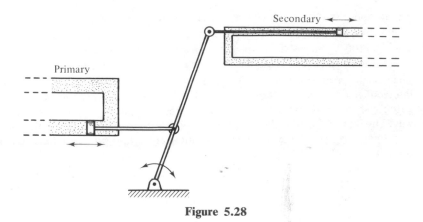

Figure 5.28

Consider the odd-looking electric transformer in Fig. 5.29(a). The winding 1 is connected to a generator. The magnetic flux divides equally between the two outer legs. One now attempts to load one of either windings 2, 3, and 4. Can this be done? Use the hydraulic analog in Fig. 5.29(b) to predict the results. Specifically, what happens with the voltages on windings 2 and 4 if you attempt to load 2?

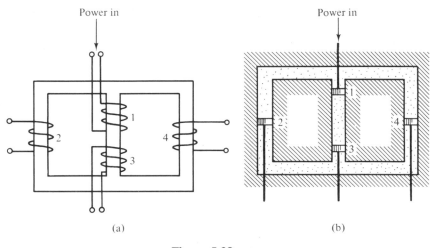

(a)   (b)

Figure 5.29

5.11 Work Example 5.11 in the text with one change: The variable tertiary load, like the fixed secondary load, is purely resistive. At what value of this tertiary load will the kVA ceilings be reached?

5.12 Three identical single-phase transformers, each having the ratings 5.500/23.20 kV, 60 Hz, 3000 kVA, are to form a Y–Y connected three-phase bank of total power rating 9 MVA. The bank will be fed from a three-phase generator of terminal voltage 9.526 kV (line value).

The load will consist of three equal resistors R, connected Δ. Use IT models exclusively in your analysis and find:

a) the value of R which will result in 9-MVA load on the transformer bank;

b) the current in each load resistor;

c) the current in each transformer, primary and secondary.

5.13 The three identical single-phase transformers in Exercise 5.12 are to be connected Δ–Y and fed from a 5.5-kV three-phase generator. The load consists this time of three identical reactors connected in Y. Each reactor has the inductance L H.

Again, use IT models and find:

a) the value of L to result in 9-MVA transformer loading.

b) the voltage across each reactor;

c) all line currents;

d) the Δ–current in each transformer primary.

5.14 Consider Exercise 5.12. Keep the load resistances R unchanged. In this exercise we will take into account the transformer series reactance which is

$$Z_g = 0.711 + j6.98 \qquad \%$$

for each single-phase unit (based upon ratings).

By how many percent must the generator voltage be increased from the previous setting of 9.526 kV in order to keep the secondary voltage at 100% exact?

5.15 In Exercise 5.3 you found by test the series impedance of a 30-kVA transformer.

a) Express the impedance in pu of rated power and voltage values.

b) Load the 200-V side of the transformer by a 1.333-Ω resistor. The transformer is fed from a 500-V source. Compute the voltage across the load!

Work the problem *exclusively* in pu values of impedances and voltages.

5.16 Rework Exercise 5.15 with one change—the load resistor is replaced by a 1.333-Ω *reactor*.

5.17 Rework Exercise 5.15 again with the load consisting of a 1.333 Ω capacitor.

Compare and contemplate the differences in voltage drops in the three cases. Does there exist a 1.333-Ω load impedance that results in zero voltage drop? If so, find it!

5.18 Two single-phase transformers are rated 2000 kVA and 4000 kVA, respectively. They have identical voltage transformation ratios and are connected in parallel, delivering a total power of 5000 + j1000 (kW and kVAr) to a common load.

a) How do the two transformers share this load assuming their impedances are $Z_{1pu} = 0.91 + j5.11$ and $Z_{2pu} = 0.81 + j7.38$ pu, respectively (based on their respective ratings)?

b) Is any of the transformers current-overloaded?

c) Prove that if we wish the transformers to divide the total kVA load in proportion to their respective kVA ratings we must require that $|Z_{1pu}| = |Z_{2pu}|$.

5.19 The auto connected three-phase transformer shown in Fig. 5.30 serves as a link between the 500 kV and 340 kV (line-to-line) portions of a power system. Determine the rms values of the currents indicated by arrows when the transformer carries a through power (three-phase) of $900 + j100$ (MW and MVAr). Model the transformer as an IT. (The "idle" Δ-winding will not carry any 60 Hz currents. It accommodates the circulating harmonic components thus preserving the sinusoidal flux.)

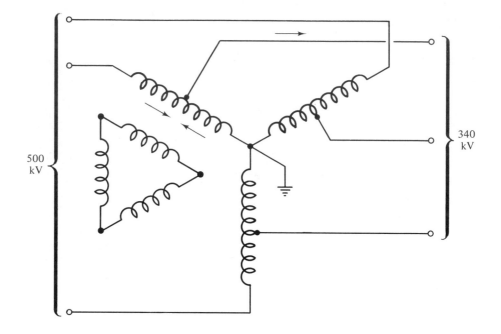

500 kV

340 kV

Figure 5.30

REFERENCES

Connelly, F. C. *Transformers.* London: Pitman, 1950

Elgerd, O. *Electric Energy Systems Theory—An Introduction.* New York: McGraw-Hill, 1971.

McNutt, W. J., et al. "Transformer Short-Circuit Strength—A State-of-the-art Paper." *IEEE Transactions,* Vol. PAS-94, no. 2 (March-April) 1975.

chapter 6

THE ELECTRIC
POWER NETWORK

*500-kV circuit breakers of compressed-gas (SF)
type. (Courtesy Westinghouse)*

It is not practically feasible with the present state of the technology to generate the electric energy at the location of its use. Local generation would be unacceptable for economic, environmental, and reliability reasons. Consequently, electricity must be generated in bulk quantities in *power stations* or *centers*. As the customer locations vary over vast geographic areas, the electric energy must be transmitted over an *electric power network* connecting the power stations with every customer.

The locations of the power centers are dictated by a number of factors. Fuel availability is one important such factor. For example, coal-fired generator units are often located in the vicinity of vast coal deposits. It is obviously more practical and economical to transport the energy to the user in electric form on electric transmission lines than in the form of coal shipments.

Environmental factors are becoming increasingly important in power-station siting. So is availability of condenser-cooling water. As this is being written, the various pros and cons of nuclear power stations and their siting problems are being hotly debated.

The size of generator units are becoming increasingly larger. Power ratings in excess of 1000 MW are not uncommon. For comparison, in 1976 in the United States, a 1000-MW unit will satisfy the varied electric power needs of a city with a population of about half a million people.

The electric power network must be designed so as to meet the need of every customer and at the same time permit energy shipments in the 1000-MW range. Immediately, one notices the analogy between an electric power network and a well-designed transportation system. The interstate superhighway system handles huge blocks of truck, bus, and long-distance auto traffic. State highways take care of medium-distance traffic. The small and lightly traveled urban and rural roads and lanes serve the local traffic needs.

In a power network the huge blocks of electric power move on the *grid* or *transmission* links. From the grid, power is being subdivided into smaller blocks and fed into the *subtransmission* portions of the power network. Finally, the individual small customers are being serviced from the *distribution* network.

6.1 POWER NETWORK STRUCTURE

From both an operational and functional point of view the power network can be divided into several substructures *based upon operating voltage level.* Highest on the voltage scale is the *transmission* or the *power grid.* The continent-spanning grid is the "interstate power highway system" over which the electric power is being "wheeled" in huge blocks. Figure 6.1 shows in a "one-line" diagram a part of our national grid. The generator, transformer, and line symbols should be self-explanatory. (Note that an autotransformer is used to intertie the 340-kV and 500-kV parts of the grid.)

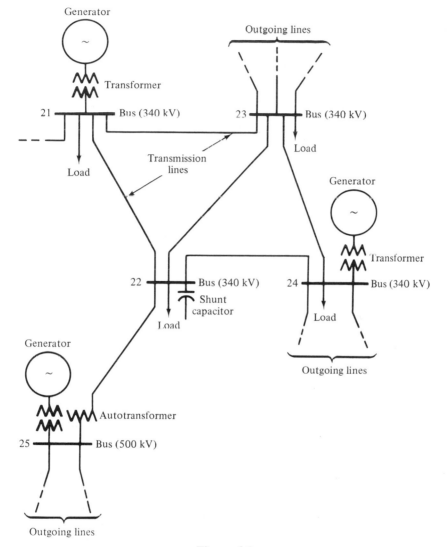

Figure 6.1

A typical feature of the power grid is its *loop structure*. Such a structure, as distinguished from a *radial type* network (Fig. 6.2) offers many alternative power paths. The transmission lines form the coarse meshes in the grid. The lines are interconnected in *switching stations* or at the *generating stations*. In power engineers' lingo, these network nodes are referred to as *buses*. The term *bus* emanates from the bus bars used for the actual intertie between the various components. Due to the physical size of the equipment these "nodes" may

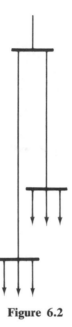

Figure 6.2

cover acres of ground. In Fig. 6.3 is sketched the physical appearance of the three-phase interconnection between a transformer and two outgoing lines; this is highly simplified. For example, no switches or curcuit breakers are indicated.†

The "load" symbols in Fig. 6.1 refer to power taps. They may symbolize service to a single huge industrial user or a whole city. Figure 6.1 also includes the symbol for a shunt capacitor. Such a device is often found on a bus which contains no generation. It is used (compare Example 6.14) to "support" the voltage of the bus in question.

The national grid is not operated by a single agency. For political, historical, and economical reasons, the grid is divided into individual *power systems.* Most of these are privately owned. Some are municipal operations. Part of the national grid is federally controlled. The sizes of these individual systems vary greatly. Many of them operate as "power pools" for mutual economic and technical benefit.

The HV power grid does not operate at a single voltage level. In the

---

† A power "switch" is a network isolation device that can be operated only under no-load condition. It is used to block power flows during repair or may also be used to sectionalize the system.

A "circuit breaker" is a device that can interrupt a circuit *under load.* Its operation is particularly crucial under fault conditions when large currents may have to be interrupted. The actual circuit interruption is performed by high-pressure gas or oil.

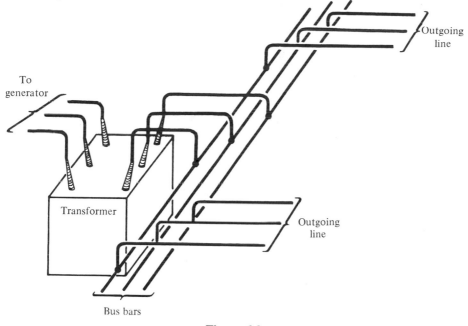

To
generator

Transformer

Bus bars

Outgoing
line

Outgoing
line

Figure 6.3

United States in 1976, the voltage levels range upward to 750 kV. The grid does not have a uniform strength. In some regions its links are quite weak. By "strength" we mean ability to transmit power (compare Section 6.5.5).

Some very large industrial customers may draw their power directly from the grid. More commonly the power is transformed in *bulk power substations* and fed into the *subtransmission system*, the voltage levels of which may vary typically from 50 to 150 kV. The subtransmission system forms an intermediate and more fine meshed link between the grid and the distribution circuits. It may be partly loop structured, partly *feeder structured*.

The power finally is fed into the fine meshes of the *distribution system* via *distribution substations*. In Fig. 6.4 we see in schematic fashion the voltage level division of the power system. It should be pointed out that no sharp demarcation lines exist between the various levels. It should also be understood that a power system, like a road system, is continually undergoing changes. What today may constitute a transmission link may be part of tomorrow's subtransmission system.

6.2 POWER SYSTEM OPERATIONS OBJECTIVES

We have pointed out that a power grid may span an entire continent. Theoretically, therefore, turning on a light switch in Florida will affect the

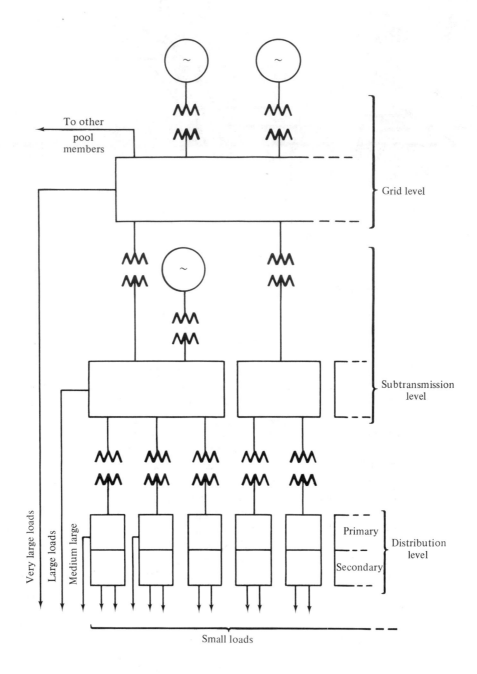

Figure 6.4

current flow in the California portion of the grid. In practice, however, the effects of network changes (like sudden steps in loads, generations, etc.) will be most strongly felt locally. The remote effects diminish rapidly with increasing distance.† This makes it possible to operate the individual power systems although interconnected, on an individual basis. However, certain functions, for example frequency control, and pooling operations must be done by mutual agreements between all power companies.

We identify now the most important objectives that must be met in *normal* operation of a power grid or the individual power systems which constitute its components:

1) Maintenance of real power balance.
2) Control of frequency.
3) Maintenance of reactive power balance.
4) Control of the voltage profile.
5) Maintaining an "optimum" generation schedule.
6) Maintaining an "optimum" power routing.

We stress that these are objectives to be met in *normal* system operation. Under *abnormal* or *fault* conditions the effects of system disturbances must be minimized; that is, we wish to operate with maximum *security*. It is not possible in this elementary treatise to cover thoroughly all aspects of system operation. We shall be content to give the reader an understanding of the most basic operational problems encountered under normal system operation.

6.3   REAL POWER BALANCE—THE LOAD-FREQUENCY CONTROL PROBLEM

The six main objectives that we stated above are not necessarily mutually exclusive. For example, the problem of keeping the frequency constant at 60 Hz is closely intertwined with the problem of real power balance. The term *load-frequency control* (LFC) describes this joint task. No doubt the LFC problem is the most basic one that confronts the power systems engineer. We shall presently turn our attention to its solution. First, however, we need to explain the term *load*.

---

† Physically this can be attributed to the "diffusion effect" that is always present in a vast power system. To demonstrate this effect consider what will happen in the case of a sudden load increase on bus 22 in Fig. 6.1. There are four incoming lines at this bus and the increased load current will thus be divided between these individual lines. The current in the line that terminates in bus 23 will again "subdivide" into five new lines, etc.

6.3.1   Load Characteristics

On several occasions in this and past chapters we have referred to system "loads." Let us now define more carefully what we mean by this term and, more importantly, study the factors that influence its magnitude.

As indicated in Fig. 6.4, power can be drained from a power network at all voltage levels. Large industrial loads tend to consist of big individual units, for example, a 300-hp mine elevator motor or a 10,000-hp drive motor for a reversible steel mill. Such huge load objects invariably are of three-phase design, that is, they represent a *balanced* three-phase loading on our system.

Domestic type loads are quite different. They are small but numerous and generally they tend to be single-phase. Figure 6.5 shows the typical connection

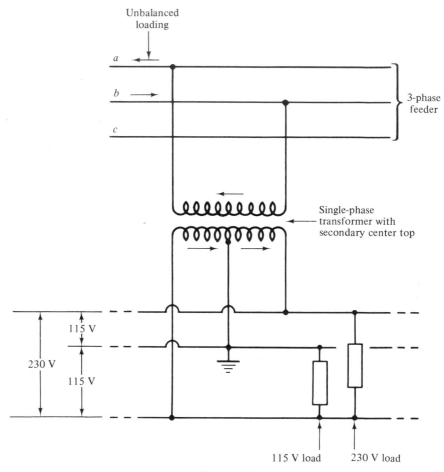

Figure 6.5

of a small single-phase distribution transformer. It has a 230-V single-phase secondary with a midpoint tap that is normally grounded. This type of secondary permits the loading of, for example, a 115-V television set or a 230-V heater—*both single-phase.*

The primary and secondary transformer currents are indicated in the figure and it is quite clear that this represents an *unbalanced* load on the three-phase network. However, the power company by distributing the individual loads between the three phases *can achieve a balancing effect.* For example, of the 60 houses in a subdivision, 20 may be connected between phases *a* and *b*, 20 between phases *b* and *c*, and the remaining 20 between phases *c* and *a*. (See exercises 6.1 and 6.2.) This balancing effect is more pronounced the further up the voltage ladder we climb. *Thus the total composite load as represented by a whole city is for practical purposes totally balanced between the three phases.*

All individual loads, whether they are toasters operated by housewives or a 10,000-hp motor run by a night-shift mill operator, are of *random* time character. However, because of the large number of individual loads we can always rely on the averaging effect of the laws of statistics. As a result, although the individual loads are entirely unpredictable, *the total load shows highly predictable time patterns.*

Figure 1.4 depicts load curves for a typical city. The power demand shifts from hour to hour in a smooth and predictable manner. The power system operator can usually with good accuracy predict from one day to the next what the load will be at a certain hour and schedule appropriate generation. What we have said about *real power* loads can also be applied to *reactive* loads. Practically all electric loads *consume*† reactive power, usually in *proportion* to the real power.

6.3.2 Load-Frequency Dynamics

Load demand curves, of the type shown in Fig. 1.4, are as we indicated characterized by a fairly smooth and predictable pattern as we view them on an hour-by-hour basis. If we look at them on a second-by-second or minute-by-minute basis they are quite different. Figure 6.6 shows a "microscopic" look at the load demand over a 15-minute interval. The system frequency is also recorded over the same time interval.

We note that the load demand averages out at about 1061 MW but is characterized by a *random fluctuation* of about ±2 MW. This is the "noise"

---

† Motors always are the most important ingredient in any type of load. A motor always contains iron cores, i.e. it will need magnetization current from the power source. (Compare the equivalent network representation of induction motors in Chapter 8.)

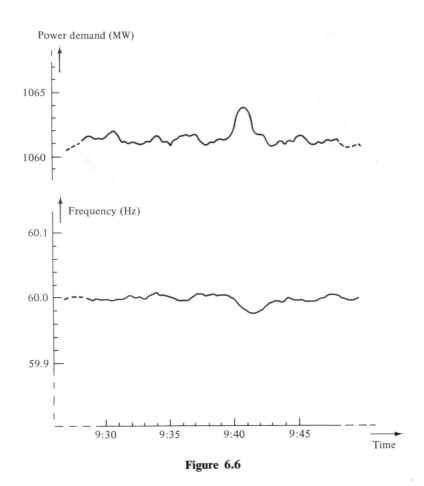

Figure 6.6

component that is entirely unpredictable. The frequency similarly averages about 60.0 Hz but on top of this average value it also has a "noise" component. There is, however, a definite correlation between the load and frequency curves. For example, a 3-MW load peak occurs around 9:40 o'clock. We can note the resulting dip in the frequency graph. Why does this happen? What are the specific connections between load demand and frequency?

6.3.3 A Mechanical Analog

The following mechanical analog illustrates the basic features of load-frequency dynamics in an interconnected electric power system.

A freight train consisting of many engines and many freight cars is supposed to travel at a constant speed of 60.0 mph. The engines represent the individual generators. The freight cars are the analogs of the electric loads. The *elastic* couplings between the engines and the freight cars represent the electric transmission lines—the "couplings" between the electric generators and loads.

Let us now see how the train analog can help us explain the load-frequency interplay in the electric power system. For example, consider what will happen if the train encounters a sudden load change in the form of an uphill grade. If the power settings of the engines remain unchanged the speed obviously will drop. Similarly, if the load in a power system suddenly increases but the generator power outputs remain fixed the frequency will drop.

In both the train analog and the power system the increase in load without a corresponding upward adjustment in generation would result in a power deficiency. The power must be taken from somewhere, and the only available energy source is the kinetic energy of the moving masses. As the kinetic energy is being consumed, the speed will drop.

Example 6.1   The total kinetic energy of the spinning rotors and turbines in a certain power system equals 600 megajoules (as measured at 60 cps). The system is running at constant frequency in perfect power balance when a sudden power-load increase of 2 megawatts sets in. What deceleration, measured in Hz/s, will the system experience if the generator turbine powers remain unchanged?

SOLUTION: As the power deficiency must equal the rate of change of kinetic energy we have the equation

$$-2 \cdot 10^6 = \frac{d}{dt} (w_{kin}) \quad \text{W.} \tag{6.1}$$

Because the kinetic energy is proportional to the square of velocity (or frequency) we can also write

$$w_{kin} = 600 \cdot 10^6 \left(\frac{f}{60}\right)^2 \quad \text{J.} \tag{6.2}$$

By substitution of (6.2) into (6.1) and upon performing the differentiation called for, we obtain for the frequency change

$$\frac{df}{dt} = -0.1 \text{ cps/s} \quad \text{(or Hz/s).} \tag{6.3}$$

6.3.4   Automatic LFC (ALFC)

In Example 6.1 if the power imbalance were sustained, the frequency would obviously drop to 59 cps in a time period of 10 seconds. A frequency dip of such magnitude would be entirely unacceptable in a modern power system, the frequency of which normally is kept within a tolerance of 60.0 ± 0.1 Hz.

There are many reasons why the frequency must be controlled to within these narrow limits of accuracy. Generally, it can be said that the tighter control we have on the frequency the better control we have over our entire system.†

Control of the frequency by *automatic* regulation of the generator output is achieved in most systems today. Figure 6.7 depicts schematically the workings of such an ALFC (automatic LFC) system. A *frequency sensor–comparator* senses the system frequency f and compares it with a reference frequency f_{ref} (60.00 Hz). A *frequency error* signal

$$\Delta f = f - f_{ref}$$

is thus generated, which is a measure of the frequency deviation. A transducer amplifies the error signal into an actuating command which is sent on to the turbine steam valve.

A positive error signal would indicate too high a frequency and the actuating signal would in this case issue a "lower" command in generator output P_G. A negative Δf would result in a "raise" command, that is, an opening of the steam valve.

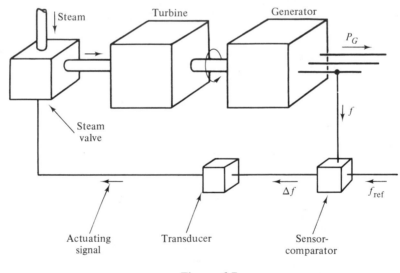

Figure 6.7

---

† There are millons of electric clocks running off today's power systems. The accuracy of these clocks depend upon the frequency constancy in the system.

A sudden shift in frequency is one of the surest signs of system trouble. The better frequency constancy we keep the easier we can detect trouble (you can detect the slightest pebble dropping in a perfectly smooth water surface).

By necessity this description is very superficial.† Many interesting questions arise in connection with the actual operation of an ALFC system, such as

1) How "responsive" should the control loop be? Clearly, it is not wise to let the generators "chase" every load excursion, however short it may be. This will cause unnecessary wear and tear on the equipment.

2) What generators should participate in the ALFC job? When the train in our previous analog encounters the uphill grade it may not be necessary to raise the power in all engines.

   Likewise, in a power system we delegate the ALFC job to those generators most suitable for the job. We have earlier noted that it is much easier to control the power level in a hydro turbine than in a steam-driven generator. Consequently, if we have a generation mix, hydro turbines are natural candidates for the ALFC job.

6.4 OPTIMUM GENERATION

The automatic LFC system maintains real power balance within the system on a second-by-second and minute-by-minute basis. This being accomplished the power system operator must make sure that the generators divide—over longer time spans—the total system load in a manner that guarantees *minimum operating costs.*

In a power system, a certain load demand can be met in an infinite number of ways. As a demonstration of this statement consider the simple two-bus system depicted in Fig. 6.8. A real-life power system never is this simple: however, it will serve well the purpose of demonstrating the principles of "optimum power dispatch."

Assume that the system is operating in the power configuration shown in Fig. 6.8(a). Sixty percent of the total load (500 MW) is tapped from bus 2, forty percent of the load exists at bus 1. Because the fuel is cheaper at bus 1, the majority of the generation occurs at this bus in generator G1. 100 MW is delivered to bus 2 from the line. Note that the line losses amount to 2 MW. Note also that power balance exists at each bus, that is, the power entering the bus equals the power leaving.

Let us now assume that the load on bus 2 increases by 50 MW. Where should this additional power be generated? The first assumption might be to let G1 handle the entire added increment because of its cheaper fuel. The *load flow* picture would then look as shown in Fig. 6.8(b). The added line power causes the line losses to increase to 5 MW—*a 3-MW increase.* (Compare loss formula (6.21).)

---

† For a detailed analysis of ALFC, see Elgerd, 1971.

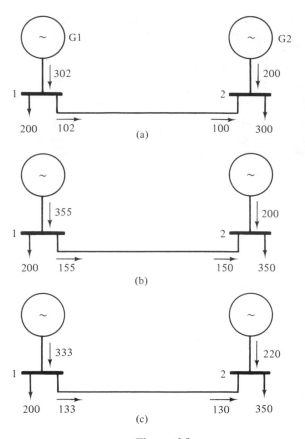

Figure 6.8

Were we to let generator G2 assume 20 MW of the added load, then the line power would be less and the losses only 3 MW—*an increase of only 1 MW.* The load flow is now shown in Fig. 6.8(c).

Which of the two generation alternatives is best? We don't know, of course, until we add up the generation costs in the two cases. In spite of the higher fuel costs at bus 2 the second alternative may prove cheaper overall because of the smaller line losses. It seems intuitively obvious that the overall costs will be minimized at some appropriately chosen load division between G1 and G2. In fact, a careful analysis confirms that *there generally exists one and only one power configuration that is cheaper than all others.* If we adjust our generators accordingly, then our system is said to be on "optimum power dispatch."

It is not particularly hard to find the optimum situation in the simple two-bus system. It is considerably more difficult in a system containing hundreds of lines and dozens of generating stations. The job would be impossible without modern computers. Most power systems are nowadays "computer dispatched."

It is beyond our scope to enter into a discussion of the mathematics of optimum dispatch. The reader will find a more detailed discussion in Elgerd, 1971.

6.5 LINE POWER AND ITS CONTROL

The transmission lines permit us to dispatch surpluses of power from one grid bus to another as was demonstrated in the previous example. They constitute the important network links that make it possible to choose alternate power flow configurations for optimum economy and security. In this section we wish to study the factors that affect the line power flows—and particularly, how we go about controlling these flows. First let us see how power lines are modeled.

6.5.1 Line Parameters

A three-phase transmission line is mostly of overhead design (Fig. 5.1). (In dense urban areas underground cables are often used when overhead lines would represent unacceptable safety hazards.) Typically, the bare stranded conductors consist of a steel core for mechanical strength and an outer current-carrying shell made of aluminum. To obtain a more flexible conductor, both the steel and aluminum portions are designed stranded. The conductors are hung on insulator strings from crossbars of the wooden, steel, or concrete transmission towers. Each of the three conductors in a three-phase line (Fig. 6.9a) is characterized by electric resistance. The current in each conductor surrounds itself with a magnetic field, resulting in a self-inductance. Finally, as shown in Fig. 6.9(a), there exists electric capacitance between each conductor. As demonstrated in Fig. 6.9(b), these are equivalent with a set of capacitances between each phase and a neutral node. (This equivalence can be verified by a so called Δ Y transformation of impedances.)

Electrically, the transmission line is thus characterized by circuit parameters in the form of both *series* and *shunt* elements. Clearly, the line resistance belongs in the former group. So does the *self-inductance* caused by the magnetic flux surrounding each conductor.

The capacitance which exists between the conductors represents a *shunt* or parallel admittance. (There is also a shunt *resistance* which represents the leakage current along the insulator strings in Fig. 5.1. For normal weather conditions this leakage current can be usually neglected.)

All of the above circuit parameters are *distributed* in nature and can be expressed in ohms-, henry-, and farad-per-meter line. If the line is designed *phase symmetrical* (like the one shown in Fig. 6.9a), all three capacitances of the Y-equivalent in Fig. 6.9(b) are equal. Consequently, the neutral node will

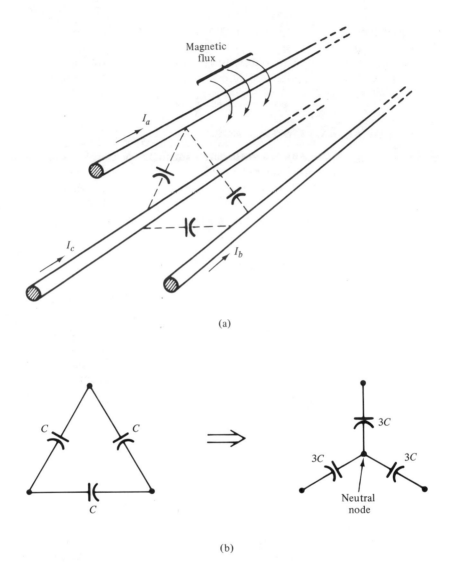

(a)

(b)

Figure 6.9

have zero (or ground) potential. It is then possible to represent the three-phase line on a per-phase basis as shown in Fig. 6.10(a).

The total effect of all these distributed parameters can be shown always to be equivalent with that of the *lumped* circuit in Fig. 6.10(b). If the line is "electrically short" (less than about 100 miles at a system frequency of 60 Hz) these lumped circuit elements are obtained from the distributed parameters by

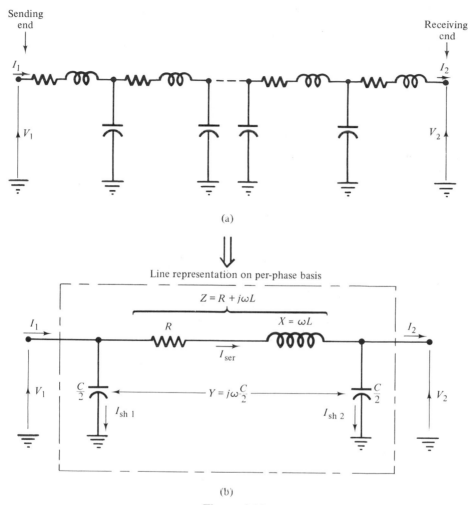

(a)

Line representation on per-phase basis

(b)

Figure 6.10

simply multiplication by the length of the line. In an "electrically long" line, the lumped elements follow more complicated formulas.† For more details, see Elgerd, 1971.

---

† The discussion given here is actually somewhat simplified. In addition to the self-inductance per phase, there is also a mutual inductance between phases. In addition to capacitance between phases, there exists also capacitance between each phase and ground. Finally (as demonstrated by the line in Fig. 5.1), a practical conductor arrangement is not always phase symmetrical.

It can be shown, however, that when all those factors are being considered a practical line can still be represented by the per-phase equivalent circuit of Fig. 6.10(b).

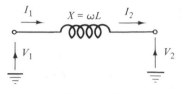

Figure 6.11

Normally the series reactance dominates over the resistance. In a typical network study, all three network parameters (R, L, and C) must be taken into account. Sometimes, for short urban lines, one may disregard both the resistance and capacitance and work with the simplified line model shown in Fig. 6.11.

Example 6.2 A three-phase 140-kV, 100-km transmission line consists of three conductors arranged in the manner shown in Fig. 5.1. By tests one has found the following line parameters:

$$\text{resistance} = 0.0910 \text{ ohms/km and phase,}$$

$$\text{inductance} = 1.34 \text{ millihenry/km and phase,}$$

$$\text{capacitance} = 8.85 \cdot 10^{-9} \text{ farads/km and phase.}$$

Compute the sending-end power if the receiving-end voltage is $|V_2| = 140$ kV rms and the receiving-end power is $S_2 = 100 + j60$ (three-phase mega-values). Assume the line to be "electrically short."

SOLUTION: We first compute the circuit parameters of the equivalent lumped circuit:

$$R = 100 \cdot 0.0910 = 9.10 \, \Omega/\text{phase,}$$

$$X = 100 \cdot 377 \cdot 1.34 \cdot 10^{-3} = 50.5 \, \Omega/\text{phase,}$$

$$Y = j50 \cdot 377 \cdot 8.85 \cdot 10^{-9} = j1.67 \cdot 10^{-4} \, \mho/\text{phase.}$$

Using these impedance and admittance values in the equivalent circuit in Fig. 6.10(b), we compute in a step-by-step fashion the sending-end power:

Step 1. Find receiving-end current!
From formula (4.28) we get

$$S_2 = V_2 I_2^* = \frac{100}{3} + j\frac{60}{3} = 33.33 + j20.00.$$

If V_2 is chosen our reference phasor we thus have

$$I_2 = \frac{33.33 - j20.00}{140/\sqrt{3}} = 0.4124 - j0.2474 \text{ kA/phase.}$$

Step 2. Find sending-end voltage!
The current† I_{sh2} (see Fig. 6.10b) is

$$I_{sh2} = V_2 Y = \frac{140}{\sqrt{3}} (j1.67 \cdot 10^{-4}) = j0.0135 \text{ kA/phase}.$$

We next compute the current† I_{ser} (see Fig. 6.10b):

$$I_{ser} = I_2 + I_{sh2} = 0.4124 - j0.2339 \text{ kA/phase}.$$

The voltage drop across the series branch is

$$\Delta V - I_{ser}(R + jX) - 15.56 + j18.70 \text{ kV/phase}.$$

The sending-end voltage is then obtained from

$$V_1 = V_2 + \Delta V = 96.39 + j18.70 \text{ kV/phase}$$

Step 3. Find sending-end current!
The shunt current† I_{sh1} (see Fig. 6.10b) is

$$I_{sh1} = V_1 Y = -0.0031 + j0.0161 \text{ kA/phase}.$$

The sending-end current thus equals

$$I_1 = I_{sh1} + I_{ser} = 0.4093 - j0.2178 \text{ kA/phase}.$$

Step 4. Finally find the sending-end power!

$$S_1 = V_1 I_1^* = (96.39 + j18.70)(0.4093 + j0.2178)$$
$$= 35.38 + j28.64 \text{ MW and MVAr/phase},$$

or

$$S_1 = 106.14 + j85.93 \text{ MW and MVAr 3-phase}.$$

Result: By comparison with the receiving-end powers we thus conclude that we lose 6.14 MW and 25.93 MVAr in transmission. We note also that the receiving-end current equals 481 A rms. The sending-end current measures 464 A. Why is the magnitude of the receiving-end current greater than that of the sending-end current?

6.5.2 Control of Line Voltage Profile

In operating a system like the one shown in Fig. 6.8 it is important to keep the voltage magnitudes $|V_1|$ and $|V_2|$ at each end *constant*, the reason being that all electric load objects are *voltage rated*, that is, they are designed to operate at a certain voltage level. For example, the light intensity from a light bulb varies strongly with voltage and we wish to avoid fluctuations in the light flux. The

---

† The currents, I_{sh1}, I_{sh2}, and I_{ser}, *cannot* be physically measured. Why?

power of an induction motor varies as the square of the voltage (Chapter 8).

This voltage control is achieved by means of *automatic excitation control* (AEC) of the individual generators. We remember from Chapter 4 that the generator emf is proportional to the rotor field current. This fact suggests that an AEC system should be designed in a manner as sketched in Fig. 6.12.

The generator bus voltage $|V|$ is sensed and compared in a *voltage sensor–comparator* with a reference voltage $|V|_{\text{ref}}$. The *error voltage*

$$\Delta|V| \equiv |V| - |V|_{\text{ref}} \tag{6.4}$$

is amplified and sent on as an actuating signal to the *field current source*. The latter will increase or decrease the field current depending upon the sign and magnitude of the error voltage.

The closed AEC loop of Fig. 6.12, together with the closed ALFC loop of Fig. 6.7, constitutes the two basic *control channels* of synchronous generators. *They are essentially noninteracting; the control of one does not significantly affect the other one.*

Because the AEC loop involves only *electrical* variables, whereas the ALFC loop also includes *mechanical* variables (steam valve settings), the former loop has much faster response than the latter.

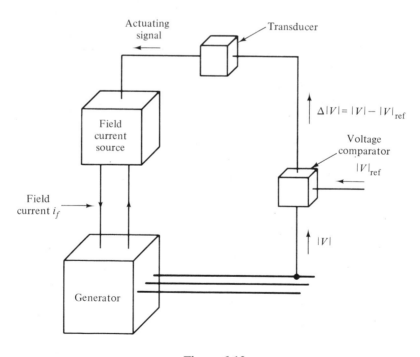

Figure 6.12

6.5.3 Control of Real Line Power

We are now in a position to understand better the important mechanism whereby the *real* power in a transmission line may be controlled in both magnitude and direction. To avoid unwieldy formulas we shall discuss a *lossless* line—we make the very reasonable and practical assumption that the line resistance $R = 0$. If we choose the simplified line representation in Fig. 6.11 the sending-end and receiving-end currents are equal. The complex powers in each end can then be computed from

$$
\begin{aligned}
S_1 &= P_1 + jQ_1 = V_1 I^*, \\
S_2 &= P_2 + jQ_2 = V_2 I^*.
\end{aligned}
\tag{6.5}
$$

These powers we already analyzed in Chapter 4 (Example 4.4). We restate the results here for easy reference:

$$
P_1 = P_2 = P = \frac{|V_1||V_2|}{X} \sin \delta \qquad \text{MW,} \tag{6.6}
$$

$$
Q_1 = \frac{|V_1|^2 - |V_1||V_2| \cos \delta}{X} \qquad \text{MVAr.} \tag{6.7}
$$

$$
Q_2 = \frac{|V_1||V_2| \cos \delta - |V_2|^2}{X} \qquad \text{MVAr.} \tag{6.8}
$$

We remind the reader that the *power angle* δ was defined as the phase angle between V_1 and V_2 (Fig. 4.3). Note that the voltages and reactance must be given in per-phase values to yield per-phase values of powers. Note also that because we neglected the line resistance, the *real* line powers at each end are equal. (For this reason we discard the subscripts and refer henceforth to the real line power P.) Here X is a *fixed* line parameter. We assume automatic excitation control of the generators in both line ends, that is, the voltage magnitudes $|V_1|$ and $|V_2|$ are both kept *constant*. Thus, *the only way in which P can be made to change is for the power angle δ to vary.*

Figure 6.13 shows how P varies as a function of δ. As δ increases in a positive sense (V_1 leading V_2) the power increases to a maximum value

$$
P_{\text{max}} = \frac{|V_1||V_2|}{X} \qquad \text{MW/phase} \tag{6.9}
$$

which occurs for $\delta = 90°$.

If we attempt to increase P by further increasing δ, the power will in fact *decrease. At this point the transmission collapses—the system steps out of synchronism.* We have reached the *transmission limit* or *static stability limit* for the line. "Stepping out of synchronism" means that G1 and the bus 1 load will run at one frequency, G2 and the bus 2 load at another.

If δ increases in a negative sense, (V_2 leading V_1) the power becomes negative. We now transmit power from right to left in Fig. 4.3.

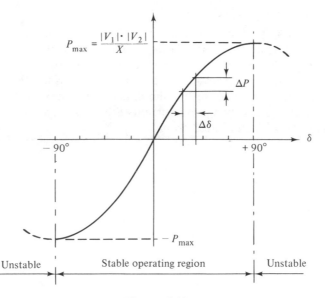

Figure 6.13

Note the difference between dc and ac transmission. In a dc system the sending-end voltage *magnitude* must exceed that of the receiving-end voltage in order for power to flow in proper direction. In an ac system the voltage magnitudes do not determine power direction but the voltage phase difference does.

Formula (6.9) tells us that the maximum transmittable power *increases* with the square of the voltage level of the line. Earlier we had found (formula 3.37) that the power loss *decreases* inversely with the square of the voltage. There is thus a double reason for use of high transmission voltages.

Example 6.3 How much power can be transmitted over the line shown in Fig. 6.8? We assume that the line is designed for 140 kV, is 50 miles long, and has a reactance of 40 ohms per phase. We neglect its resistance.

SOLUTION: Assuming that the line voltage is kept at 140 kV in both ends we get directly from Eq. (6.9) for the total three-phase maximum power

$$P_{max} = 3 \frac{\dfrac{140}{\sqrt{3}} \cdot \dfrac{140}{\sqrt{3}}}{40} = \frac{140^2}{40} = 490 \text{ MW}.$$

(Note that the formula yields the answer in *three*-phase *mega*-values of power if we use *line* voltage expressed in *kilo*-volts.)

Example 6.4 In the introductory part of Chapter 5 we concluded that we would encounter difficulties in transmitting 1000 MW along a 20-km line at a transmission voltage of 20 kV. Let us see, in light of formula (6.9), how much power this line in fact will transmit.

SOLUTION: In Chapter 5 we computed the line reactance $X = 8.72 \, \Omega/\text{phase}$. If we assume 20 kV in both ends and if we neglect resistance, we have

$$P_{max} = \frac{20^2}{8.72} = 45.9 \, \text{MW} \qquad \text{3-phase.}$$

(No wonder we had difficulties squeezing 1000 MW through this line!)

6.5.4 A Mechanical Analog

We can again make good use of the spring-coupled mechanical transmission depicted in Fig. 4.31. We remind ourselves that the power (or torque) transmitted through the coupling can be written in the form

$$P \sim R_1 R_2 \sin \delta. \tag{6.10}$$

In the case of an electric transmission line the voltages at *both* ends generally can be voltage controlled. For this reason, for the analog to make sense, both rods in Fig. 4.31 must be length controlled, that is, both rods should be telescoping.

Example 6.5 100 kW is transmitted via the "rod coupling" in Fig. 4.31. The angle δ between the rods equals 76°. Assume now that each rod is lengthened by 10%. What will be the new power angle? It is assumed that the transmitted power remains fixed.

SOLUTION: From Eq. (6.10) we get for the two cases

$$\begin{aligned} 100 &\sim R_1 R_2 \sin 76°, \\ 100 &\sim (1.1 R_1)(1.1 R_2) \sin \delta. \end{aligned} \tag{6.11}$$

Upon division we obtain:

$$1 = \frac{\sin 76°}{1.21 \sin \delta}; \tag{6.12}$$

$$\therefore \quad \delta = 53.3°.$$

How would the lesson you learn from this example be applicable to the operation of a power line?

6.5.5 Synchronizing Coefficient

The "strength" or "stiffness" of the mechanical rod coupling in Fig. 4.31 depends on the stiffness coefficient of the spring and on the lengths of the rods. The "electrical stiffness" or "synchronizing coefficient", T_{sync}, of a power transmission line is defined as

$$T_{\text{sync}} \equiv \frac{\Delta P}{\Delta \delta} \quad \text{MW/rad.} \tag{6.13}$$

It is a measure (Fig. 6.13) of the incremental increase in power, ΔP, resulting from an incremental increase in the power angle, $\Delta \delta$. By noting that the definition of T_{sync} is identical (in the limit $\Delta \delta \to 0$) to the derivative $dP/d\delta$ we get

$$T_{\text{sync}} = \frac{dP}{d\delta} = \frac{|V_1||V_2|}{X} \cos \delta \quad \text{MW/rad.} \tag{6.14}$$

Note that the stiffness approaches zero when we get close to the stability limit ($\delta = 90°$). *For this reason one never operates a line close to its power limit.*

Example 6.6   Consider again the 140-kV lossless transmission line treated in Example 6.3. We operate it with "flat" rated voltage profile—140 kV at both ends. The real power flow is 200 MW. Find power angle δ and also T_{sync}!

SOLUTION: This time we shall work our example in *per-unit* values. We must thus first choose a power and voltage base. We make the following selection:

Voltage base: 1 pu $\equiv$ 140 kV line to line or $140/\sqrt{3}$ phase to ground,

Power base: 1 pu $\equiv$ 100 MVA 3-phase or 33.3 MVA 1-phase.

Formula (5.59) then gives the impedance base:

$$|Z_b| = \frac{(140/\sqrt{3})^2}{33.3} = \frac{140^2}{100} = 196 \ \Omega.$$

The line reactance, 40 Ω, thus becomes

$$X = \frac{40}{196} = 0.2041 \ \text{pu.}$$

The transmitted power, 100 MW, is

$$P = \frac{100}{100} = 1 \ \text{pu.}$$

Formula (6.6) now reads in pu values

$$1.00 = \frac{1.00 \cdot 1.00}{0.2041} \sin \delta;$$

$$\therefore \delta = 11.78°.$$

Formula (6.14) gives the "electrical stiffness":

$$T_{\text{sync}} = \frac{1.00 \cdot 1.00}{0.2041} \cdot \cos 11.78° = 4.796 \text{ pu MW/rad},$$

or

$$\frac{4.796}{57.30} = 0.0837 \text{ pu MW/degree.}$$

This result means that if we increase the power angle from 11.78° to 12.78° the 3-phase power will approximately increase by 0.0837 pu or 8.37 MW.

6.5.6 Control of Reactive Line Power

Previous analysis shows the strong interrelation between real line power and the difference in *phase angles* between the sending-end and receiving-end voltages. The reactive line power shows an equally strong relationship with the difference in voltage *magnitudes*.

We see this most clearly from a study of formulas (6.7) and (6.8), which can be written

$$Q_1 = \frac{|V_1|}{X}(|V_1| - |V_2| \cos \delta) \qquad \text{MVAr/phase,}$$

$$Q_2 = \frac{|V_2|}{X}(|V_1| \cos \delta - |V_2|) \qquad \text{MVAr/phase.}$$

(6.15)

In normal operations $\cos \delta$ is fairly close to unity (compare Example 6.6). The factor inside the parenthesis in both Q-formulas thus tends to be proportional to the difference $(|V_1| - |V_2|)$.

Differently expressed:

The reactive line power flow tends to be in the direction of lowest voltage; the greater the voltage difference, the stronger the flow.

This reactive flow tendency is still more clearly seen if we look at the *average* reactive power flow:

$$Q_{\text{ave}} \equiv \frac{Q_1 + Q_2}{2} = \frac{|V_1|^2 - |V_2|^2}{2X} \qquad \text{MVAr/phase.} \qquad (6.16)$$

Example 6.7   Let us study the reactive power flow in the transmission line shown in Fig. 6.8. We shall assume the same real power flow (100 MW) as in Example 6.6. We will treat the following three cases:

1) $|V_1| = |V_2| = 1.00$ pu ("flat" voltage profile),
2) $|V_1| = 1.20$ pu, $|V_2| = 1.00$ pu,
3) $|V_1| = 1.00$ pu, $|V_2| = 1.20$ pu.

SOLUTION:

1) We had computed δ earlier (11.78°). We thus get directly from Eq. (6.15)

$$Q_1 = \frac{1.00}{0.2041} (1.00 - 1.00 \cos 11.78°) = 0.103 \text{ pu MVAr},$$

$$Q_2 = \frac{1.00}{0.2041} (1.00 \cos 11.78° - 1) = -0.103 \text{ pu MVAr}.$$

Note that when the voltage profile is kept flat, reactive power flows *into* the line from both directions. Clearly, the line (i.e., its reactance) consumes 0.206 pu MVAr in this case.

2) We first recompute δ:

$$1.00 = \frac{1.20 \cdot 1.00}{0.2041} \sin \delta;$$

$$\therefore \ \delta = 9.79°.$$

Thus

$$Q_1 = \frac{1.20}{0.2041} (1.20 - 1.00 \cos 9.79°) = 1.262 \text{ pu MVAr},$$

$$Q_2 = \frac{1.00}{0.2041} (1.20 \cos 9.79° - 1.00) = 0.894 \text{ pu MVAr}.$$

3) The δ angle will be the same as in (2). Using Eq. (6.15) we compute

$$Q_1 = -0.894 \text{ pu MVAr},$$
$$Q_2 = -1.262 \text{ pu MVAr}.$$

Figure 6.14 shows the three cases in graphical form.

It is interesting to note that by changing the voltage levels of the two buses, via AEC, a strong effect is noted in the reactive power flow *but no effect whichever on the real power flow*. There will, however, be a slight effect on the power angle δ.

6.5.7   Real Power Losses

Figure 6.14 shows a rather substantial *reactive power loss* on the line. In case (1) in the previous example the line absorbs a reactive loss of 0.206 pu MVAr (0.103 + 0.103). In cases (2) and (3) the loss has increased to 0.368 pu MVAr (1.262 − 0.894). This loss is, of course, consumed by the series line reactance. Formulas (6.6), (6.7), and (6.8) were derived on the assumption of zero resistance, and the *real power losses* will thus be zero.

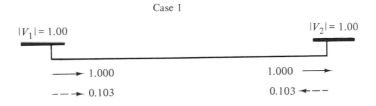

Case 1

$|V_1| = 1.00$ $|V_2| = 1.00$

→ 1.000   1.000 →

--→ 0.103   0.103 ←--

Case 2

$|V_1| = 1.20$ $|V_2| = 1.00$

→ 1.000   1.000 →

--→ 1.262   0.894 --→

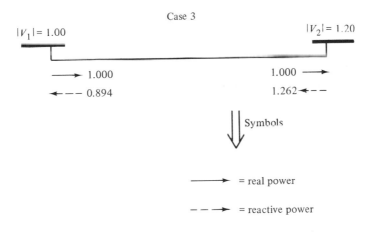

Case 3

$|V_1| = 1.00$ $|V_2| = 1.20$

→ 1.000   1.000 →

←-- 0.894   1.262 ←--

⇓ Symbols

——→ = real power

--→ = reactive power

Figure 6.14

A real line will, of course, have a certain series resistance R which will cause a certain ohmic power loss P_Ω. This power loss which can be written in the form

$$P_\Omega = R\,|I|^2 \quad \text{W/phase} \tag{6.17}$$

is in practice of greater importance than the reactive loss. We shall presently study how the real losses vary with the line flows.

We have earlier noted that voltage, current, and powers vary along the line. However, let V, I, P, and Q represent *average* values measured, for example, midline. The following relation exists between these variables:

$$P + jQ \approx VI^*. \tag{6.18}$$

We thus have

and

$$\left.\begin{array}{c} I^* \cong \dfrac{P + jQ}{V}, \\[2em] I \cong \dfrac{P - jQ}{V^*}. \end{array}\right\} \tag{6.19}$$

Multiplication of Eqs. (6.19) results in

$$I \cdot I^* = |I|^2 \cong \frac{P - jQ}{V} \cdot \frac{P + jQ}{V^*} \cong \frac{P^2 + Q^2}{|V|^2}. \tag{6.20}$$

By substitution of $|I|^2$ into Eq. (6.17) we obtain the following approximate loss formula:

$$P_\Omega \approx R|I|^2 \approx R\frac{P^2 + Q^2}{|V|^2} \quad \text{MW/phase.} \tag{6.21}$$

This formula is important because it tells us *that the real and reactive line powers contribute equally to the real power losses*. From the point of view of power loss one should therefore minimize the reactive line flow.

In practice this is accomplished by generation of the reactive power at the bus where it is needed. If no generator is available (remember that an overexcited generator generates reactive power) one often installs shunt capacitors for this purpose.

Example 6.8   Use the simple but approximate loss formula (6.21) to find the real line losses in Example 6.2. We remember that an exact (but tedious) computation had yielded the *exact* loss power $P_\Omega = 6.14\,\text{MW}$.

SOLUTION: From Example 6.2 we compute the following average values for line voltage and line power flows:

$$|V|_{\text{ave}} = \frac{170 + 140}{2} = 155 \text{ kV} \qquad \text{(line to line)},$$

$$P_{\text{ave}} = \frac{106.1 + 100}{2} = 103 \text{ MW} \qquad \text{(3-phase)},$$

$$Q_{\text{ave}} = \frac{85.9 + 60}{2} = 73.0 \text{ MVAr} \qquad \text{(3-phase)}.$$

From formula (6.21) we thus get

$$P_\Omega \approx 9.10 \cdot \frac{103^2 + 73^2}{155^2} = 6.04 \text{ MW} \qquad \text{(3-phase)}.$$

The approximate loss formula yields results with an error less than 2%!

Example 6.9   In Example 6.2, if the 60 MVAr needed at the receiving end of the line were produced locally what would be the real transmission loss?

SOLUTION: Formula (6.21) gives in this case

$$P_\Omega \approx 9.10 \frac{103^2}{155^2} = 4.0 \text{ MW} \qquad \text{(3-phase)}.$$

Result: The real losses would decrease from 6 MW to about 4 MW, that is, a 33% reduction. (Note that in using the formula (6.21) we assumed the same average voltage as in Example 6.8. This is not quite correct, of course, but we must remember that the formula is approximate.)

Example 6.10   Three equal shunt capacitors, C, are to be connected phase to ground at the receiving-end bus of the 140-kV line in Example 6.2 to produce locally exactly the 60 MVAr needed at this bus. Find the capacitor size!

SOLUTION: The voltage across each capacitor is $140/\sqrt{3}$ kV. From Table 4.1, item 3, we thus get

$$Q_{\text{produced}} = \frac{60}{3} = 20 = \omega C |V^2|;$$

$$\therefore \quad C = \frac{20}{377 \cdot (140/\sqrt{3})^2} = 8.12 \cdot 10^{-6} = 8.12 \ \mu\text{F/phase}.$$

6.5.8 Summary Observations

Let us at this time take stock of the most important findings of the previous sections and put them in proper perspective.

Observation 1.   An interconnected, synchronously run power system is operated at a uniform frequency, which is kept constant by automatic load-frequency control. ALFC implies a continuous process of real power balance within the system, using the frequency deviations as the indicator that the balance is achieved.

Observation 2.   The load as it is felt at the transmission level is of symmetrical three-phase type, fairly predictable and slowly changing throughout the day.

Observation 3.   The *real* power flow on a transmission line depends in magnitude and direction on the difference in phase angles between the end-point voltage phasors. The power magnitude increases with phase difference. The flow direction is from leading to lagging voltage.

Observation 4.   The *reactive* power flow on a line depends in magnitude and direction on the difference in magnitudes of the end-point voltages. The magnitude of the reactive power flow increases with increased difference. The flow takes place from higher to lower voltage.

Observation 5.   The real line power losses are proportional to the sum of the squares of real and reactive line flows and inversely proportional to voltage magnitude square.

Observation 6.   If an individual generator in an interconnected system is made to increase its real or reactive power output, the changes are felt most strongly locally, with the effects fast diminishing at increasing distance.

This is such an important observation that we find it appropriate to look at the situation more carefully. For this purpose, consider the generator feeding power into bus 21 in Fig. 6.1. Assume that we increase its real power output by, say, 50 MW by increasing correspondingly the steam flow through its turbine. What happens?

If the local load were to absorb the added 50 MW *nothing would be felt beyond bus 21*. Assuming that the local load remains constant, then the 50-MW surplus would be forced out on the system. Let us look at the mechanism with which this will happen, by considering the mechanical rod–spring analog in Fig. 6.15. With the relative rod positions given in the figure, rod 21 is pulling the rods 22 and 23 in the direction of rotation. This corresponds to electric line power in the directions $21 \rightarrow 23$ and $21 \rightarrow 22$. The line between buses 22 and 23 carries power in the direction $22 \rightarrow 23$.

Rod 23 via four springs is pulling on four other rods, etc. This means that electric power flows outward from bus 23 via the corresponding lines.

The added 50 MW at bus 21 advances the rotor of that generator by the angular increment $\Delta\delta_{21}$ (shown dashed in Fig. 6.15). As a result the other rods in Fig. 6.15 will advance by certain angular increments, *but those increments will be smaller since there are several restraining springs pulling the other way.*

Had we shown additional rod positions in the figure (corresponding to more distant buses in the electric network) these would have shown still smaller angular deviations.

The interpretations of these observations is as follows: The added 50 MW will first divide into two incremental line flows, maybe 25 MW in each line. The 25 MW reaching bus 23 will divide into four increments, of about maybe 6 MW each, which will continue on from bus 23, etc. We note that the real power increment will thus rapidly "dissipate" into the network.

We could make an analogous conclusion in regard to a change in the *reactive* output from the same generator. Now, of course, the result would not be changes in the angular position of the bus voltages but voltage *magnitude increases.* In the case of the reactive power flow, a large portion of it will dissipate in the line reactances (compare Example 6.7), the reason being that the series line reactance is much greater than the resistance.

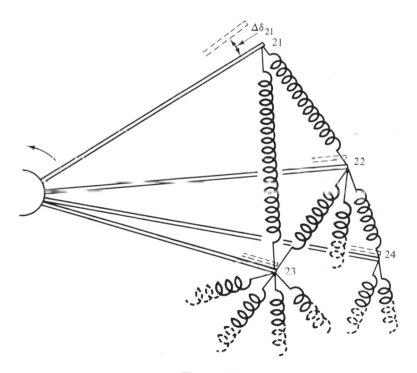

Figure 6.15

Observation 7. Three equal shunt capacitors connected to the bus will have the same effect on voltage and Q-flow as an overexcited generator. Three equal shunt reactors will have the effect of an underexcited machine.

Observation 8. A bus lacking both generator and shunt capacitors or/and reactors has an "uncontrollable" voltage. Its voltage (magnitude or phase) is determined solely by the effects of ALFC and AEC at the other network buses.

6.6 LOAD FLOW ANALYSIS

In the previous sections we analyzed the factors which influence the real and reactive power flows on an individual single transmission line. It is considerably more difficult to analyze the power flows in interconnected systems. For example, assume that 200 MW are demanded by the load on bus 22 of the system shown in Fig. 6.1. This power will be delivered via the four incoming lines. How will these lines share the load? What portion of the load will be supplied by the various generators in the system?

"Load flow analysis" (LFA) is the collective term for a number of computer-aided analysis procedures aiming at determining the actual power flow patterns in a given system and—more importantly—how to control these patterns.

Here are some of the most important objectives of LFA:

1) Determination of the real and reactive power flows in the transmission lines of a system based on certain *a priori* assumptions regarding loads and generations;

2) Computation of the voltages at all system buses;

3) Checking that no transmission line is overloaded. "Overload" can mean operation too close to its transmission limit or (in the case of UG cables) overheating;

4) Rerouting of power in case of emergencies;

5) Determination of the specific load flow pattern that results in "optimum dispatch."

LFA of power systems containing hundreds of buses and transmission lines is a rather complex procedure far beyond the objectives of this elementary treatise.† It is possible however, to demonstrate some of the basic features of LFA by considering very simple but instructive networks.

We shall choose for our analysis purposes the simple two-bus system that we have referred to earlier in this chapter. We simplify the system further by

---

† For a detailed presentation of LFA as applied to large-scale systems see Elgerd, 1971 or Stevenson, 1975.

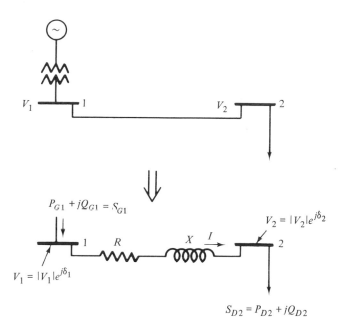

Figure 6.16

assuming that generation is available only at bus 1. Load demand exists at bus 2 only. In Fig. 6.16 we have depicted this simplified two-bus system. We have also shown its network model plus the power symbols that we shall be using in our analysis. (The subscript "D" refers to load "demand.")

6.6.1 LFA—Not a "Standard" Circuits Problem

An electrical engineer immediately identifies the LFA problem as an electric circuits problem. The "standard" procedure in solving such problems is to first represent the active sources as either *voltage* or *current sources*. Network equilibrium equations are then written in which either the network voltages or currents take on the role of unknowns. The "loads" are invariably represented by impedances, and if these and other circuit impedances are assumed known and constant, the resultant network equations turn out to be linear.

For example, assume for a moment that the load in Fig. 6.16 could be specified in terms of a *load impedance* Z_D. We could then model our two-bus system in the manner shown in Fig. 6.17. (For simplicity we have neglected the shunt impedance elements for the transformer and the transmission line). The current in the circuit would equal

$$I = \frac{E}{Z_G + Z_T + Z_L + Z_D} \quad \text{A.} \qquad (6.22)$$

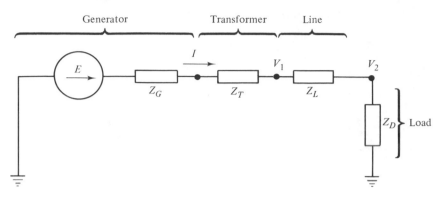

Figure 6.17

If E were known, the formula would yield the current. With a knowledge of the current we then could easily compute the bus voltages. From a knowledge of the bus voltages it would then be easy to compute all powers of interest.

In summary, the analysis would be straightforward, simple, and *linear*.† The linearity feature would still prevail if we were to extend the analysis to a multi-bus system with many generators and loads. Instead of the single linear current equation (6.22) we would now have a *linear system* of equations.

LFA in a power system can *never* be performed in the above simple manner for these reasons:

1) A power systems load never behaves in a manner as to make it possible to represent it by a constant impedance.

2) In a real situation the generator emf E is never explicitly known.

Instead we must write our network equations in terms of variables that can be easily measured and have practical significance. In power systems work these variables are (1) real and reactive *powers*, and (2) bus voltage *magnitudes.*

A typical LFA thus involves network equations written in terms of voltages and powers, *not* voltages and currents as exemplified by Eq. (6.22), a difference that will (as will be demonstrated in several examples) cause our LFA equations to be *nonlinear*. This fact will eliminate the possibility of an *analytical* solution of the load flow equations in most cases. We can, however, always arrive at *numerical* solutions by computer assistance.

---

† Linearity implies that I is proportional to E (compare formula 6.22).

6.6.2 LFA By Means of Analytical Approach

We noted in Section 6.5.6 how the reactive power flow in a system to a great extent depends on the voltage profile. Reactive power tends to flow from higher voltages to lower voltages. In review, if we wish to elevate the voltage level of a particular bus we should inject reactive power into the bus from appropriate sources. A Q-source connected to a bus thus in effect becomes a means for voltage control of that particular bus.

In Fig. 6.16, bus 2 lacks a reactive source. Therefore, we have no direct means for voltage control of that bus. However, bus 1 can obtain reactive power from the generator. An increase in the reactive generator output will result in a voltage elevation of bus 1.

By manipulation of the excitation control of the generator we can thus control the voltages of both buses, *but only in unison—not independently.*

Example 6.11   We demonstrate this type of situation by performing a load flow study of the two-bus system under the following set of assumptions:

1) The voltage of bus 2 must be kept at a magnitude $|V_2| = 1.00$ pu.

2) The load demand from bus 2 amounts to $S_{D2} = 8 + j5$ pu. The line impedance equals $Z_L = 0.005 + j0.030$ pu/phase.

SOLUTION: We have the following relationship between line current I, bus voltage V_2, and load power S_{D2}:

$$V_2 I^* = S_{D2}. \tag{6.23}$$

For the current we thus have

$$I = \frac{S_{D2}^*}{V_2^*} \quad \text{A/phase.} \tag{6.24}$$

The two bus voltages are related in the following way:

$$V_1 = V_2 + Z_L I \quad \text{V/phase.} \tag{6.25}$$

If we substitute for the current we thus obtain

$$V_1 = V_2 + Z_L \frac{S_{D2}^*}{V_2^*} \quad \text{V/phase.} \tag{6.26}$$

All variables on the right-hand side of Eq. (6.26) are specified. If we choose V_2 as our reference phasor we can thus solve for V_1:

$$V_1 = 1.00 + (0.005 + j0.030)\frac{8 - j5}{1.00} = 1.19 + j0.215 = 1.209\underline{/10.24°}.$$

Once we know V_1 we can compute the power† S_{G1} injected into bus 1:

$$S_{G1} = V_1 I^*$$ (6.27)

or numerically,

$$S_{G1} = (1.19 + j0.215)(8 - j5) = 8.44 + j7.67 \text{ pu.}$$

We have summarized the results in the power flow graph in Fig. 6.18. We note that the voltage drop (expressed as a difference between the rms values of the two bus voltages) equals 0.209 pu. Note that this voltage drop $|V_1| - |V_2|$ does not equal the drop $|V_1 - V_2| = |Z_L I| = 0.287$ pu.

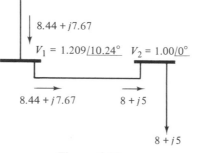

Figure 6.18

We see also that the power loss in the line is

$$S_{\text{loss}} = 0.44 + j2.67 \text{ pu.}$$

Most importantly, the analysis procedure is straightforward and the value sought for V_1 is obtained from a simple formula.

Once the two bus voltages are known, all other variables of interest are found most conveniently. (This is true in all types of LFA.)

6.6.3 LFA By Means of Iterative Computation

As a rule the LFA problem is not as simple as the previous example leads us to believe. Consider the following example.

Example 6.12 Rework Example 6.11 under these slightly changed specifications:

1) The load demand is unchanged, that is

$$S_{D2} = 8 + j5 \text{ pu};$$

2) The voltage of bus 1 must be kept at magnitude $|V_1| = 1.00$ pu.

Compared with Example 6.11 the voltages V_1 and V_2 have changed roles as knowns and unknowns.

---

† This power is measured (see Fig. 6.16) at the transformer HV terminals and represents thus the generator power minus the transformer losses.

SOLUTION: The solution procedure presented in the previous example does not work well in this case. In formula (6.26) V_1 is now known, and V_2 is an unknown. The equation is, in fact, a *nonlinear, second order, complex* equation in V_2. It looks fairly innocent but if the reader attempts to solve for V_2 he will find it surprisingly laborious.

We shall choose therefore a simpler approach utilizing what is referred to as an *iterative* computation procedure. It works as follows:

We make an initial reasonable *guess* at V_2. Call this value $V_2^{(0)}$. Using Eq. (6.24) we can now solve for I. We obtain

$$I^{(0)} = \frac{S_{D2}^*}{(V_2^{(0)})^*} .$$

From Eq. (6.25) we are then able to compute a *new* value for V_2. Call it $V_2^{(1)}$:

$$V_2^{(1)} = V_1 - Z_L I^{(0)} = V_1 - Z_L \frac{S_{D2}^*}{(V_2^{(0)})^*} \qquad \text{pu.} \qquad (6.28)$$

This new value for V_2 must be more accurate than the originally guessed value.

Starting with this updated value for V_2 we repeat the total process and obtain a still better value $V_2^{(2)}$, etc.

After *k iterations* this repetitive or *iterative* process obviously yields the V_2-value

$$V_2^{(k)} = V_1 - Z_L \frac{S_{D2}^*}{(V_2^{(k-1)})^*} \qquad \text{pu.} \qquad (6.29)$$

This equation represents a "computational rule" or *algorithm*. How many iterations need to be made? To find out, we made the initial guess

$$V_2^{(0)} = V_1 = 1.000 + j0.$$

The algorithm (6.29) yielded the following sequence of computed values:

Number of Iterations	Bus Voltage V_2
0	$1.000 + j0$
1	$0.810 - j0.215$
2	$0.715 - j0.190$
3	$0.677 - j0.215$
.	.
.	.
.	.
12	$0.619 - j0.215$
13	$0.618 - j0.215$
14	$0.618 - j0.215$

Within 3-decimal accuracy, the 14th iteration does not improve the computed V_2-value. The iterative process has numerically *converged on the solution.*

The above iteration scheme is named after Gauss. It is characterized by slow convergence but the algorithm is simple. There are other algorithms that converge much faster.† Clearly, this type of repetitive computation should be delegated to a digital computer.

We summarize the results:

$$V_2 = 0.618 - j0.215 = 0.654 \underline{/-19.18°} \text{ pu,}$$

$$I = \frac{8 - j5}{0.618 + j0.215} = 14.43 \underline{/-51.19°} \text{ pu,}$$

$$S_{G1} = V_1 I^* = 1.00 \cdot 14.43 e^{j51.19°} = 9.04 + j11.24 \text{ pu.}$$

The load flow picture is shown in Fig. 6.19.

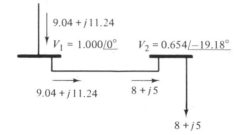

Figure 6.19

Note that the voltage drop is now 0.346 pu and the loss power has increased to

$$S_{\text{loss}} = 1.04 + j6.24 \text{ pu.}$$

These figures are considerably higher than those in the previous example.

6.6.4 Nonconvergence of the Iterative Process

The above examples, although performed on a very simple system, contain features which characterize LFA of large-scale grids. First, an *analytical* approach proves very unpractical—in fact, impossible in the general case. The equations simply do not lend themselves to analytical solutions. Second, by means of an iterative computing process one tries to obtain a *numerical* solution of sufficient accuracy. Sometimes, however, the iterative computations do *not converge.* Let us demonstrate.

---

† For a discussion of different iterative methods and their convergence properties, consult Traub, 1964.

Example 6.13 Work the previous example with one single change—the load demand has increased from $8 + j5$ to

$$S_{D2} = 10 + j5 \text{ pu.}$$

SOLUTION. Using the identical algorithm that we used in Example 6.12 we find that the iterations now look as follows:

Number of Iterations	Bus Voltage V_2
0	$1.000 + j0$
1	$0.800 - j0.275$
.	.
.	.
.	.
.	.
23	$0.123 - j0.275$
24	$-0.105 + j0.234$
.	.
.	.
.	.
39	$0.203 - j0.275$
40	$0.712 - j0.217$
.	.
.	.
.	.

....

No convergence! What is the explanation? The computer tries to tell us that *there is no solution.* We have actually asked *for the impossible!*

It is simply not physically feasible to deliver the power $10 + j5$ via a line with the impedance $0.005 + j0.030$ if we simultaneously specify that the bus voltage $|V_1|$ must equal 1 pu.

Whenever an iterative computation process does not converge on a solution one is wise to recheck the power and voltage specifications in order to verify that *physical laws have not been violated.* It should be added that convergence may not be achieved for other reasons as well. For example, if the initial guess is too far removed from the solution the computations may diverge *although the problem does possess a physical solution.* For example, the reader may try to solve Example 6.12 using the initial poor guess $V_2^{(0)} = 0.2$ pu.

6.6.5 LFA Involving Reactive Bus Power Injection

Example 6.12 teaches us an additional important lesson. If we compare the flowcharts in Figs. 6.18 and 6.19 we note that the voltage drops in the two cases are 0.209 and 0.346 pu, respectively. *As we lower the voltage on the generator bus it is getting increasingly difficult to keep up the voltage on the load bus and simultaneously maintain the specified load power on the latter.*

The physical explanation for this phenomenon is as follows: As we lower the voltage profile the line current must be correspondingly higher in order to give the *same* specified load power. (Note that the current increased from 9.43 pu in Example 6.11 to 14.43 pu in Example 6.12.) As a result the voltage drop along the line increases drastically. In fact, as we attempt to increase the load demand to $10 + j5$, as we tried in Example 6.13, *the load bus voltage simply collapses.*

It is illustrative to look at the situation graphically. We have done so in Fig. 6.20(a) where the bus voltages V_1 and V_2 in Example 6.12 are shown to scale. The shaded, voltage-drop triangle represents the line drop. Because the line impedance is predominantly reactive note how the voltage drop $|V_1| - |V_2|$ increases as the phase angle φ_2 increases. This phase angle is determined by the

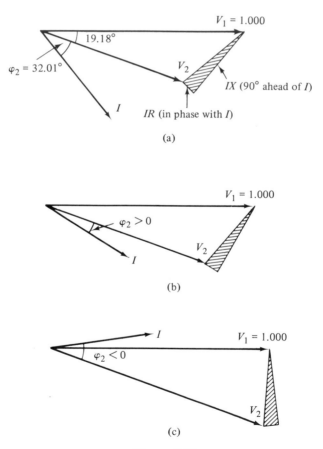

(a)

(b)

(c)

Figure 6.20

load S_{D2} and increases with the reactive part, Q_{D2}, according to the formula

$$\varphi_2 = \tan^{-1}\left(\frac{Q_{D2}}{P_{D2}}\right). \tag{6.30}$$

In our case we have

$$\varphi_2 = \tan^{-1}\left(\frac{5}{8}\right) = 32.01°.$$

In Fig. 6.20(b) we see how the voltage drop can be reduced if φ_2 could be reduced. If we actually make φ_2 negative (I leading V_2 as shown in Fig 6.20(c)) we can make $|V_2|$ not only equal to $|V_1|$ *but even larger than* $|V_1|$.

Thus Fig. 6.20 shows graphically what formulas (6.15) tell us *mathematically*—we need to inject reactive power into bus 2 if we want to increase ("support") its voltage. As we have noted before, if we lack a generator at bus 2 we can obtain the needed reactive generation from a bank of shunt capacitors.

The latter arrangement is shown in Fig. 6.21. The capacitor delivers Q_{G2} pu MVAr to bus 2. Clearly, if Q_{G2} exceeds the load demand of 5 pu MVAr, then the surplus reactive power will be flowing into the line. This means then that the current I will lead the voltage V_2 and we will have achieved the situation depicted in Fig. 6.20(c).

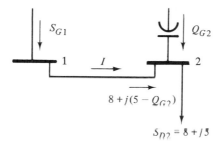

Figure 6.21

Example 6.14 In this example we want to determine the exact reactive generation Q_{G2} needed to maintain a "flat" unity voltage profile. That is, we specify $|V_1| = |V_2| = 1.00$ pu.

The load demand shall remain unchanged at the previous level:

$$S_{D2} = 8 + j5 \text{ pu.}$$

SOLUTION: Since the line power, as measured at the receiveng end, now equals $S_{D2} - jQ_{G2}$, we obtain for the line current

$$V_2 I^* = S_{D2} - jQ_{G2}. \tag{6.31}$$

Thus

$$I = \frac{S_{D2}^* + jQ_{G2}}{V_2^*}. \tag{6.32}$$

The voltage relation (6.25) is still valid. Substitution of the above I value into (6.25) yields

$$V_1 = V_2 + Z_L \frac{S_{D2}^* + jQ_{G2}}{V_2^*}. \tag{6.33}$$

We now choose V_2 as our reference phasor, that is, we set $V_2 \equiv 1.00$. Upon substitution of numerical values into Eq. (6.33) we obtain

$$V_1 = 1.00 + (0.005 + j0.030)\frac{8 - j5 + jQ_{G2}}{1.00}. \tag{6.34}$$

The specifications call for $|V_1|$ to equal unity. Thus we must require that

$$|1.00 + (0.005 + j0.030)(8 + j(Q_{G2} - 5))| = 1. \tag{6.35}$$

Upon separation of real and imaginary parts we obtain

$$|1.040 - 0.030(Q_{G2} - 5) + j[0.240 + 0.005(Q_{G2} - 5)]| = 1.$$

Using the formula

$$|a + jb| = \sqrt{a^2 + b^2}, \tag{6.36}$$

we are thus led to the equation

$$[1.040 - 0.030(Q_{G2} - 5)]^2 + [0.240 + 0.005(Q_{G2} - 5)]^2 = 1. \tag{6.37}$$

This is a second-order equation in Q_{G2} which can be readily solved. Solution yields

$$Q_{G2} = 7.410 \text{ pu MVAr.}$$

(Note that although we used an *analytic* solution in this simple case we could have used an iterative procedure instead. In a large-scale system this would *invariably* be the choice.)

Having found Q_{G2} the completion of the analysis is simple. First we find the line current from (6.32):

$$I = \frac{8 + j2.410}{1.00} = 8 + j2.410 = 8.355\underline{/16.76°} \text{ pu.}$$

(Note that current is leading V_2.)

Next we compute V_1 from Eq. (6.33):

$$V_1 = 1.000 + (0.005 + j0.030)(8 + j2.410) = 1.000\underline{/14.60°} \text{ pu.}$$

(Note that $|V_1| = 1$ pu as could be expected.)

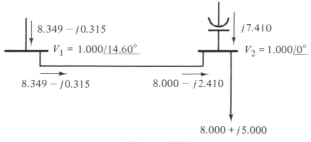

Figure 6.22

Finally we have the sending-end power:

$$S_{G1} = V_1 I^* = 1.00 \cdot 8.355\underline{/-2.16°} = 8.349 - j0.315 \text{ pu.}$$

We have summarized our findings in the flowchart in Fig. 6.22.

6.6.6 Some Comparisons of the Three LFA Examples

It is appropriate at this time to make some comparative observations of the three load-flow examples that we have analyzed above and the results of which have been summarized in the flowcharts in Figs. 6.18, 6.19, and 6.22.

Observation 1. The power angles vary between a maximum of 19.18° to a minimum of 10.24°. It should be noted that the real power delivered from the line is the same in all three cases.

This agrees well with the approximate† formula (6.6) which indicates an inverse relation between power angle and voltage (assuming constant line power). For the case in Fig. 6.18 the formula yields

$$\delta = \sin^{-1}\left(\frac{0.030 \cdot 8.22}{1.209 \cdot 1.00}\right) = 11.77°$$

which value should be compared with the *exact* value 10.24°. (The *I*-value 8.22 is the average as obtained from Fig. 6.18.)

Observation 2. In the last case (Fig. 6.22) all of the reactive power demand at bus 2 is supplied locally by the shunt capacitor. As a result the reactive power flow on the line has been drastically reduced (in fact, reversed). According to loss-formula (6.21) we should therefore expect a considerable loss reduction.

This is also confirmed from the flowcharts. The real power loss is only 0.349 pu compared with 0.44 and 1.04 pu, respectively, in the other two cases.

---

† Remember that the formula was derived on the assumption of zero line resistance.

Observation 3. The examples also give a very vivid confirmation of the *independence* between the *real* and *reactive* control channels. As we raise the voltage of bus 2 from the value 0.654 pu to 1.00 pu by the insertion of the capacitor the *reactive power flow changes drastically.*

However, the *real* power flow is almost invariant. (We assume of course that the load objects do not change their power demand as the voltage level increases.) The only change is a slight reduction of the real generator output resulting from the reduced line losses. This reduction is brought about automatically by the actions of the ALFC system. (If this readjustment would not take place we would be stuck with a slight real power surplus—and an increasing frequency.)

6.7 SUMMARY

The main function of an electric power network is to connect the generating stations with the individual customers. It must be designed so that the smallest blocks and largest blocks of power can be properly transmitted. We have learned that the transmission capacity of a line grows as the square of the line voltage, inversely with the magnitude of the line reactance.

The primary concern of the power systems engineer is to maintain a constant operating frequency. This job is normally assigned to an automatic control system that maintains at all times real power balance within the system. A mismatch in real power results in a frequency deviation.

The next most important job is to maintain a proper voltage profile throughout the system. This is accomplished by proper flows of *reactive* power on the various lines. One may express the problem thusly: If proper reactive power balance can be maintained, the voltage profile remains controlled. If the reactive power balance is not maintained, the voltage profile will drift (just like the frequency will drift if real power balance is not maintained).

This is particularly noticeable during night hours. The reactive generations in the shunt capacitors of the lines tend to provide a reactive power surplus (during the day hours this reactive power is consumed in the motor loads). Consequently, the bus voltages tend to increase.

We have also presented, in simplest of terms, the important load-flow analysis problem. By controlling the flow of both real and reactive power on the electric grid it is possible to control losses and thereby affect the economic operation of the system.

EXERCISES

6.1 Consider the transformer shown in Fig. 6.5. The 3-phase feeder voltage measures 11.5 kV between lines. The 115-V load consists of a total of 0.95 kW of single-phase inductive motor load of power factor $\cos \varphi = 0.8$. The 230-V load consists of 3.1 kW heaters at $\cos \varphi = 1.0$.

Model the transformer as an IT and compute all currents indicated by arrows. [*Hint*: The total mmf on the transformer core must be zero.]

6.2 The single-phase load in the previous exercise totals 4.05 kW connected between phases *a* and *b*. Assume now that we have three identical single-phase loads of this type. The remaining two are connected between phases *b* and *c* and *c* and *a*, respectively. Show that the total set of currents in the 3-phase feeder constitutes a symmetrical 3-phase set! Find the rms value of the current in each phase!

6.3 A power company does not normally try to exert any control over the amount of power that its customers drain from its network. However, in times of energy crisis the load may exceed the generating capacity of the company. The load must therefore be reduced. If voluntary means fail, the company may in the end have to disconnect its customers on some priority basis. Before this "final solution" is adopted, the company can reduce the customer load gradually by reducing the voltage.

Consider an industrial heating load consisting of three identical 12-ohm resistors connected in Y to a 12-kV, 3-phase feeder.

a) What MW load do these resistors represent if the 12-kV bus measures 12 kV?

b) What load do the same resistors represent if the power company lowers the voltage of the 12-kV bus by 5%?

If you work the problem correctly you will find that the voltage reduction causes a 9.75% reduction in the MW load. This would seem a saving for the customer. Why is he not happy?

6.4 The previous exercise demonstrates the dependency of the MW load upon the voltage. The load may also depend upon the frequency. Consider a load consisting of three identical impedances again connected in Y on the 12-kV, 3-phase bus in the previous example. Each impedance consists of a 20-Ω resistor in series with a 40-mH reactor. Prove that the MW load drawn by this set of impedances will *increase* if the frequency drops. In particular, by how many percent will the MW load increase if the frequency drops from 60.0 Hz to 59.5 Hz?

6.5 In Example 6.1 in the text the total kinetic energy of a certain power system amounts to 600 MJ. To get a feel for magnitudes, place all this kinetic energy in an equivalent cylinder made of solid steel, having a diameter that equals its length, and rotating at 1000 rpm. How big will this cylinder be? The density of steel is 7800 kg/m^3.

6.6 Consider the 3-phase, 140-kV, 100-km transmission line, the line parameters of which were given in text Example 6.2. The sending-end voltage is kept at 145 kV. The 3-phase sending-end power equals 120 MW at a 0.8 power factor (voltage leading current); $f = 60$ Hz.

a) Compute the current in each end of the line.

b) Compute the voltage in the receiving end.

c) Compute the 3-phase power at the receiving end.

d) Find transmission efficiency.

6.7 Consider again the transmission line in the previous exercise. The load at the receiving end consists of three equal 200-Ω resistors connected in Y.

    a) It is required to keep the receiving-end voltage $|V_2|$ at 140 kV exactly. What voltage $|V_1|$ must be maintained at the sending end to make this possible? $f = 60$ Hz.

    b) If the voltage condition in part (a) is maintained, what will be the sending-end and receiving-end powers, real and reactive?

6.8 The 100-km line in the previous two exercises is kept at a sending-end voltage of $|V_1| = 140$ kV. The opposite end of the line is open-circuited; $f = 60$ Hz.

    a) What will be the sending-end current?

    b) Show that the line *consumes* real power (how much?) but *generates* reactive power (how much?). Explain this physically.

    c) What is the voltage $|V_2|$ at the open end?

6.9 The line in the previous exercise has been de-energized for repair. A sleet storm has put a layer of ice on the line. Before taking the line into operation we wish to melt the ice by sending an estimated 100-A current into each phase.

For this purpose we short-circuit all three phases in one end and apply a symmetric 3-phase voltage in the other. How much voltage must be applied in order to inject 100 A into each phase? $f = 60$ Hz.

6.10 We represent a 100-mile line by the simplified model in Fig. 6.11. The voltage in each end is 100 kV. The line reactance is $X = 60$ Ω/phase. According to formula (6.6) this line can at most transmit

$$P_{max} = 167 \text{ MW}.$$

By putting three equal *series* capacitors C midpoint in each phase, we seek to reduce the series reactance from 60 Ω to 35 Ω and thus increase the transmittable power to

$$P_{max} = 286 \text{ MW}.$$

    a) What size capacitors are needed? $f = 60$ Hz.

    b) Compute the voltage across each capacitor if 200 MW is being sent over the line. The voltages in both ends are kept at 100 kV.

6.11 Although we would never do this in reality, assume that we would operate the 140-kV line in the text Example 6.6 at a power angle of 75°. (The voltages are assumed to equal 140 kV at both ends.)

    a) Compute the synchronizing coefficient, expressed in MW/degree.

    b) If the line power increases by 8.37 MW, by what amount will the power angle change? Compare with results in Example 6.6!

6.12 Assume that you operate the line in the previous exercise at a power angle $\delta = 89.9°$ (theoretically, of course).

Prove that the addition to the line load of one single horsepower (0.746 kW) will bring about collapse of the transmission!

6.13 Consider the 2-bus system shown in Fig. 6.23. The line connecting the two buses is modeled according to Fig. 6.11. The line reactance X equals 0.10 pu/phase.

There are synchronous generators at each bus. The generated powers indicated in the figure are measured at the HV terminals of the step-up transformers. A particular load case appears as follows:

$$P_{D1} + jQ_{D1} = 5 + j3 \text{ pu},$$
$$P_{D2} + jQ_{D2} = 1 + j1 \text{ pu}.$$

The two bus voltages must be maintained at 1 pu each, that is, $|V_1| = |V_2| - 1$ pu. The generators divide the *real* load equally; $P_{G1} = P_{G2} = 3$ pu MW. This means that 2 pu MW must be transmitted from bus 2 to bus 1.

Find the reactive load flows in each end of the line and also the reactive generations Q_{G1} and Q_{G2}! What will be the line power angle δ?

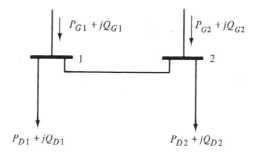

Figure 6.23

6.14 Consider again the system in Fig. 6.23. All *power* specifications are unchanged from the previous exercise. The line voltages must now be

$$|V_1| = 1.10 \text{ pu},$$
$$|V_2| = 1.00 \text{ pu}.$$

a) Find real and reactive line flows in each end!

b) Find reactive generations!

c) Find power angle, δ, of the line!

(Note from your results how the higher line voltage $|V_1|$ now requires a higher reactive generation at bus 1.)

6.15 Consider again the system in Fig. 6.23. We specify the following powers and voltages:

$$P_{D1} + jQ_{D1} = 5 + j3 \text{ pu},$$
$$P_{D2} + jQ_{D2} = 1 + j1 \text{ pu},$$
$$P_{G1} = 0,$$
$$P_{G2} = 6 \text{ pu},$$
$$|V_1| = 1.10 \text{ pu},$$
$$|V_2| = 1.0 \text{ pu}.$$

a) Find real and reactive line flows!

b) Find reactive generations!

c) Find power-angle, δ, of the line!

6.16 Consider Example 6.12 in the text. Write a computer program that will repetitively compute the voltage V_2 using the algorithm (6.29). The computer should stop when the computations have converged to within an accuracy of 0.0001 in *both* the real and imaginary parts of V_2.

6.17 This exercise is relatively difficult and requires a digital computer for its solution.

Consider the 3-bus system in Fig. 6.24. At the buses 2 and 3 loads are supplied in the amounts indicated. The load powers are given in pu MW and MVAr. The voltage $|V_1|$ of the generator bus 1 must be kept at 1 pu. The two lines are identical and each can be represented by a series impedance of $0.002 + j0.010$ pu/phase.

Develop a load-flow algorithm and write a computer program that will yield both the voltages V_2 and V_3 of buses 2 and 3, respectively. Start your iterations by assuming initially that $V_2 = V_3 = V_1 = 1.0$ pu. Stop the computer when your iterations have converged to within three-decimal accuracy in both the *real* and *imaginary* components of *both* voltages. Then compute the power flows in both ends of both lines.

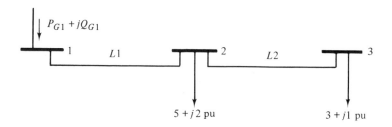

Figure 6.24

6.18 Example 6.13 in the text demonstrated the impossibility of delivering a power of $10 + j5$ pu at bus 2, assuming no voltage control of the latter. Prove that no such difficulties exist if voltage control is available.

a) Specifically, rework the text Example 6.14 with the specified load power increased to $S_{D2} = 10 + j5$.

b) What is in fact the largest real load power, P_{D2}, this line can accommodate, assuming $|V_1| = |V_2|$ and no limits on Q_{G2}? Q_{D2} is assumed to be fixed at 5 pu.

REFERENCES

Byerly, Richard T., and Edward W. Kimbark. *Stability of Large Power Systems.* New York: IEEE, 1974.

Elgerd, O. *Control Systems Theory.* New York: McGraw-Hill, 1967.

————. *Electric Energy Systems Theory—An Introduction.* New York: McGraw-Hill, 1971.

Neuenswander, J. *Modern Power Systems.* Scranton, Pennsylvania: International Textbook, 1971.

Stevenson, William T. *Elements of Power Systems Analysis.* New York: McGraw-Hill, 1975.

Traub, J. F. *Iterative Methods for the Solution of Equations.* Englewood Cliffs, N.J.: Prentice-Hall, 1964.

chapter 7

THE DC MOTOR

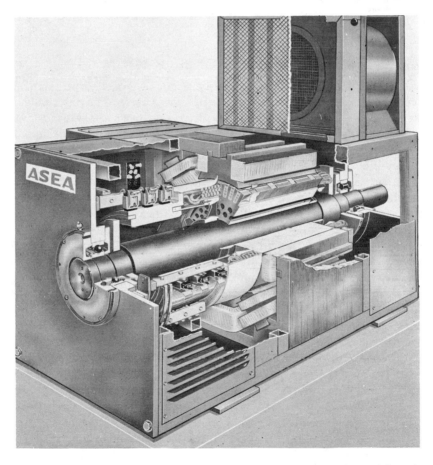

Artist's sketch of a dc machine of forced-ventilated design. (Courtesy ASEA)

This chapter and the next will deal with electric motors and their controls. In the present chapter we cover dc motor theory; Chapter 8 is devoted to ac motors.

The system loads that we discussed in the previous chapters consist, in general, of a multitude of electrical devices ranging from electric bread toasters and lighting equipment in domestic-type loads to huge motors and lift magnets in the industrial sector. As we survey this vast and varied field of devices we find that electric motors constitute typically the majority of the overall load, measured in MW.

The vast majority of the electric motors are of ac design (Chap. 8). Dc motors constitute a distinct minority. Not only is a dc motor more expensive than an ac motor of equivalent size, but it also requires a special dc supply system. The reason why the dc motor still survives in the competition is that it has some very special characteristics to which no other motor can lay claim.

7.1 TORQUE-SPEED REQUIREMENTS OF MOTORS

There is a large variety of motors, electrical and nonelectrical, in use in today's technology. They turn the wheels of our society. The internal combustion engine (ICE) is the most important drive system in the transportation sector. The hydroturbine—the descendant from the waterwheel—still finds important use, particularly as a prime mover in hydropower plants. The steam turbine drives the majority of the world's electric generators.

The electric motor is, however, by far the most versatile powering device available. It completely dominates the industrial, commercial, and domestic application areas. Electric motors come in sizes varying from a few watts to thousands of horsepower. The ac induction motor has a simple and rugged design, and it meets the requirements of a vast spectrum of applications. The dc motor does not find its match in terms of versatility and ease of control. It is used when no other motor can do the job. Important dc motor uses are in the areas of hoist and crane applications, electric traction, and in a number of industrial controls.

The single most important requirement that a motor—any motor—must meet is matching the torque–speed (TS) characteristics of the load it pulls. Consider for example the TS demands put on an automobile engine.

Figure 7.1(a) shows a work duty cycle which incorporates most of the requirements of an auto, including start, run, stop, and reverse. During those periods of the duty cycle when the speed must be changed the torque requirements are most severe. For example during startup the motor drive torque T_m must equal the sum of the dominating inertia torque T_i and the friction torque T_f:

$$T_m = T_i + T_f = ma + T_f. \tag{7.1}$$

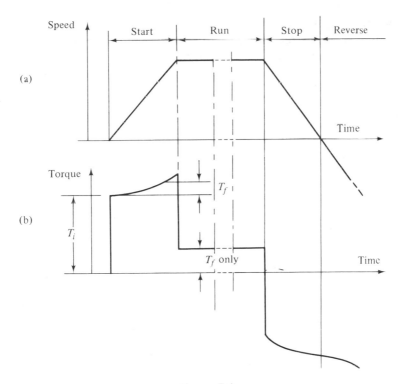

Figure 7.1

If the acceleration a is assumed constant, then T_i will also be constant. Usually T_f increases about quadratically with speed. We have indicated the motor torque requirement, T_m, in Fig. 7.1(b). The ICE poorly meets these requirements. It has zero starting torque—in fact, it takes a dc starter motor to get the ICE itself going. It cannot deliver the smooth torque needed from zero to full speed. Either a manual gearshift or a complicated torque converter (automatic transmission) must be installed between the engine and the drive wheels to obtain the desired TS characteristics called for.

The ICE easily delivers the torque needed during "run," but during "stop" most of the negative torque needed is obtained from mechanical brakes. During the "reverse" part of the duty cycle, the torque and speed both are negative. The ICE engine can run in only one direction, so again we must call upon the torque converter to give proper TS characteristics.

In contrast, a dc motor could provide easily the required TS demands—without assistance of either torque converter or (to some extent) even brakes. Furthermore, a dc motor is quiet, nonpolluting, cheap, and has only one moving part—the rotor.

Why, then, is the ICE still dominating the auto market?

Clearly, the single reason is the unique ability of the ICE to run 200–300 miles on a single tank of gasoline. When the day arrives when we finally find an electric battery that contains—per weight unit—as much energy as a tank full of gasoline, we shall most probably see the "electrics" rapidly capture the auto market. In the meantime we shall see them winning inroads in low-speed, short-range, heavy-duty urban applications.

7.2 A DC MOTOR PROTOTYPE

The simple arrangement in Fig. 7.2 depicts a "linear" dc motor. In its simplicity it demonstrates most of the typical characteristics of a normal rotating dc motor. A rod of length L can freely move along two supporting rails, perpendicular to a uniform, vertical magnetic field B. A dc source of constant terminal voltage, V, supplies current to the rod via a "starting resistor."

If the circuit breaker CB is closed the source will cause a current i to flow in the indicated direction. According to formula (3.68) the rod will be subject

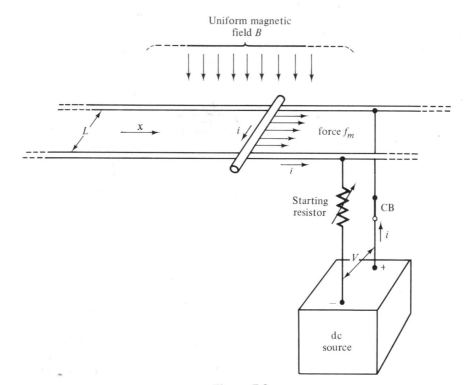

Figure 7.2

to an electromechanical or *motor* force f_m of magnitude

$$f_m = BiL \quad \text{N} \tag{7.2}$$

acting in the positive x-direction.

The motor force can be used to accelerate the rod, that is, to overcome the inertia force f_i. It can also pull along a cart that we may hook onto the rod and which represents a load force, f_L.

Mathematically this can be stated

$$f_m = f_i + f_L = m\dot{s} + f_L \quad \text{N}, \tag{7.3}$$

where

$$s = \text{rod velocity, m/s;}$$

$$m = \text{rod mass, kg.}$$

As the rod picks up speed, an emf, e, will be induced which according to formula (3.58) will be of magnitude

$$e = BLs \quad \text{V} \tag{7.4}$$

According to Lenz's law, the emf polarity will be such as to oppose the current, that is, the cause of the motion. The total loop resistance R is made up of the rod resistance, the contact† resistance between rod and rails and the external current limiting "starting" resistor. By application of Ohm's law the current will assume the value

$$i = \frac{V - e}{R} = \frac{V}{R} - \frac{BL}{R}s \quad \text{A.} \tag{7.5}$$

If we combine the Eqs. (7.2), (7.3), and (7.5) we obtain the following first-order, linear differential equation for the speed s:

$$\dot{s} + \frac{B^2L^2}{mR}s - \frac{BLV}{mR} + \frac{f_L}{m} = 0. \tag{7.6}$$

Also included in the load force f_L are all friction forces acting on the rod.

7.2.1 Steady-State Speed in No-Load Case

In the absence of friction or other restraining forces the load force f_L is zero. This is the "no-load" case. The rod will now accelerate until the emf equals the source voltage at which time the current reaches zero value. The force f_m and thus the acceleration $\dot{s}$ will then also be zero, and the rod velocity will remain constant.

---

† This contact resistance is not constant but we neglect this in our analysis. It is small in comparison with the resistance of the starting resistor anyway.

This steady-state *no-load* velocity, s_0, can be readily derived from Eq. (7.6) by setting both $\dot{s}$ and f_L equal to zero. We get

$$s \equiv s_0 = \frac{V}{BL} \quad \text{m/s.} \tag{7.7}$$

Note that this velocity *increases* as the magnetic field density *decreases*. This is a somewhat surprising situation and one that always confuses the novice in his first encounter with dc motors.

7.2.2 Start-up Speed and Current—The No-Load Case

It is of interest to study how the speed and current vary during start-up. We can do so by solving the differential Eq. (7.6). We proceed to do so for the no-load case ($f_L = 0$).

It is easy† to prove that the solution of Eq. (7.6) is in this case

$$s = \frac{V}{BL}(1 - e^{-t/T}) = s_0(1 - e^{-t/T}) \quad \text{m/s,} \tag{7.8}$$

where for brevity we have introduced

$$T \equiv \frac{mR}{B^2L^2} \quad \text{s.} \tag{7.9}$$

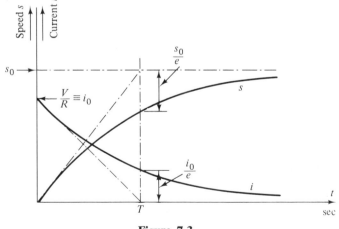

Figure 7.3

---

† The most popular solution method is by Laplace transform techniques. If the reader has no familiarity with differential equations, he can still prove that (7.8) represents the correct solution simply by "back substitution" into (7.6).

The current solution is readily found by substitution of (7.8) into (7.5). We obtain

$$i = \frac{V}{R} e^{-t/T} \quad \text{A.} \tag{7.10}$$

We have plotted the start-up current and speed versus time in Fig. 7.3.

7.2.3   Start-up Time Constant

During the start-up period the speed grows and the current decays at the same exponential rate. Mathematically, steady-state velocity is reached after infinite time. However, in practice, steady-state values are reached in a finite time which is determined by the "time constant" T, defined by Eq. (7.9).

Equation (7.8) informs us that T seconds after closing the switch the rod velocity will have attained a value that is $100/2.718 = 36.8\%$ from the steady-state value. After the passage of five time constants, the figure has dropped to 0.67%. For all practical purposes, we have now arrived at "steady-state speed."

Example 7.1   Find the start-up time constant and distance for the following numerical case:

$$m = 0.1 \text{ kg,}$$

$$L = 1 \text{ m,}$$

$$B = 1 \text{ T,}$$

$$R = 10 \ \Omega,$$

$$V = 10 \text{ V.}$$

SOLUTION: From Eq. (7.9) we get

$$T = \frac{0.1 \cdot 10}{1 \cdot 1} = 1 \text{ s}$$

If by "starting distance" D_{st}, we mean the distance covered in the first $5T$ seconds, we have

$$D_{st} \equiv \int_0^{5T} s \, dt = \frac{V}{BL} \int_0^{5T} (1 - e^{-t/T}) \, dt \approx 4 \frac{VT}{BL} \quad \text{m.}$$

For our numerical case we have

$$D_{st} \approx 4 \frac{10 \cdot 1}{1 \cdot 1} = 40 \text{ m.}$$

7.2.4   Energy Transformations in Motors

The linear motor converts electric power drawn from the source into ohmic loss power in the loop resistance plus useful motor power. The latter goes for adding to the kinetic energy of the rod plus performing useful work. We can obtain a mathematical statement of this mechanism by considering Eq. (7.5) which can be written

$$V = Ri + e \quad \text{V.} \tag{7.11}$$

If we multiply by i we have

$$Vi = Ri^2 + ei \quad \text{W.} \tag{7.12}$$

The left-hand side of Eq. (7.12) represents the power delivered by the source. The first term on the right-hand side is the ohmic power loss. The last term thus must represent the *motor power, p_m*;

$$p_m = ei \quad \text{W.} \tag{7.13}$$

Another expression can be obtained for the motor power by multiplying Eq. (7.3) by the speed s;

$$p_m = f_m s = f_i s + f_L s = ms\dot{s} + f_L s \quad \text{W.} \tag{7.14}$$

By combining the three Eqs. (7.12), (7.13), and (7.14) we thus have the *power balance equation* within the motor

$$Vi = Ri^2 + ms\dot{s} + f_L s \quad \text{W.} \tag{7.15}$$

Upon multiplication by dt and integration we obtain the motor *energy balance equation*:

$$\int Vi \, dt = \int Ri^2 \, dt + \int ms\dot{s} \, dt + \int f_L s \, dt \quad \text{J.} \tag{7.16}$$

The second and third terms on the right-hand side of this equation can be rewritten, respectively, as

$$\int ms\dot{s} \, dt = \int ms \frac{ds}{dt} \, dt = \int ms \, ds \tag{7.17}$$

and

$$\int f_L s \, dt = \int f_L \frac{dx}{dt} \, dt = \int f_L \, dx. \tag{7.18}$$

The expression (7.17) is recognized (compare Eq. 2.39) as the increase in kinetic energy. The Eq. (7.18) represents the energy required to overcome the load force (compare Eq. 2.17).

Equation (7.16) thus tells us the following important story. The energy delivered by the source, $\int Vi\, dt$, is used for three purposes:

1) ohmic heat dissipation, $\int Ri^2\, dt$,
2) increase in system kinetic energy, $\int ms\, ds$,
3) "useful" work, $\int f_L\, dx$.

7.2.5   Starting Efficiency

An important relation exists between the first two terms on the right-hand side in the energy equation. To find this relation consider the no-load case discussed earlier. The steady-state no-load speed was given in Eq. (7.7). For the increase in *kinetic* energy during start up we thus have

$$w_{kin} = \int_0^{s_0} ms\, ds = \tfrac{1}{2}ms_0^2 = \tfrac{1}{2}m\frac{V^2}{B^2 L^2} \qquad \text{J.} \qquad (7.19)$$

For the ohmic *heat dissipation* during start we have

$$w_\Omega = \int_0^\infty Ri^2\, dt = \int_0^\infty R\frac{V^2}{R^2}e^{-2(t/T)}\, dt = \frac{T}{2}\cdot\frac{V^2}{R} \qquad \text{J.} \qquad (7.20)$$

In view of Eq. (7.9) the ohmic loss energy can be written

$$w_\Omega = \frac{1}{2}\cdot\frac{mR}{B^2 L^2}\cdot\frac{V^2}{R} = \tfrac{1}{2}m\frac{V^2}{B^2 L^2} \qquad \text{J.}$$

We thus make the important conclusion:

$$w_{kin} = w_\Omega. \qquad (7.21)$$

In words:

If the motor is started from a constant voltage source, the energy lost in ohmic heat exactly equals the kinetic energy imparted to the moving mass.

Differently stated, the energy efficiency during start up equals 50 percent.
This has important practical consequences. If a high-inertia motor is to be started, the heat energy lost in the starting resistors can pose a severe disposal problem.

Example 7.2   Find the energy supplied by the battery during start-up of the linear motor in Example 7.1.

SOLUTION: We have

$$w_{kin} = \tfrac{1}{2}\cdot 0.1\cdot\frac{10^2}{1^2\cdot 1^2} = 5\,\text{J.}$$

Total start-up energy supplied by the battery is thus 10 J.

7.2.6 The Linear DC Motor under Loading Conditions

In previous sections the linear dc motor was analyzed under the assumption of zero mechanical load. During start-up we had to overcome only the inertia of the moving mass. Once the steady-state velocity is reached the current will be zero and from this time on the electric source will deliver zero power.

In reality, of course, for a motor to serve a useful purpose it must pull a load. In the case of our linear motor this load simply could consist of frictional drag or some additional cart or similar arrangement, representing a load force f_L. In the presence of this load force, the motor upon starting will no longer accelerate to the no-load speed given by Eq. (7.7). The speed will reach a lower value determined by the fact that the motor force f_m must equal the load force f_L.

The former is obtained by substitution of Eq. (7.5) into (7.2);

$$f_m = BLi = BL\frac{V - BLs}{R} \qquad \text{N.} \qquad (7.22)$$

If we solve this equation for s we obtain

$$s = s_0 - \frac{R}{B^2L^2}f_m \qquad \text{m/s.} \qquad (7.23)$$

In a speed–force graph this is the equation for a straight line of negative slope. (See Fig. 7.4.)

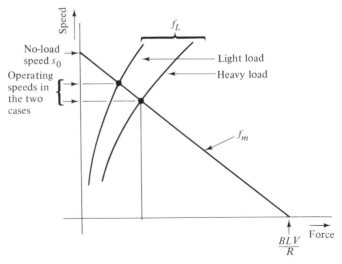

Figure 7.4

We plot also in the same speed–force graph the load force f_L. Typically the load force always increases with speed. (For example, wind drag increases about quadratically with speed.) Actually, we have plotted two load cases, representing "light load" and "heavy load," respectively. The intersection of the f_L and f_m curves indicates that balance exists between the two forces. The intersection point thus gives the operating speed of the motor.

Note that the heavy load results in a lower speed. The reason for this is quite obvious. The heavy load requires a higher value for the motor force. This higher force can be attained only by a higher current. A higher current can occur only if the motor emf is reduced. A reduced emf results from lower speed. (Note that the current–speed relationship (7.5) looks similar to the motor force–speed Eq. (7.23). The only difference is a scale factor. This means therefore that the sloping f_m line in Fig. 7.4 also represents the motor current, in a different scale.)

7.2.7   Motor Rating

Too heavy a motor load may result in such a high current that the ohmic losses in the circuit will become of intolerable magnitude. Overheating of the motor thus sets a limit to its load-pulling capacity. Were we to *stall* the motor, the motor emf e would be zero. The motor current would now be limited only by the loop resistance R. Not only would the current attain its maximum value (V/R) but the stalled motor has lost all its cooling capacity and would heat up very rapidly.

It is logical, therefore, to rate the motor in terms of its maximum load force. We may also multiply this force by the speed and obtain a maximum *power* that the motor can deliver.

7.2.8   Motor Turning Into Generator

Assume that the rod is traveling at the no-load velocity s_0. Its generated emf e equals the source voltage V and the current, as we have noted, is zero.

Assume now that we reverse the load force f_L, that is, we make it act in the direction of speed thus pulling the rod along at a speed *in excess of* s_0. The emf e will now exceed V. As a result a current will now be generated *in the opposite direction*, thus feeding into the positive terminal of the source.

The linear motor has thus *turned into a generator* feeding energy into the "source," which now turns into an energy "sink." Of course, a reversal of the load force f_L means that the "load" has turned into a "prime mover" needed to pull the rod along. With i and f_L thus assuming *negative* values the first and fourth terms of the energy Eq. (7.16) become negative. By letting these terms change places we keep them positive, and the equation now takes the form

$$\int (-f_L)s\, dt = \int Ri^2\, dt + \int ms\dot{s}\, dt + \int V(-i)\, dt \qquad \text{J.} \qquad (7.24)$$

This equation, valid for the *generator case*, states in words: The (positive) energy, $\int(-f_L)s\, dt$, delivered by the prime mover is used for these three purposes:

1) ohmic losses, $\int Ri^2\, dt$,
2) changes in kinetic energy, $\int ms\dot{s}\, dt$,
3) energy delivered to sink, $\int V(-i)\, dt$.

7.2.9 Equivalent Circuits

As viewed from the source (or sink) the dc machine† behaves like a variable but unidirectional emf e behind the resistance R. The latter consists of an "external" part (the current-limiting "starting" resistor) and an "internal" part (contact resistance between fixed and moving parts plus conductor resistance).

We have separated in Fig. 7.5 the "motor" and "generator" cases. Note

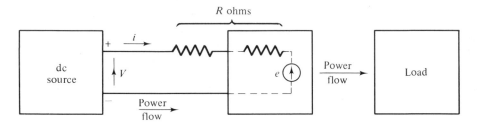

(a)

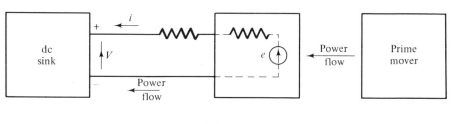

(b)

Figure 7.5

---

† Calling the device "machine" rather than "motor" implies that we now realize its potentialities also as a "generator."

the reversal of current and power flow but the unidirectionality of machine emf
e.

7.3 PHYSICAL MOTOR DESIGN

The linear motor shown in Fig. 7.2 would clearly have only a very limited
practical usefulness. Moreover, it would be very hard to design. How, for
example, would we get the strong magnetic field needed? Let us proceed now
to turn this linear unpractical concept into a very practical rotating design.

7.3.1 The Homo-Polar Machine

An obvious first improvement is to add strength to the linear motor by adding
parallel conductors. We thus obtain the ladder-like contraption depicted in Fig.
7.6. If the "ladder" contains n step conductors each carrying the current i, the
total motor force will equal

$$f_m = nBiL \qquad \text{N.} \qquad (7.25)$$

Next we bend the linear "racetrack" of Figs. 7.2 and 7.6 into a circular
shape. We can do this in several ways. By arranging the current-carrying
conductors *radially* into a spoke-like "rotor" as shown in Fig. 7.7 the motor
forces f_m acting on each spoke will result in a *motor torque* T_m. A practical
variant of this design is the *homo-polar* machine depicted in Fig. 7.8.

The rotor consists here of a circular aluminum disc. Current is fed radially
through the disc via the two ring-shaped carbon brushes. The magnetic flux is

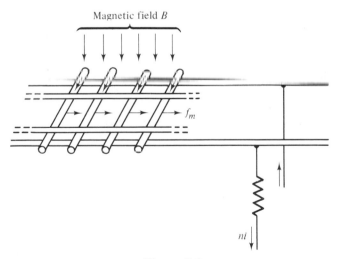

Figure 7.6

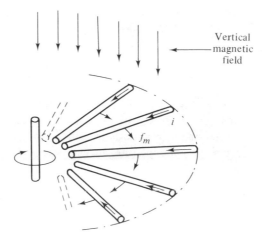

Figure 7.7

created by the *field current* i_f in the *field winding*. The flux follows the indicated path through the stator iron. It penetrates the disc vertically, to give the flux-current geometry of Fig. 7.7.

However, the homo-polar machine is not very practical. As will be demonstrated in Example 7.3, this type of motor is *a high-current/low-voltage* device. This is not a good combination. We remember that low voltage levels esult in high power losses.

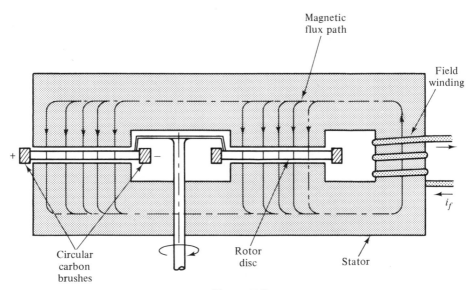

Figure 7.8

Example 7.3 The aluminum disc in a homo-polar motor runs at a speed of $n = 3000$ rpm. The disc (see Fig. 7.9) has outer and inner radii $R = 0.2$ m and $r = 0.05$ m, respectively. The flux density is $B = 0.5$ T. A total current of i A flows between the brushes. Find the magnitude of the emf e generated in the disc and the motor power p_m.

SOLUTION: Consider the emf de generated in the small shaded radial element of length dx in Fig. 7.9. According to Eq. (7.4) the differential emf will have the magnitude

$$de = Bs\,dx \quad V \tag{7.26}$$

where

$$s = \omega x = \frac{n\pi}{30} x \quad \text{m/s} \tag{7.27}$$

and ω is the rotating speed of the disc measured in rad/s.

Upon substitution of (7.27) into (7.26) and integrating from $x = r$ to $x = R$ we obtain the total emf e generated in the "spoke" between the two brushes. We get

$$e = B\frac{n\pi}{30}\int_{x=r}^{x=R} x\,dx = \frac{Bn\pi}{60}(R^2 - r^2) \quad V. \tag{7.28}$$

By inserting the given numerical values we obtain

$$e = 2.945 \text{ V.}$$

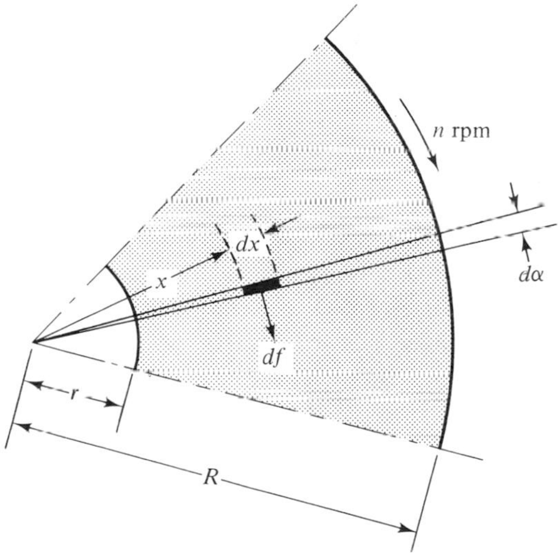

Figure 7.9

The current in the "spoke" element dx will interact with the magnetic flux to give the tangential motor force df and a corresponding motor torque dT_m. By a simple analysis (see exercise 3.16) we find the total motor torque to be

$$T_m = \frac{Bi}{2}(R^2 - r^2) \qquad \text{N} \cdot \text{m,} \qquad (7.29)$$

where i is the total current fed radially into the disc.

For the *motor power* we then get

$$p_m = \omega T_m = \frac{n\pi Bi}{60}(R^2 - r^2) \qquad \text{W.} \qquad (7.30)$$

In view of formula (7.28) we can thus write

$$p_m = ei \qquad \text{W}$$

in full agreement with formula (7.13). For example, assume $i = 500\,\text{A}$. We then have

$$p_m = 2.945 \cdot 500 = 1473\,\text{W} \approx 2\,\text{hp.}$$

A motor that would supply only 2 hp and need a current supply of 500 amps at a terminal voltage of about 3 volts would not find many buyers. However, there are application areas where electric power of low voltage–high current variety is obtained (solar electric panels and magnetohydrodynamics generators, for example). Homopolar machines represent, at least in principle, suitable electromechanical converters in such cases.

7.3.2 Cylindrical Conductor Arrangement

A considerably more practical motor design is obtained by arranging the current-carrying rotor conductors in a cylindrical geometry as shown in Fig. 7.10. The conductor currents now have an axial orientation. The magnetic flux is everywhere radially directed. The electromechanical forces are therefore tangential again, resulting in a motor torque.

A practical design version based upon this geometry is shown in Fig. 7.11. Practically all dc machines in use today are variants of this design. Figure 7.11 depicts a two-pole machine with only twelve torque-creating rotor conductors. Normal dc machines are not quite this simple, but their workings can be fully explained in terms of this simple example.

The magnetic flux is created by the dc field current i_f in the stator field winding. The flux passes through the air gap radially, exiting from the magnetic N-pole and entering the S-pole.

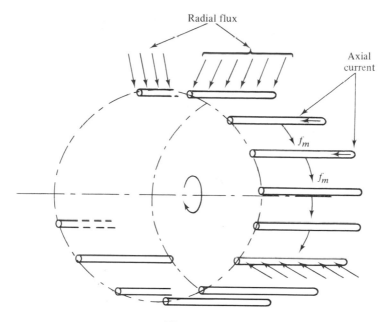

Figure 7.10

The cylindrical rotor winding is placed in slots in the rotor surface. We refer to it as the *armature* winding. The armature current i_a is supplied by an external dc source via the *carbon brushes* and the *commutator*, the actions of which will be explained in the next section. As noted from Fig. 7.11, the conductor currents are of different direction under the two poles. Because the magnetic field, as earlier observed, also is of opposite direction, the resulting motor forces and torque will be of *equal* direction around the total rotor surface. The motor torque T_m will have a clockwise direction in Fig. 7.11.

Note that whereas the stator iron experiences a constant or dc magnetic field, the rotor (when running) is subject to a changing or ac field.† For this reason, in order to reduce iron losses, the rotor is made laminated.

7.3.3 The Commutator Action

It is most important that the current direction in all rotor conductors under each poleface remains fixed *as the rotor turns*. Otherwise we would not preserve the constant directionality of the electromechanical torque T_m. Clearly, to *accomplish this, we must reverse the current direction in every conductor as it passes the interpole regions.*

---

† Remember the opposite situation prevailing in the synchronous machine.

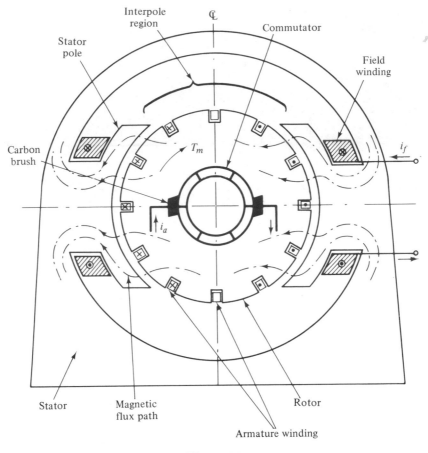

Figure 7.11

This continuous reversal of the current in the armature conductors is accomplished by the commutator. This device consists of a number (six in our case) of identical copper segments, insulated from each other, forming a ring fixed to the rotor. Each segment is connected to the armature conductors in a symmetrical pattern as shown in Fig. 7.12. This connection can be performed in many ways—the pattern of Fig. 7.12 is called a "lap winding." Another representation of this winding is shown (Fig. 7.15) where the winding is exposed as an endless coil, each turn of which is connected to a commutator segment.

The reader can readily trace the current paths from the positive to the negative brushes in Fig. 7.12. With the indicated brush position (on segments c and f) half of the armature conductors (#3 through 8) carry currents in the same direction. The other half have the opposite current polarity.

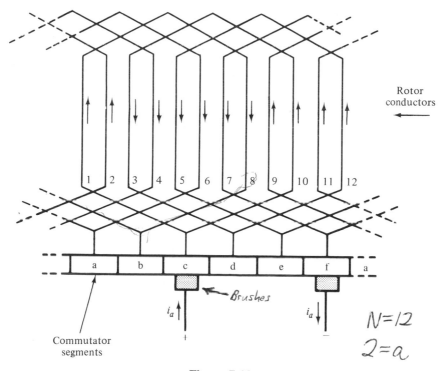

Rotor
conductors

$N=12$

$2=a$

Brushes

i_a i_a

Commutator
segments

Figure 7.12

As the rotor turns and the brush position changes to the next segment pair (b and e), the currents in the conductors 1, 2 and 7, 8 will reverse direction. Note that as the brushes change position, the commutator segment pairs b–c and e–f will be short-circuited for a short period. During this interval, the conductors 1, 2 and 7, 8 will all be short-circuited. The current reversal or current *commutation* will take place during this commutation period.

The current in the individual rotor conductors will look somewhat like the sketch in Fig. 7.13. For one full turn of the rotor, the current will complete a highly nonsinusoidal ac cycle. During the major portion of the cycle, the current will have a constant value. During the relatively short commutation interval, the current will fully reverse.

As the total armature current i_a is split into two equal parallel paths (see Fig. 7.12) the individual conductor currents will alternate between the peak values $\pm\frac{1}{2}i_a$.

7.3.4 The Motor Torque (T_m)

Of fundamental importance in study of motors is the electromechanical torque or motor torque, T_m. We proceed now to find a formula for this torque. We

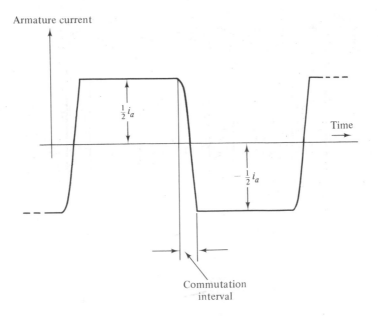

Figure 7.13

shall do so under the following general assumptions:

1) The armature winding consists of N conductors, each occupying on the average a rotor space $\pi D/N$ meters, measured tangentially, as shown in Fig. 7.14. (In the machine previously discussed, N equalled 12.)

2) The armature winding consists of a parallel paths. (The specific lap winding in Fig. 7.12 was characterized by $a = 2$.)

3) The machine has p poles, where p of course is an even integer. (In the previous case, $p = 2$.)

In Fig. 7.14 are shown the magnetic flux distribution and the rotor conductors under one adjoining pole pair of the machine. The individual forces on each conductor have been identified. The force f_ν on the individual conductor ν will be of magnitude

$$f_\nu = L \frac{i_a}{a} B_\nu \qquad \text{N} \tag{7.31}$$

where B_ν is the magnetic flux density as measured at the conductor in question, i_a/a is the conductor current, and L the conductor length (as measured in the magnetic field).

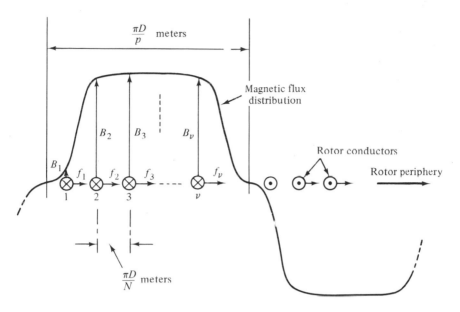

$$\frac{\pi D}{p} \text{ meters}$$

Magnetic flux distribution

B_2 B_3 B_ν

Rotor conductors

Rotor periphery

B_1 f_1 f_2 f_3 ---- f_ν

$$\frac{\pi D}{N} \text{ meters}$$

Figure 7.14

The sum of the forces on all N/p conductors located under one pole will thus equal†

$$\sum_{\nu=1}^{N/p} f_\nu = L \frac{i_a}{a} \sum_{\nu=1}^{N/p} B_\nu \quad \text{N.} \tag{7.32}$$

For the total motor force f_m acting on all rotor armature conductors under the p poles we thus have

$$f_m = p \sum_{\nu=1}^{N/p} f_\nu = pL \frac{i_a}{a} \sum_{\nu=1}^{N/p} B_\nu \quad \text{N.} \tag{7.33}$$

We had earlier concluded (compare Fig. 7.14) that each rotor conductor occupies on the average a peripheral rotor space of $\pi D/N$ meter. Through this small rotor surface "window" passes a magnetic flux of magnitude

$$L \frac{\pi D}{N} B_\nu \quad \text{Wb.} \tag{7.34}$$

---

† Note that the formula is slightly incorrect because the conductors which are commutating do not carry the current i_a/a. However, these conductors are located in the interpole regions where the flux density is very small. Therefore, the contribution which the commutating conductors make to the total force is negligible.

The total pole flux Φ_p entering (or leaving) one pole will thus equal the sum

$$\Phi_p = L\frac{\pi D}{N}\sum_{\nu=1}^{N/p} B_\nu \qquad \text{Wb.} \qquad (7.35)$$

Upon substitution of (7.35) into (7.33), we thus obtain

$$f_m = \frac{pN}{\pi a D}\Phi_p i_a \qquad \text{N.} \qquad (7.36)$$

Finally, the motor torque T_m is obtained as the product of force times radius; that is,

$$T_m = f_m\frac{D}{2} = \frac{pN}{2\pi a}\Phi_p i_a = k_T\Phi_p i_a \qquad N\cdot m \qquad (7.37)$$

where the torque constant

$$k_T \equiv \frac{pN}{2\pi a} \qquad (7.38)$$

is a machine design parameter.

Example 7.4   Find the torque delivered by the machine in Fig. 7.11 if the pole flux equals $\Phi_p = 0.25$ Wb and the armature current is $i_a = 25$ A.

SOLUTION: Formula (7.38) gives directly

$$k_T = \frac{2\cdot 12}{2\cdot\pi\cdot 2} = \frac{6}{\pi}.$$

From Eq. (7.37) we then have

$$T_m = \frac{6}{\pi}\cdot 0.25\cdot 25 = 11.94 \qquad N\cdot m.$$

....

7.3.5 The Induced emf (e)

As the rotor conductors cut the magnetic flux lines, emf's will be generated in each conductor. The magnitudes of these emf's follow from formula (7.4). If we assume that the rotor rpm n is constant then the tangential conductor speed s is likewise constant, and the instantaneous value of the emf will be proportional to the flux density B.

Consequently, as the conductor travels the total distance of one pole pair, the generated conductor emf will complete a full ac cycle of the same waveshape as the flux wave in Fig. 7.14. This waveshape is highly nonsinusoidal. However, due to the action of the commutator, the emf, *as felt between the brushes*, will be a dc emf. Let us see why this must be so.

As we traced the path between the two brushes in Fig. 7.12, we noted earlier that the twelve conductors constitute two parallel circuits, each containing six armature conductors. One such parallel circuit is shown schematically in Fig. 7.15. At each instant the six emf's (identified by arrows in the figure) add up to the total emf e which can be measured between the brushes. (Note however, that during the commutation interval some conductors are shorted out. These "commutating conductors" are, however, located midways between the poles where the flux, and thus the emf, is negligible.)

The total dc emf which can thus be picked off from the N/p conductors under one pole is the sum of the instantaneous† emf's existing in the conductors in question, written as

$$\sum_{\nu=1}^{N/p} e_\nu = \sum_{\nu=1}^{N/p} sLB_\nu = sL \sum_{\nu=1}^{N/p} B_\nu \quad \text{V.} \tag{7.39}$$

As we have p poles and a parallel paths the total emf measured between the brushes will thus be

$$e = \frac{p}{a} \sum_{\nu=1}^{N/p} e_\nu = \frac{psL}{a} \sum_{\nu=1}^{N/p} B_\nu \quad \text{V.} \tag{7.40}$$

In view of Eq. (7.35) the e formula can be written in the form

$$e = \frac{pNs}{\pi Da} \Phi_p \quad \text{V.} \tag{7.41}$$

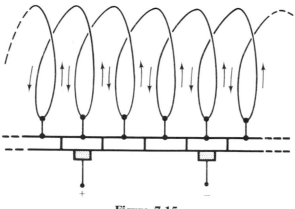

Figure 7.15

---

† This refers to the emf existing at the instant the brush makes contact with the commutator segment in question.

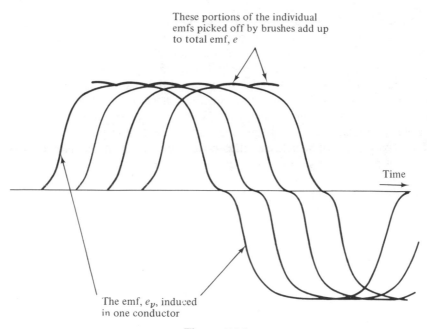

These portions of the individual
emfs picked off by brushes add up
to total emf, e

Time

The emf, e_p, induced
in one conductor

Figure 7.16

The tangential conductor speed s is related to the angular motor speed ω_m of the rotor by the formula

$$s = \frac{D}{2}\,\omega_m \qquad \text{m/s.} \tag{7.42}$$

By substitution of this expression into (7.41) and by making use of formula (7.38) we obtain the following important formula for the machine emf:

$$e = k_T\omega_m\Phi_p \qquad \text{V.} \tag{7.43}$$

The emf e is not perfectly constant. It will contain a slight ripple which usually is of negligible magnitude. This ripple is the result of the commutator action. Each commutator segment has a finite width and the brushes will therefore, in effect, pick off a finite "slice" of the armature coil emf.

Because the coil emf varies with time, this "emf slice" will not be of constant magnitude. The situation is graphically depicted in Fig. 7.16.

Example 7.5   Find the emf generated in the dc machine discussed in Example 7.4. The machine is running at its rated speed $n = 3000$ rpm.

SOLUTION: We first compute ω_m

$$\omega_m = \frac{n}{60}\cdot 2\pi = \frac{3000}{60}\cdot 2\pi = 314.2 \text{ rad/s.}$$

From Example 7.4 we borrow

$$k_T = \frac{6}{\pi} = 1.910.$$

Formula (7.43) then gives directly

$$e = 1.910 \cdot 314.2 \cdot 0.25 = 150 \, \text{V}.$$

7.3.6 The Motor Power (p_m)

In view of formula (2.24) we can express the motor power p_m in terms of T_m and ω_m thus:

$$p_m = T_m \omega_m \quad \text{W.} \tag{7.44}$$

By making use of the torque relation (7.37) we obtain

$$p_m = k_T \Phi_p i_a \omega_m \quad \text{W.} \tag{7.45}$$

By using the emf formula (7.43) we can also express the motor power in the alternate form

$$p_m = e i_a \quad \text{W.} \tag{7.46}$$

Note that this last formula is identical with the linear dc motor formula (7.13).

Example 7.6 Compute the motor power delivered by the motor discussed in Examples 7.4 and 7.5.

SOLUTION: The armature current was specified to be $i_a = 25$ A. The emf was earlier computed to be $e = 150$ V.

Formula (7.46) then gives directly

$$p_m = 150 \cdot 25 = 3.75 \, \text{kW}.$$

Note that this is not the power delivered to the motor shaft. The shaft power is somewhat smaller due to rotational losses in the motor (see discussion in Section 7.3.9).

7.3.7   The Motor Equivalent Circuit

Figure 7.17 shows the equivalent circuit of a dc motor. (Compare also Fig. 7.5.) The motor is fed from an armature voltage source supplying a terminal voltage V_a. The field coils are separately fed from an excitation or field voltage source V_f. (This type of motor is often described as "separately excited." The field coils could, of course, be fed from the same source as the armature. We then have a so-called *shunt* motor. Shunt motors and separately excited motors have slightly different characteristics (compare Section 7.4.4).)

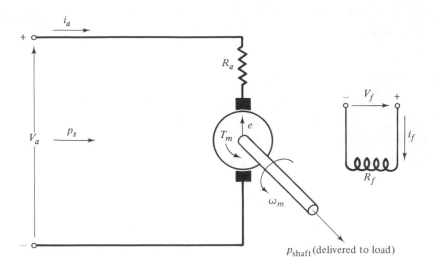

Figure 7.17

The total resistance of the armature winding including the brush contact resistance is lumped into the *armature resistance* R_a.

The difference between V_a and the generated emf e must equal the voltage drop across the armature resistance, that is,

$$V_a - e = R_a i_a \qquad \text{V.} \qquad (7.47)$$

Example 7.7 Find the source voltage V_a required to supply the 25-A armature current needed in Example 7.4. The armature resistance of the machine is $R_a = 0.41 \ \Omega$.

SOLUTION: The emf e had earlier been found to be 150 V. From Eq. (7.47) we thus get

$$V_a = 150 + 0.41 \cdot 25 = 160.25 \ \text{V.}$$

....

7.3.8 Energy Balance in a DC Motor

The motor torque T_m is used to overcome the inertia torque $I(d\omega_m/dt)$ plus the load torque T_L of the motor; that is,

$$T_m = I \frac{d\omega_m}{dt} + T_L \qquad \text{N} \cdot \text{m} \qquad (7.48)$$

where I equals the moment of inertia of rotor plus load, kgm^2, and ω_m equals the mechanical angular speed, rad/s. The load torque also includes windage and friction torques.

By multiplication of Eq. (7.48) by ω_m and by subsequent use of the relations (7.44), (7.46), and (7.47) we obtain the following *energy balance equation* for the dc motor:

$$\int V_a i_a \, dt = \int R_a i_a^2 \, dt + \int I\omega_m \, d\omega_m + \int T_L \omega_m \, dt \qquad \text{J.} \qquad (7.49)$$

Compare this with Eq. (7.16)!

Equation (7.49) states the following: the energy $\int V_a i_a \, dt$ delivered by the armature voltage source is used for these purposes:

1) ohmic heat dissipation in the armature, $\omega_\Omega = \int R_a i_a^2 \, dt$,

2) increase of the kinetic energy of the rotating masses, $\int I\omega_m \, d\omega_m = \frac{1}{2} I\omega_m^2$,

3) "useful work," $\int T_L \omega_m \, dt$.

7.3.9 Additional Losses

Assume that the motor has reached a steady-state, constant speed ω_m. Its kinetic energy has thus assumed a constant value. Thus Eq. (7.49) tells us that the power delivered by the source $V_a i_a$ is used to overcome the load torque T_L plus supplying the ohmic losses in the armature.

A. Rotational Losses (p_{rot}). There are, however, additional losses that are not revealed in the above analysis. Consider first the load torque T_L. A small portion of this torque consists of windage and brush and bearing friction torques. In addition, as the motor spins the rotor flux as we noted earlier will be of ac type. There will therefore arise eddy current and hysteresis, that is, *core losses*, in the rotor. These must be supplied by the armature source. The sum of windage, friction, and magnetic core losses are referred to as *rotational losses*, p_{rot}. The useful load power, also called *shaft power*, p_{shaft}, thus is obtained by deducting the rotational loss power from the motor power:

$$p_{shaft} - p_m - p_{rot} \qquad \text{W.} \qquad (7.50)$$

B. Field Losses (p_{field}). The field coil (see Fig. 7.17) in steady state consumes the power

$$p_{field} - V_f i_f - R_f i_f^2 \qquad \text{W.}$$

As this power does not reach the load, it must be considered a loss power.

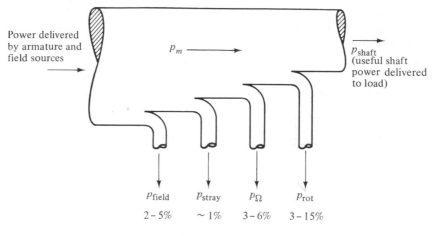

Power delivered
by armature and
field sources

p_m

p_{shaft}
(useful shaft
power delivered
to load)

p_{field} p_{stray} p_Ω p_{rot}

2–5%   ~1%   3–6%   3–15%

Figure 7.18

C. Stray Load Losses. A loss component, very hard to determine by either analysis or test, is the so-called *stray load loss*, p_{stray}. It is caused by nonuniform current distribution in windings and nonuniform magnetic densities in the stator teeth (that is, in the spaces between slots). It is usually estimated to amount to about one percent of the output power of the motor. This loss in effect reduces the motor power p_m. Figure 7.18 shows in schematic fashion the power flow within a dc motor. The figure also informs of the typical ranges for the loss powers.

Example 7.8 Consider the dc motor discussed in Examples 7.4 through 7.7. In order to measure the rotational losses of the motor, it is run at rated speed ($n = 3000$ rpm) and unchanged emf (150 V) but disconnected from its load. The motor draws 2.21 A from the source in this no-load test.

Find rotational losses and also the motor efficiency when operated at full rated current (25 A) and full speed (3000 rpm). When operated under these rated conditions the field coil consumes 173 W and the stray load loss is estimated to equal 40 W.

SOLUTION: Since the motor is disconnected from the load during the no-load test the delivered power p_{shaft} is zero. The measured loss power† thus equals p_{rot}. We therefore have

$$p_{rot} = V_a i_a = 150 \cdot 2.21 = 332 \text{ W}.$$

The armature loss at rated load current is

$$p_\Omega = 0.41 \cdot 25^2 = 256 \text{ W}.$$

In Example 7.6 we had computed the motor power at rated armature current to equal 3750 W. We now adjust this value for stray load losses:

$$p_m = 3750 - 40 = 3710 \text{ W}.$$

From Eq. (7.50) we thus have for the shaft (or output) power

$$p_{\text{shaft}} = 3710 - 332 = 3378 \text{ W}.$$

The motor efficiency when operated at rated current and speed is thus

$$\eta = \frac{p_{\text{shaft}}}{p_{\text{shaft}} + \text{total losses}}$$

$$= \frac{3378}{3378 + 332 + 256 + 173 + 40}$$

$$= 0.808$$

7.4 DC MACHINE OPERATING CHARACTERISTICS

We have now derived the most important formulas, relating to motor torque, power, and emf. A more detailed analysis of the dc machine would include the study of phenomena related to the following:

a) Commutation voltages and the tendency for "sparkover" of the commutator. In an actual dc machine these voltages must be compensated for by means of "compensation windings."

b) The "armature reaction," a term which describes the magnetizing effect of the armature current. The armature current represents, of course, an mmf that will superimpose upon the mmf of the field winding, and will have a second order effect upon both the emf and the torque. (It also is part of the cause of the stray load losses discussed in the previous section.)

c) Nonlinear effects due to magnetic saturation.

---

† We know that p_{rot} varies with speed and rotor flux. If we wish the test to render correct value of the loss, we must thus make sure that we measure it (as we did) at proper speed and emf values. Note also that the ohmic armature loss during the no-load test amounts to only

$$p_\Omega = R_a i_a^2 = 0.41 \cdot 2.21^2 = 2 \text{ W}.$$

Therefore, for all practical purposes, the power consumed during the no-load test goes into rotational losses.

These effects, although of great importance for the proper functioning of the motor, are not of overriding significance to the motor user who wishes to learn about the basic operating features of the dc machine. They will therefore not be further discussed.

Instead, we shall now proceed to explain the operating characteristics of the dc machine based upon previously derived formulas. We focus our attention first on the separately excited machine—the machine where the total pole flux is obtained from a separate dc source.

7.4.1 Starting the DC Motor

At standstill, zero emf is generated in the dc motor. Were we to apply full source voltage to the armature winding, the current would be limited only by the armature resistance R_a. The result would be a current that could easily damage the machine. The torque would likewise be very large and the corresponding sharp acceleration might damage the load. Actually, one of the very attractive features of the dc motor is its high starting torque (compare the internal combustion engine). As the machine picks up speed, an emf will be generated and the current will drop. The starting current would actually experience an exponential decrease of the type shown in Fig. 7.3.

Obviously, we need to control the magnitude of the starting current. The simplest way of doing this is by means of starting resistors as shown in Fig. 7.19. Initially, all three sections of the 3-step starting resistor are in series with the armature. As the motor picks up speed, the three resistor sections are successively shorted out. As a result, the starting current will have a sawtooth shape as shown in the figure.

Of course, the lower current resulting from the insertion of the starting resistors results in a longer start-up period. The energy dissipated in the resistor equals the kinetic energy imparted to rotor plus load. (Compare formula 7.21.) But this power dissipation takes place *outside* the motor. Without the starting resistor the same energy would be dissipated *inside* the machine.

Example 7.9   The total moment of inertia of a dc motor rotor plus its load amounts to

$$I = 770 \text{ kgm}^2.$$

(This is the equivalent inertia of a solid steel cylinder having a diameter and length of 1 m.)

Find the energy dissipated in the starting resistor as this load is being accelerated to the speed of $n = 3000$ rpm.

SOLUTION: The kinetic energy (formula 2.41) equals

$$\omega_{kin} = \tfrac{1}{2}I\omega_m^2 = \tfrac{1}{2} \cdot 770 \cdot 314.2^2 = 38.0 \cdot 10^6 \text{ J} = 10.6 \text{ kWh}.$$

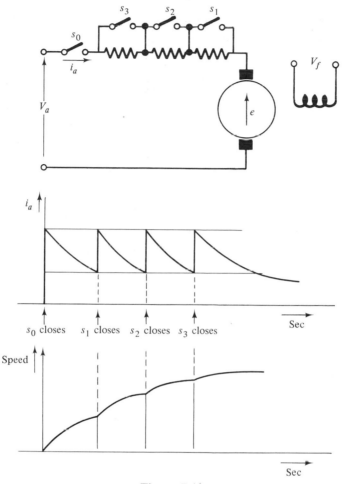

Figure 7.19

An equal amount of energy is dissipated in the starting resistor, which obviously must be designed to withstand the heat. The total energy thus taken from the source during start-up is

$$2 \cdot 10.6 - 21.2 \text{ kWh.}$$

(In addition energy is needed to overcome the windage and friction.)

7.4.2 The Separately Excited DC Machine Operated as Generator

For the separately excited machine, the poleflux is constant if the field source current i_f (see Fig. 7.17) is kept fixed. According to formula (7.43) the emf e will now be proportional to the rotor speed.

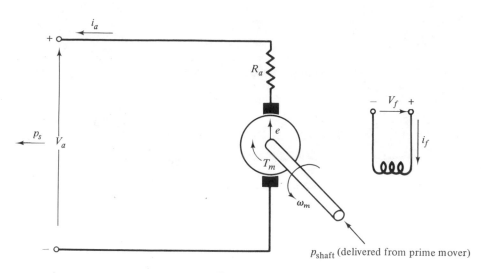

Figure 7.20

Assume now that we reverse the polarity of the shaft power p_{shaft}. This requires that the load be replaced by a "prime mover." The prime-mover torque will tend to accelerate the rotor. Assume that the torque is of sufficient magnitude to speed up the rotor to the point where the emf e exceeds the source voltage V_a. The current i_a will now reverse direction, and the Eq. (7.47) will change to

$$e - V_a = R_a i_a \qquad \text{V.} \qquad (7.51)$$

Since the current is fed into the source at its positive terminal the machine evidently delivers energy to the source—it operates as a generator. The power flows in the machine are depicted in Fig. 7.20. In comparing the figures 7.17 and 7.20 note that current, torque, and power all have reversed polarity, but emf and speed have not.

Example 7.10   The dc machine in Example 7.7 was fed from a dc source of voltage $V_a = 160.3$ V. The source voltages V_f and V_a are kept constant. The load is replaced by a prime mover which will speed up the rotor to $n = 3400$ rpm.

Find the emf, armature current, and power delivered by the machine.

SOLUTION: With the motor running at 3000 rpm, the emf was 150 V (Example 7.5). At 3400 rpm the emf will thus be

$$e = \frac{3400}{3000} \cdot 150 = 170 \text{ V.}$$

Equation (7.51) then yields the current†

$$i_a = \frac{170 - 160.3}{0.41} = 23.7 \text{ A}$$

(entering the source at its positive terminal).

Finally, the power delivered to the source equals

$$p_s = 160.3 \cdot 23.7 = 3799 \text{ W}.$$

The ability of a dc machine to smoothly change from a motor to a generator, and thus equally smoothly change the polarity of its motor torque T_m is often used for what is called *dynamic braking*. For example, decelerating trains and descending elevators require negative or braking torques. Instead of using brakes which would waste the energy in heat, the dc motor may be operated as a generator, thus producing the negative torque needed and delivering the energy back into the source. In all such cases the required current reversal is accomplished by field-current control of the dc machine. In other words, one accomplishes the required increase in the emf by an increase in the field current. (For more details see Section 7.4.9.)

7.4.3 The Torque–Speed Characteristics of the Separately Excited DC Machine

The torque–speed characteristics of the separately excited dc machine are very similar to the force–speed characteristics of the linear motor as given by Eq. (7.23) and depicted graphically in Fig. 7.4.

When the motor load is disconnected (no-load), the speed increases to its no-load value, at which speed the emf equals the source voltage V_a. The motor torque and armature current are now both zero.‡ As the motor is loaded, the speed will drop, resulting in a lower emf and a higher current. When the motor torque balances out the load torque, the rotor speed settles down to its "operating" value, ω_m.

We can display this speed–torque relation speed by combining Eq. (7.37), (7.43), and (7.47) into the following formula:

$$\omega_m = \frac{V_a}{k_T \Phi_p} - \frac{R_a}{k_T^2 \Phi_p^2} T_m \qquad \text{rad/s.} \tag{7.52}$$

---

† We have tacitly assumed that the source voltage V_a is "stiff," that is, it will not vary as the current changes polarity.

‡ In reality as demonstrated in Example 7.8 both are nonzero but small because a small torque is needed to handle the no-load losses.

We introduce the no-load speed

$$\omega_0 \equiv \frac{V_a}{k_T \Phi_p} \quad \text{rad/s} \tag{7.53}$$

and can now write Eq. (7.52) as follows:

$$\omega_m = \omega_0 - \frac{R_a}{k_T^2 \Phi_p^2} T_m \quad \text{rad/s.} \tag{7.54}$$

Note the similarity with formula (7.23)!

On a graph of ω_m vs. T_m this is the equation for a negatively sloping straight line. We have drawn several of these lines in Fig. 7.21, each corresponding to a different value of V_a. Note that increased V_a results in an upward parallel shift of the torque–speed curves. If we extend these lines into the negative torque region we obtain the torque–speed characteristics of the machine operated as generator.

Shown also in the diagram is a "load–torque" curve. This line represents the torque characteristics of the load (including the rotational losses of the motor itself). The point of intersection between this load curve and the torque–speed curve of the motor indicates balance between the driving and loading torques of the motor. This point thus yields the operating speed of the motor for the particular operating voltage in question. Compare Fig. 7.4.

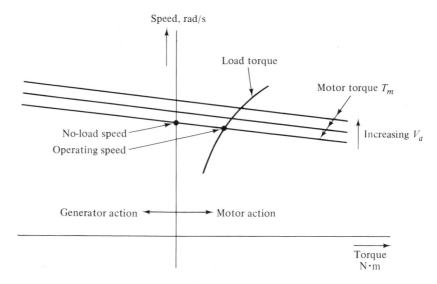

Figure 7.21

Example 7.11 Find the no-load speed of the motor in Example 7.5.

SOLUTION: We had earlier computed

$$V_a = 160.25 \text{ V} \qquad \text{(Example 7.7)},$$

$$k_T = 1.910 \qquad \text{(Example 7.5)},$$

$$\Phi_p = 0.25 \text{ Wb} \qquad \text{(Example 7.4)}.$$

From Eq. (7.53) we now get directly

$$\omega_0 = \frac{160.25}{1.910 \cdot 0.25} = 335.6 \text{ rad/s}$$

which corresponds to a no-load speed of $n_0 = 3205$ rpm.

In other words, if the motor is running at its rated speed of 3000 rpm and drops its load, it will accelerate to 3205 rpm, that is, a 6.83% speed jump.

Example 7.12 What would be the no-load speed of the motor in the previous example if we increase the excitation current by 10%?

SOLUTION: Let us assume that the pole flux Φ_p is proportional to the field current i_f. (In reality this is not quite correct, as a more correct analysis must consider the magnetic saturation effect in the machine iron.)

A 10% increase in i_f will thus mean a 10% increase in Φ_p. Our earlier computed no-load speed of 3205 rpm will then, according to formula (7.53), *decrease* to

$$n_0 = \frac{3205}{1.1} = 2914 \text{ rpm}.$$

Compare: An increase in the *armature* voltage will *increase* the speed.

7.4.4 The Torque–Speed Characteristics of the DC Shunt Motor

In a shunt-connected dc motor the field coils are fed directly from the motor terminals, usually via a variable field rheostat, as shown in Fig. 7.22. By means of the rheostat we can vary the field current i_f and thus the poleflux Φ_p and the emf e.

In a shunt motor, the source current i_s is the sum of armature and field currents:

$$i_s = i_a + i_f \qquad \text{A}. \tag{7.55}$$

Usually i_f is only a few percent of i_a.

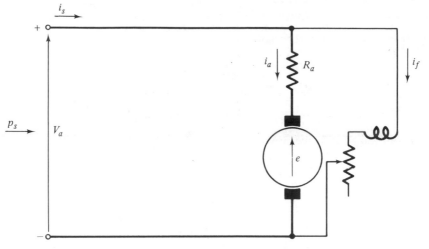

Figure 7.22

With a fixed setting of the field rheostat we have

$$i_f = \frac{V_a}{R_f} \tag{7.56}$$

where R_f is total field circuit resistance. If we further make the assumption of linearity between flux and field current, we obviously also have linearity between pole flux and motor voltage V_a; that is,

$$\Phi_p = k_1 V_a \qquad \text{Wb} \tag{7.57}$$

where k_1 is a parameter that varies with rheostat setting.

Substituting this expression for Φ_p in the speed–torque formula (7.52) yields the following torque–speed relation:

$$\omega_m = \frac{1}{k_1 k_T} - \frac{R_a}{k_T^2 k_1^2 V_a^2} T_m \qquad \text{rad/s.} \tag{7.58}$$

We note the following dissimilarities between the shunt-connected and the separately excited motors:

1) The no-load speed, $\omega_0 = 1/k_1 k_T$, is constant and independent of the voltage V_a, but dependent upon rheostat setting.

2) The slope of the torque speed curves decreases with increasing V_a.

We summarize these findings into the family of torque–speed curves drawn in Fig. 7.23.

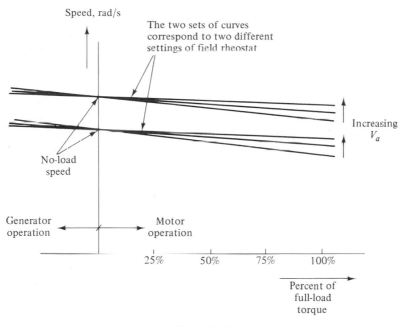

Figure 7.23

The magnitude of the motor torque T_m is directly proportional to the armature current i_a (Eq. 7.37). As the current is a measure of the degree of motor loading, it is practical to plot the speed versus "percentage torque" as has been done in Fig. 7.23. The "full-load torque" is the torque corresponding to rated armature current.

7.4.5 Speed Control of a DC Motor

The graphs in Figs. 7.21 and 7.23 clearly indicate a very good speed constancy of the dc motor as a function of the load torque. Typically, the speed will drop not more than five to ten percent between zero and full-load torque for a shunt-connected or separately excited motor. (A series excited motor behaves quite differently, as we shall discover in Section 7.4.7.) Of course, the dc motor does not compare to the synchronous motor in terms of speed constancy. We remember from Chapter 4 that the latter will experience zero speed drop as a result of load changes.

In many applications—for example, vehicular drive systems—a very important requirement is easy speed control. No electric motor surpasses the dc motor in this regard. There are two basic ways in which the speed of the dc motor can be controlled. Both of them can be best understood from the expression for no-load speed, Eq. (7.53). We treat them separately.

A. Speed Control by Variation of Armature Voltage V_a. Formula (7.53) informs us that the motor no-load speed is directly proportional to the armature voltage V_a. A change in this voltage therefore results in a proportional parallel shift in the speed curves as we already indicated in Fig. 7.21. We offer the following comments about this speed control method:

1) Although it works well for the separately excited machine, it does *not* work in the case of a shunt motor. In a shunt motor, the field current, and thus the pole flux Φ_p is proportional to V_a. Consequently the ratio V_a/Φ_p will be unaffected by the voltage change. (Note that the no-load speed, $\omega_0 = 1/k_1 k_T$, is independent of V_a.)

2) *By varying the armature voltage V_a* throughout the range $\pm 100\%$, the speed will vary from full forward to full reverse (Fig. 7.24a). Note that zero V_a corresponds to zero speed, a fact that permits smooth speed reversal. This type of speed control is thus very useful for the many industrial and transportation applications *where speed reversal is required.*

3) This type of control action (in contrast to the alternate method discussed below) will not weaken the pole flux Φ_p and will therefore not weaken the motor torque.

B. Speed Control by Variation of Field Resistance.   Formula (7.53) indicates that the magnitude of the pole flux Φ_p greatly affects the speed. The pole flux is roughly proportional to the field current i_f—i.e., the speed can be conveniently controlled by variation of the field rheostat.

Because Φ_p appears in the denominator in Eq. (7.53), the speed–field current graphs have the hyperbolic shape shown in Fig. 7.24(b). We make the following observations about this control method:

1) It works both for separately excited and shunt-excited motors.

2) Higher speeds are attained by a lowering of Φ_p. This reduces the magnitude of the torque (compare Eq. 7.37), that is, it causes a weakening of the motor.

3) The formula (7.53) indicates that the speed approaches infinity when Φ_p goes to zero. Therefore care must be exercised not to run the motor at excessive speeds. (A classic danger is to "open circuit" the field circuit accidentally.)

4) As clearly evidenced by Fig. 7.24(b), there is no smooth way of going from positive to negative speeds—one must stop the motor, disconnect it from the source, and reverse field polarity in order to reverse speed. This control method therefore *cannot* be used when speed reversal is required.

Example 7.13   A separately excited dc motor is being operated from an armature voltage of 300 V. Its no-load speed is 1200 rpm. When fully loaded,

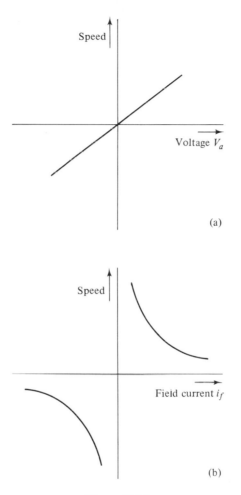

Speed

Voltage V_a

(a)

Speed

Field current i_f

(b)

Figure 7.24

it delivers a motor torque of 400 N · m (295 lbf · ft) and the speed drops to 1100 rpm. Find full-load motor torque, power, and speed if operated with an armature voltage of 600 V. Excitation is assumed unchanged.

SOLUTION: According to Eq. (7.53) the no-load speed will increase to 2400 rpm as a result of the voltage doubling. Because Φ_p is unchanged the rated current† will give rise to unchanged torque (400 N · m) at the higher speed.

---

† Because of better cooling at the higher speed the motor can actually accept even larger armature current without excessive heating.

At the lower speed (1100 rpm) the motor power p_m is

$$p_m = \omega_m T_m = \frac{1100}{60} \cdot 2\pi \cdot 400 = 46.08 \text{ kW.}$$

According to Fig. 7.21, at the higher speed and full torque the speed will drop from 2400 to 2300 rpm. The motor power will thus increase to

$$p_m = \frac{2300}{1100} \cdot 46.08 = 96.35 \text{ kW.}$$

....

Example 7.14   Consider again the motor in the previous example. We now wish to increase the speed from 1200 to 2400 rpm by keeping the armature voltage V_a constant at 300 V but decreasing the pole flux to one-half its original value. Find the full-load torque and power at the higher speed.

SOLUTION: According to formula (7.37), the full-load torque will now reduce to one-half the original value, 200 N · m. (Note that we are not permitted to compensate the reduced Φ_p by a higher armature current i_a. This would cause excessive heat losses.)

With speed doubled but the torque at half value, the motor power will be unchanged at 46 kW. Result: A weakened motor! (Compare with the 96 kW figure in the previous example!).

....

7.4.6   The Shunt Machine Operated as a Generator onto a Resistive Load

When a shunt machine is operated as a generator feeding energy into the armature voltage source, its torque–speed characteristics are directly obtained from Fig. 7.23 by extending them into the negative torque region. The situation would be quite different were we to feed the energy into a "passive" load resistance R_L as shown in Fig. 7.25. Under certain conditions the generator would now be unable to maintain its own voltage.

To appreciate the problem, consider the graphs in Fig. 7.25. One graph depicts the generated emf as a function of the field current i_f.[†] We note that because of magnetic saturation the emf graph is linear only in its lower range.

The two other graphs show "field resistance lines." They give the linear relationship between the voltage V_a, across the field circuit, and the current i_f for two different values of field resistance, R_f' and R_f''.

Assume that the field rheostat is set to correspond to the field resistance $R_f' = 100\,\Omega$. The graphs indicate that for a certain field current i_f' to be

---

† The emf is proportional to the rotor rpm. The emf graph therefore refers to one particular speed—for example the rated speed of the machine.

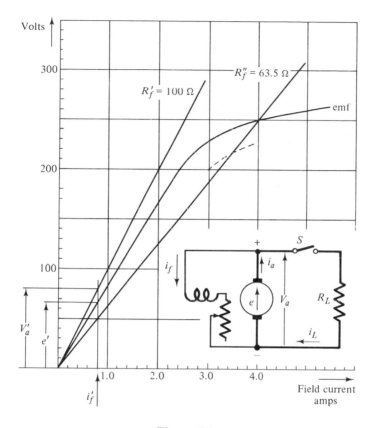

Figure 7.25

maintained, a terminal voltage V_a' would be needed. However, the available emf is only of value e' and because $e' < V_a'$, the voltage equilibrium obviously cannot be maintained. *The generator would never build up its own voltage.*

Were we to set the rheostat to the lower resistance value R_f'', then the right conditions for voltage buildup would be present. The emf would grow until balance would be achieved between emf and terminal voltage. If the internal voltage drop $R_a i_a$ is neglected, this balance evidently exists at the intersection between the curves.

Example 7.15   What would be the largest value of R_f for which the machine could build up its own voltage assuming the speed is constant?

SOLUTION: This would obviously be the resistance value for which the field-resistance line would coincide with the linear portion of the emf curve. From the graph in Fig. 7.25 we read that an emf $e = 200\,\text{V}$ is obtained for a field current of i_f = 2.4 A. This corresponds to a field resistance value of 200/2.4 = 83.3 Ω. This would then be the critical value above which the machine would be unable to maintain its own excitation.

Example 7.16   Consider the generator in Fig. 7.25 with R_f set at 63.5 Ω. What will be the terminal voltage V_a before and after closing the switch S to the load resistance R_L = 5.0 Ω? It is assumed that the prime mover maintains a constant speed. The armature resistance R_a of the generator equals 0.5 ohms.

SOLUTION:

Switch Open: We now have i_a = i_f. Were we to neglect the voltage drop $R_a i_a$, voltage equilibrium would occur according to the graph for e = V_a = 250 V and i_f = 4.0 A. The voltage drop $R_a i_a$ will obviously be about 2.0 V so if we compensate for it the correct answer would thus be

$$V_a \approx 248 \text{ volts.}$$

Switch Closed: By making the very rough and unjustified assumption that the insertion of the 5-Ω load resistance will not affect the terminal voltage, we get for the load current

$$i_L \approx \frac{248}{5.0} = 49.60 \text{ A.}$$

The armature current would therefore be

$$i_a = 49.60 + 4.0 = 53.60 \text{ A.}$$

The armature voltage drop can then be computed:

$$R_a i_a = 0.5 \cdot 53.60 = 26.8 \text{ V.}$$

This drop is too large to be neglected, so we deduct it from the emf curve as shown by the dashed line in Fig. 7.25. The intersection of this dashed line and the 63.5 = Ω resistance line corresponds to i_f = 3.4 A, and V_a = 212 V.

With this corrected value for the terminal voltage we recompute a corrected value for the load current:

$$i_L = \frac{212}{5.0} = 42.40 \text{ A.}$$

The corrected value for the armature current would thus be

$$i_a = 42.40 + 3.4 = 45.8 \text{ A,}$$

and the corrected value for the armature drop becomes

$$i_a R_a = 45.8 \cdot 0.5 = 22.9 \text{ V.}$$

We can now go back and readjust all values once more, following the above procedure. In summary the finally adjusted current and voltage values will be

$$i_f = 3.5 \text{ A,}$$
$$V_a = 220 \text{ V,}$$
$$i_L = 44.0 \text{ A,}$$
$$i_a = 47.5 \text{ A,}$$
$$e = 244 \text{ V.}$$

Delivered power $= V_a i_L = 10.45$ kW.

7.4.7 The Series-Excited Machine

Very special torque–speed features are obtained if the dc motor is series excited, that is, its field winding is connected in series with the armature as depicted in Fig. 7.26. We have also included a variable series control rheostat R_s.

The pole flux Φ_p is now controlled by the armature current rather than by the armature voltage as is the case in a shunt motor. As the magnitudes of both the emf and torque are directly proportional to Φ_p, both can be expected to vary greatly with armature current.

If we disregard magnetic saturation, we can assume the pole flux to be proportional to the current that produces it, that is, i_a, and we can therefore

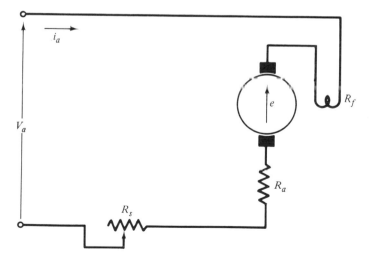

Figure 7.26

write

$$\Phi_p = k_I i_a \qquad \text{Wb} \qquad\qquad (7.59)$$

where k_I is a machine constant.

By using Eqs. (7.37) and (7.43) we then obtain for the motor torque and emf

$$T_m = k_T k_I i_a^2 \qquad \text{N} \cdot \text{m}, \qquad\qquad (7.60)$$

$$e = k_T k_I \omega_m i_a \qquad \text{V}. \qquad\qquad (7.61)$$

If the field winding has the resistance,[†] R_f, the armature voltage equation will read

$$V_a = e + i_a(R_a + R_f + R_s). \qquad\qquad (7.62)$$

By elimination of e between Eqs. (7.61) and (7.62) we obtain

$$i_a = \frac{V_a}{R_a + R_f + R_s + k_T k_I \omega_m} \qquad \text{A}. \qquad\qquad (7.63)$$

Finally, by substituting this expression for i_a in formula (7.60) we have for the motor torque

$$T_m = k_I k_T \frac{V_a^2}{(R_a + R_f + R_s + k_T k_I \omega_m)^2} \qquad \text{N} \cdot \text{m}. \qquad\qquad (7.64)$$

Using this formula we plot the torque–speed curves in Fig. 7.27. The most prominent features of the torque–speed characteristics of the series motor are as follow:

1) For zero load the motor has a "runaway" tendency (compare the earlier motor types where the no-load speed is just a few percent above the full-load speed).

2) The speed drops sharply with increasing torque. This means that a sharp torque spike results in a sharp speed drop and thus only a modest peak in the power. (A shunt motor which is "speed stiff" would respond with an equally sharp power peak.) The speed may drop all the way to zero (stall) and the motor may not be damaged (if R_s is not too small).

3) Good starting torque.

The series motor will thus "cushion" the source against power peaks during severe torque overloads. It can also withstand severe starting duties. For these reasons the series motor is used extensively in hoists and cranes, and in traction and similar fields.

---

† A shunt-connected field winding consists of many turns of light wire—thus its resistance is high. A series-connected winding consists of a few turns of heavy wire—thus its resistance is low.

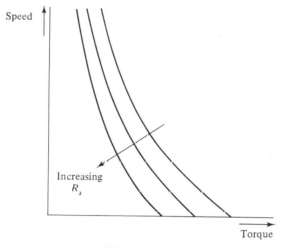

Figure 7.27

7.4.8 The "Universal" Motor

As seen from the expression for torque (Eq. 7.60) the series-motor torque will be positive for positive *or* negative currents. (The reason being, of course, that reversal of the current also reverses the flux, thus preserving the torque direction.) *Thus, the series dc motor will preserve torque direction if the source voltage changes polarity—it will operate on ac.*†

As a consequence of its unique ability to run on both ac and dc, the series dc motor is referred to as *universal motor*. When intended for ac, however, it must have both rotor and stator laminated, to avoid heating the core excessively.

When operating on ac the motor torque will pulsate. (Why?) Its average value will be approximately equal to the dc torque. The torque–speed characteristics are also about the same.

Universal motors can be designed for high speed, often in the range 5000 to 15,000 rpm. We know from earlier discussions that high-speed motors have a high power-to-weight ratio. This type of motor is therefore often found in portable equipment like vacuum cleaners, hand drills, etc.

7.4.9 A Drive-System Design Example

We shall find it instructive to present in some detail the design of a specific drive system incorporating dc machines.

---

† The shunt motor will also change both flux and current directions thus preserving the unidirectionality of torque. Why will the shunt motor *not* operate on ac?

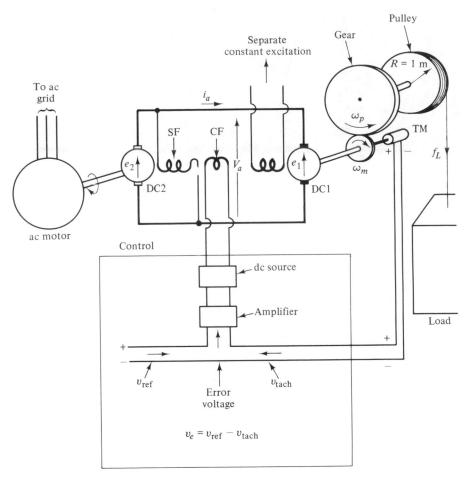

Figure 7.28

A. *System Description.* Consider the mine elevator discussed in Example 2.3. We are asked to develop the preliminary design for a drive system for the elevator. A possible solution is sketched in Fig. 7.28. The elevator is raised and lowered by means of a cable pulley driven by a separately excited dc motor DC1. In order for the motor not to be too low-speed† and thus too bulky (and expensive), a gear train with ratio 20:1 is provided between the pulley and motor.

---

† The size of a motor increases with decreasing speed. (Compare comments on page 171.)

DC1 is connected terminal to terminal to another dc machine, DC2, which will serve as a generator. (This type of arrangement is referred to as a "Ward–Leonard" system.) DC2 is driven by an ac motor which in turn receives its power from the ac power grid.

Speed control of DC1 is accomplished by controlling its terminal voltage V_a according to the description in Section 7.4.5A. The voltage V_a, which is being delivered by DC2, is controlled by field-current manipulation of this machine. The total speed control task of DC1 can be handled by the circuit marked "control" in Fig. 7.28, and will be discussed in section G below.

B. Load Data and Operating Specifications. The elevator load is five tons and the lift distance is 200 m. The speed profile is the same as assumed in Example 2.3, that is, the load is being accelerated and decelerated in five seconds and run at a constant speed of 5 m/s for the rest of the Raise cycle. The lift force f_L for the Raise cycle was computed in Example 2.3 and plotted in Fig. 2.7. We replot those force graphs for easy reference in Fig. 7.29(a).

We add here the additional requirement that the empty elevator (empty weight = 1 ton) must be lowered in 25 s at a top speed of 10 m/s, and top acceleration of ± 2 m/s^2.

The lift force during the steady-state portion of the Lower cycle is

$$f_L = mg_0 = 1000 \cdot 9.81 = 9810 \text{ N}.$$

During the acceleration and deceleration portion of the Lower cycle the lift force is

$$f_L = m(g_0 \mp 2) = 1000 \cdot (9.81 \mp 2);$$

or, 7810 and 11,810 N, respectively. These forces have been plotted in Fig. 7.29(a).

C. Computation of Required Motor Speed and Torque. Because the pulley radius is 1 m, we can compute its steady-state speeds ω_p, during the Raise and Lower cycles in the following manner:

$$\omega_p = \frac{5}{1} = 5 \text{ rad/s} \quad \text{(Raise)},$$

$$\omega_p = \frac{-10}{1} = -10 \text{ rad/s} \quad \text{(Lower)}.$$

The gear ratio is 20:1 and we therefore have for the required steady-state motor speeds

$$\omega_m = 20 \cdot 5 = 100 \text{ rad/s} \quad \text{(Raise)},$$
$$\omega_m = -20 \cdot 10 = -200 \text{ rad/s} \quad \text{(Lower)}.$$

These speeds have been plotted in Fig. 7.29(b).

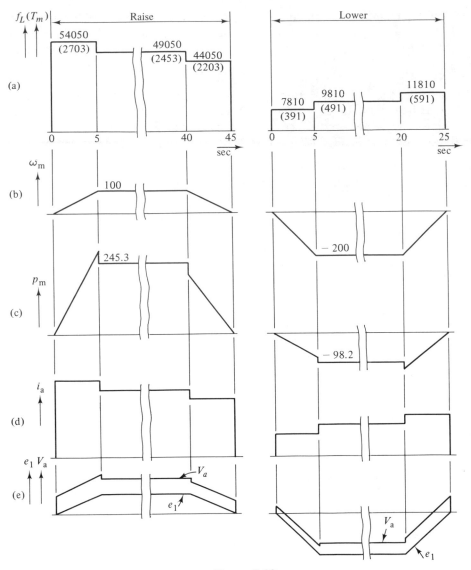

Figure 7.29

The required motor torque T_m is obtained from the knowledge of the load force f_L which exerts a torque on the pulley of magnitude $f_L \times 1 \, \text{N} \cdot \text{m}$. Since the gear ratio is 20:1, the load torque, as felt on the high-speed side, is reduced in the same ratio. If, for simplicity, we neglect all rotational losses the motor torque must equal the load torque and we get

$$T_m = T_L = \frac{f_L \cdot 1}{20} = \frac{f_L}{20} \qquad \text{N} \cdot \text{m}. \tag{7.65}$$

From this formula we derive the required motor torques that have been indicated within parenthesis in Fig. 7.29(a).

D. Required Motor Power (p_m). At this point we can calculate the required motor power. As the power p_m is the product of ω_m and T_m, we obtain p_m simply by multiplying the graphs in Figs. 7.29(a) and (b). The power graph in Fig. 7.29(c) results.

We make the following observations:

1) The power during the steady-state portion of the Raise cycle is constant and equals 245.3 kW. (This figure agrees with the earlier computed value in Example 2.3.)

2) Note that the negative motor power during the Lower cycle means that the dc machine DC1 now is being driven as a generator (by the elevator) and delivers power to DC2. The latter machine in turn drives the ac machine which becomes a generator feeding energy back into the ac network.

3) The power and torque requirement during one complete duty cycle (Raise and Lower) is highly fluctuating.

4) We have disregarded all losses in computing the above powers and torques. Were we to take the losses into account we would have to adjust T_m and p_m correspondingly.

E. Required Armature Current (i_a). We now turn our attention to armature current requirements of DC1. As already mentioned, DC1 will be operated separately excited with *constant* field current and thus *constant pole flux*, Φ_p. According to formula (7.37) its armature current must therefore be proportional to the torque; that is,

$$ i_a = \frac{1}{k_T \Phi_p} T_m \sim T_m \qquad \text{A.} \qquad (7.66) $$

The graph (see Fig. 7.29d) of the armature current thus looks identical to the torque graph.

The required torque is highest at the lowest speeds (Fig. 7.29a). The machine must therefore tolerate the highest currents at speeds which give the poorest cooling. This fact must be taken into account when motor type and size is settled.

F. Required Armature Voltage (V_a). The required armature current i_a must be supplied by the application of proper armature voltage V_a. How shall we select the proper V_a throughout the work cycle? According to Eq. (7.47) the voltage V_a must be

$$ V_a = e_1 + R_a i_a \qquad \text{V} \qquad (7.67) $$

where e_1 is the emf induced in DC1 and R_a is its armature resistance.

According to formula (7.43) the emf e_1 is

$$e_1 = k_T \Phi_p \omega_m. \tag{7.68}$$

Because k_T and Φ_p are constant, the emf e_1 will thus be proportional to the motor speed ω_m. We have plotted e_1 in Fig. 7.29(e).

Thus as both e_1 and i_a are now known throughout the workcycle, we can readily compute V_a from formula (7.67). The result is shown in Fig. 7.29(e). (Note that because the voltage drop $R_a i_a$ is relatively small, V_a and e_1 are *not* drawn to scale.)

During the Raise portion of the work cycle the magnitude of V_a exceeds the magnitude of e_1. The current thus flows *against* e_1 indicating *motor action*.

During the Lower portion of the work cycle the magnitude of e_1 exceeds the magnitude of V_a, thus making i_a flow in the direction of e_1. This is *generator action*—DC1 acts like a generator feeding energy into DC2. (Recall the discussion of dynamic braking in Section 7.4.2.)

G. Automatic Speed Control. One obvious way of making the elevator drive motor follow the desired speed-versus-time curve shown in Fig. 7.29(b) would be to make sure that the generator, DC2, outputs a voltage equalling the one shown in Fig. 7.29(e). This would be achieved by controlling the field of DC2.

This would be an example of "open-loop" control and its success would be dependent upon the accuracy of the equations that led to the V_a curve in the first place.

A better way of accomplishing the speed control job is shown in Fig. 7.28. The suggested scheme is of automatic "closed-loop" type. We give a brief description of this more practical approach.

A *tachometer*, TM, is coupled to the drive motor shaft. This device is a small dc generator, the pole flux of which is provided by a constant permanent magnet. Its output voltage v_{tach} will thus be proportional to the motor speed—*it is a replica of the actual elevator speed*. The voltage v_{tach} is compared with a *voltage* v_{ref}. The latter voltage may be derived, in a simple case, from a manually operated rheostat or, in a more sophisticated design, from a computer. In any case, the voltage v_{ref} must portrait the *desired* speed, that is, it should be a replica of the speed-versus-time graph in Fig. 7.29(b).

The difference voltage v_e between v_{ref} and v_{tach} is referred to as the "error" signal, as it obviously is a measure of the error between desired and actual speed. This error voltage is first amplified and then fed as an *actuating* signal to a controllable DC source which feeds the control field (CF) of the generator DC2. This machine is also equipped with a shunt field (SF). This latter field has a "field resistance line" (see Fig. 7.25) that is adjusted to coincide with the emf curve as shown in Fig. 7.30.

Due to this "critical adjustment" of SF the emf e_2 and thus the voltage V_a *will be very sensitive* to a small mmf increment from the control field, CF. In an

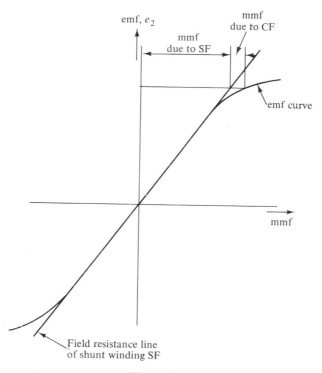

Figure 7.30

automatic, continuous manner the emf e_2 and thus the terminal voltage V_a will be kept at such values as to minimize the error voltage. (It is assumed, of course, that the DC source is "polarity sensitive.")

This is the basic feature of an automatic closed-loop control system. A more detailed analysis of the dynamic response of this system is beyond our scope. It suffices to say that the basic requirements of a control loop of this type are

1) stability,
2) speed of response, and
3) accuracy.

7.5 DC SUPPLY SYSTEMS

Essentially all electric energy is being distributed to the consumer in the form of ac. Before this energy can be utilized for driving dc motors (or any other device requiring dc) it must be transformed into dc form. In this section, we shall briefly discuss those devices and circuits that make this ac → dc transformation possible.

Let it first be said that there are many ways in which this transformation (or the opposite one, from dc → ac) can be performed. We remind the reader of the arrangement in Fig. 7.28 consisting of an ac motor driving a dc generator. This type of arrangement is finding increasingly less use. It is comparatively expensive, involves rotating equipment, and is characterized by relatively poor efficiency (the losses of *two* rotating machines must be supplied). *Solid-state* equipment such as *diodes, thyristors,* and *transistors* are today dominating the dc motor control field and also finding increased uses in the ac motor control field as well.

7.5.1 Basic Rectifier Elements

Diodes and thyristors are the basic elements used in the *rectifier circuits* to be discussed. Space does not permit us to give a detailed physical account of these devices. For a more detailed presentation see Mazda, 1973. We shall limit ourselves to a description of the input–output behaviors of the diode and the thyristor.

A. The Diode. The diode constitutes the basic rectifier element. The circuit symbol for a diode is shown in Fig. 7.31 along with its voltage–current

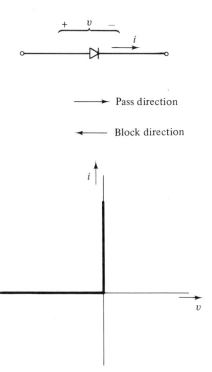

Figure 7.31

characteristics. When the voltage v across the diode element assumes positive values the element permits current flow essentially without voltage drop. We may express the situation differently by saying that the resistance in the "open" direction of the element is very low, approaching zero. As the voltage v changes polarity the resistance takes on very high values and the current flow through the element is now blocked.

An "ideal diode" would be described as having zero resistance for $v > 0$ and infinite resistance for $v < 0$. These extreme values are not, or course, measured in reality. The diode has a small but nonzero resistance in its "open" direction. This resistance results in a small ohmic heat loss which in effect sets a current limit of the device. In the "close" direction, the resistance is large but not infinite. A small current leakage will thus take place when $v < 0$.

The diode characteristics are evidently highly *nonlinear*, a fact that introduces considerable complexity into the analysis of circuits containing this type of device. It is helpful to think of the diode as a switch which, depending upon the polarity of the voltage v, alternately and without inertia opens and closes the circuit.

B. The Thyristor.   The polarity of the voltage v solely determines the "open" or "closed" state of a diode. A thyristor is a *controlled-rectifier* element, the "open" and "close" characteristics of which can be controlled via a third terminal, the *gate*. "Silicon-controlled rectifier" (SCR) is an alternate name of the device. The thyristor symbol is shown in Fig. 7.32. The figure also depicts its voltage–current characteristics.

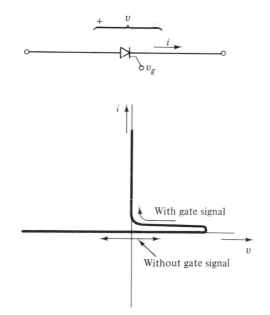

Figure 7.32

With no gate signal ($v_g = 0$) applied, the thyristor blocks current in *both* directions. When a gate signal is applied (usually a pulse of a few volts amplitude) the thyristor "fires," that is, it becomes conductive like the diode. It remains conductive for as long as $v > 0$.

7.5.2   Single-phase, Half-wave Rectifier Circuits

The simplest type of dc supply circuit is obtained by using a single rectifier element in series with the load.

A. Diode Circuit.   The diode case is shown in Fig. 7.33. As only the positive voltage waves are passed through the diode, the arrangement is referred to as "half-wave." The load† voltage v_L will be *pulsating* at the rate of 60 Hz, having an average value $v_{L\,ave}$ of magnitude

$$v_{L\,ave} = \frac{1}{T}\int_0^{T/2} v_{max} \sin \omega t \, dt = \frac{v_{max}}{\pi} \quad \text{V.} \tag{7.69}$$

(The formula assumes zero voltage drop in the diode element.)

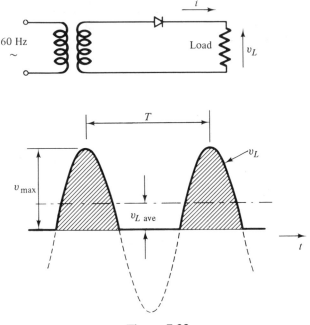

Figure 7.33

---

† Here for simplicity assumed purely resistive.

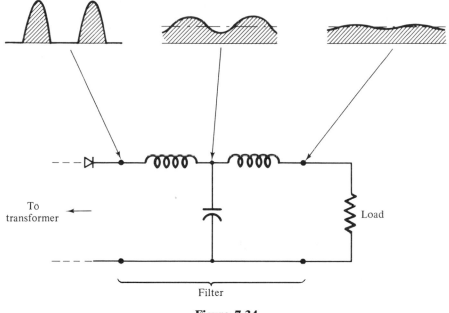

Figure 7.34

If the load cannot tolerate the highly pulsating dc voltage, a *wave filter* can be placed between the diode and the load as shown in Fig. 7.34. The series-inductor sections of the filter represent *high* impedance for the ac harmonics in the diode voltage. The shunt capacitor represents a *low* impedance. Consequently, only a slight ac ripple remains at the output terminals of the filter.

B. Thyristor Circuit. A single-phase, half-wave thyristor rectifier circuit is shown in Fig. 7.35. By a *control circuit* (not shown) the gate pulse is delayed a certain delay period T_d. Clearly, this results in a reduction of the dc load voltage $v_{L\ ave}$. By control of T_d, we can thus exert control over the dc load voltage, the magnitude of which is computed from

$$v_{L\ ave} = \frac{1}{T}\int_{T_d}^{T/2} V_{max} \sin \omega t\, dt$$

$$= \frac{v_{max}}{2\pi}\left[1 + \cos\left(2\pi\frac{T_d}{T}\right)\right] \quad \text{V.} \tag{7.70}$$

Note that the dc average becomes zero for $T_d = T/2$. For $T_d = 0$, the dc average equals that of the diode circuit (Eq. 7.69).

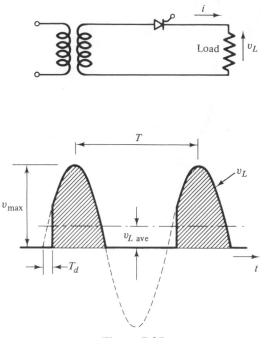

Figure 7.35

7.5.3   Single-phase, Full-wave Rectifier Circuits

By employment of additional diode elements, one can rectify both the positive and negative voltage waves and obtain *full-wave* rectification. A typical rectifier *bridge* circuit is shown in Fig. 7.36. As both half-waves are now rectified, the dc average value will be of twice the magnitude given by formula (7.69); that is, we have for the dc component in Fig. 7.36

$$v_{L\ ave} = \frac{2}{\pi} v_{max} \quad \text{V.} \tag{7.71}$$

Note that we have also doubled the lowest ripple frequency from 60 to 120 Hz.

7.5.4   Three-phase Rectifier Circuits

If dc power in excess of about 5 kW is required, one must usually employ a three-phase source. Figure 7.37 shows one of the simplest three-phase rectifier arrangements. Figure 7.38(a) depicts the rectified voltage wave assuming zero control (diodes). If thyristors are employed, the dc voltage magnitude may be controlled as shown in Fig. 7.38(b).

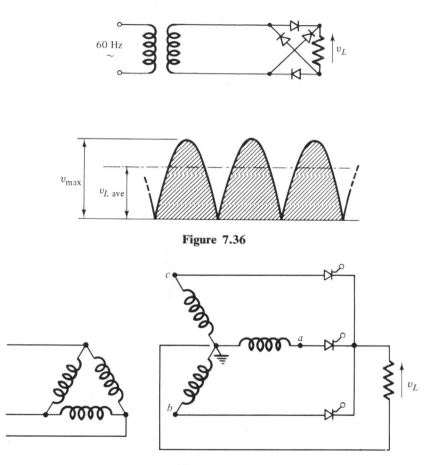

Figure 7.36

Figure 7.37

7.6 SUMMARY

This chapter treats the dc machine and, briefly, methods for transforming ac to dc power. The dc motor (including its rectifier supply circuitry) is a comparatively expensive motor type. Its use is therefore limited to those applications where the torque–speed requirements are so severe and exacting that no other motor can do the job.

The operating principles and characteristics of the dc motor are presented by the use of a linear motor prototype. Upon explaining the formation of motor forces and force-speed characteristics of this simple (but unpractical)

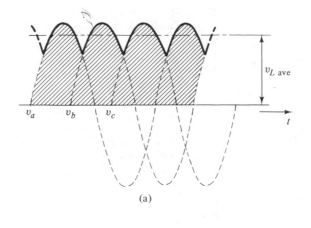

(a)

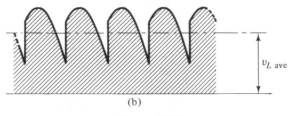

(b)

Figure 7.38

device, we develop by a step-by-step procedure a rotating and more practical motor design. The need for current commutation is explained and the principles of the commutator are presented. The most important aspects of any motor type are its torque–speed features. Formulas are derived for torque and emf and torque–speed relationships are then presented.

Various methods for speed control are discussed and compared in terms of torque and power ratings.

EXERCISES

7.1 The "linear" motor discussed in the text develops a motor *force* the magnitude of which depends upon the speed. What is the maximum value of the force? At what speed is this maximum force developed?

7.2 What maximum value of the motor *power* is the linear motor in Exercise 7.1 able to deliver? At what speed (in terms of the no-load speed s_0) is this maximum power being delivered? What is the power delivered by the source? What is the ohmic loss power? What is the motor efficiency?

7.3 What is the no-load speed s_0 of the linear motor in the text Example 7.1?

7.4 If the linear motor in text Example 7.1 were to pull a load representing a constant load force of 0.3 N what would be its steady-state speed? What power would the motor develop? What would be the ohmic power loss?

7.5 Consider a separately excited dc motor. Prove that it will deliver its maximum power at a speed of one-half its no-load speed. (Compare also Exercise 7.2.) Would a motor ever be operated like this? Explain why not!

7.6 In this and several of the following exercises, we shall study the operating characteristics of a six-pole dc machine that is characterized by the following design data:

1) It is to be operated from a 500-V dc armature voltage source.

2) It must tolerate a maximum armature current of 200 A.

3) The no-load speed is 2500 rpm (when operated from 500 V).

4) The field winding is separately excited from a 500-V dc excitation source. When the machine is fed from the 500-V armature source and running at a no-load speed of 2500 rpm, the field current measures 5.05 A. We shall call this the *nominal excitation level.*

Laboratory tests yield the following data:

R_a = 0.211 Ω

p_{stray} = 0.9 kW at fully loaded motor

p_{rot} = 4.65 kW, when motor is running at a no-load speed of 2500 rpm and fed from a 500-V voltage source.

In order to determine how the rotational losses vary with speed the no-load test is performed at three different armature voltage levels. At all three tests the excitation current is of nominal value. Test results are tabulated below:

Armature voltage, V	Speed rpm	Rotational losses, kW
450	2250	3.69
500	2500	4.65
550	2750	5.73

If we assume that the rotational losses increase as the xth power of speed, n, we can express them in the simple formula

$$p_{rot} = 4.65 \left(\frac{n}{2500}\right)^x. \tag{7.72}$$

Find x from the above test data! Why do the rotational losses increase with speed? Why will the speed increase as the armature voltage is raised?

7.7 Compute the emf and armature current for the machine in the previous exercise if it is being run at no-load from an armature voltage source of

a) 500 V,

b) 525 V.

Assume nominal excitation level!

7.8 Assume the above machine to be running at no-load at nominal excitation level. The armature source is 500 V. A load is now applied and increased slowly until the motor draws 200 A from the armature source. The motor is now fully loaded. To what value will the speed drop? How many kW are drained from the armature source? How many kW are drained from the excitation source?

7.9 From the previous exercise you will find that the fully loaded motor will draw a total of 102.53 kW from the armature and excitation sources. How many kW will it deliver to the load? Therefore, what will be its operating efficiency? What is the magnitude of the shaft torque? What would be the hp rating of this motor?

7.10 Consider the fully loaded machine in exercises 7.8 and 7.9. The armature voltage is decreased by 7%. By how many percent will the speed drop?

Assumption: The load torque varies as the square of the speed.

7.11 Consider again the fully loaded machine in exercises 7.8 and 7.9. Now we decrease the excitation voltage by 7%. By how many percent will the speed increase?

Assumptions: (1) The pole flux is directly proportional to field current. (2) The load torque has the same speed characteristics as assumed in previous exercise.

7.12 The dc machine in exercise 7.6 is being operated as a generator. It is being run at 2500 rpm driven by a diesel motor. The excitation voltage is adjusted until the voltage across the open-circuited armature terminals equals 500 V.

A load resistance R is connected across the armature terminals. To what value can R be lowered if rated armature current (200 A) must not be exceeded? When the armature current is 200 A, compute also the generated power (that is, the power dissipated in the load) and the power delivered by the diesel!

Assumption: The armature current will give rise to an electromechanical torque that will tend to decrease the speed. We assume that the diesel motor can maintain a *constant* speed.

7.13 Same as in the previous exercise but assume now that the diesel cannot maintain a constant speed. Assume a speed drop of 1% for each added armature current increment of 35 A.

7.14 As the load current in exercise 7.12 is being increased the terminal voltage decreases from its initial value (500 V) due to the drop across R_a. For example, an armature current of 200 A will result in a voltage drop of 42.2 V, or 8.4%. This is a distinct drawback.

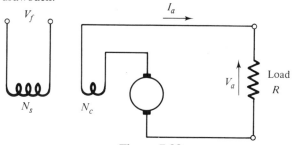

Figure 7.39

By adding a "compound" field winding (see Fig. 7.39) the above voltage drop can be compensated for. The compound winding having N_c turns adds an mmf to that caused by the separate excitation winding N_s.

Explain how this added field winding works and also compute the turns ratio $N_c:N_s$ if we want to maintain a terminal voltage V_a that is totally independent of load current I_a.

REFERENCES

Kusko, A. *Solid-State DC Motor Drives*. Cambridge, Mass.: M.I.T., 1969.

Langsdorf, A. *Principles of DC Machines*. New York: McGraw-Hill, 1959.

Mazda, F. F. *Thyristor Control*. London: Newnes-Butterworth, 1973.

Siskind, C. S. *Electrical Machines*. New York: McGraw-Hill, 1959.

chapter 8

AC INDUCTION MOTORS

Three-phase induction motor of the squirrel-cage type. Note the cooling ducts in both the rotor and the stator. Note also the similarity of the stator winding with that of the synchronous machine shown at the opening of the Chapter 4. (Courtesy General Electric)

357

8.1 WHY INDUCTION MOTORS?

There are large numbers of application areas for which the dc motor is definitely "overqualified." Its outstanding torque–speed characteristics and unmatched controllability come at a price. A dc motor always requires a dc supply and this entails extra costs. Numerous industrial, domestic, and commercial uses of motors require fairly simple torque–speed features. For example, an air compressor motor is required to operate at a constant speed and deliver a constant torque. In most of these applications ruggedness of design and ability to be operated directly off the ac network are much more sought after characteristics.

The ac induction motor fits very well into this picture. In fact, this motor is today the most widely used powering device in every technological society. It comes in sizes ranging from a fraction of to several thousand horsepowers. Large ac induction motors (usually above 5 hp) are invariably designed for three-phase operation, the reason being that one desires a constant torque and symmetrical network loading. Small fractional horsepower motors are often of single-phase design. The chapter will begin with a treatment of three-phase motors. The final section shall be devoted to single-phase induction motor design.

8.2 BASIC DESIGN FEATURES

Typical design features of a three-phase induction motor are depicted in Fig. 8.1.

The stator contains a distributed three-phase winding essentially identical with that found in the stator of synchronous machines as described in Chapter 4 (Fig. 4.14). The speed of the induction motor like that of the synchronous motor depends upon the number of poles. Figure 8.1 shows a two-pole machine. The rotor winding design varies depending upon the need for torque control. The most common design is shown in Fig. 8.1, where a "squirrel-cage" winding is depicted. It consists of solid copper or aluminum bars embedded in rotor slots, each bar short-circuited by end rings.

The electromechanical torque of the induction motor is obtained by the interaction between a stator-bound magnetic flux wave and a rotor-bound current wave. The latter is obtained from the stator flux wave through magnetic induction—thus giving the motor its name.

8.3 THE ROTATING STATOR FLUX WAVE

All induction motors, three-phase as well as single-phase, base their operation upon the existence of a traveling or revolving magnetic flux wave. We analyze in this section the conditions which must be satisfied for the creation of such a wave.

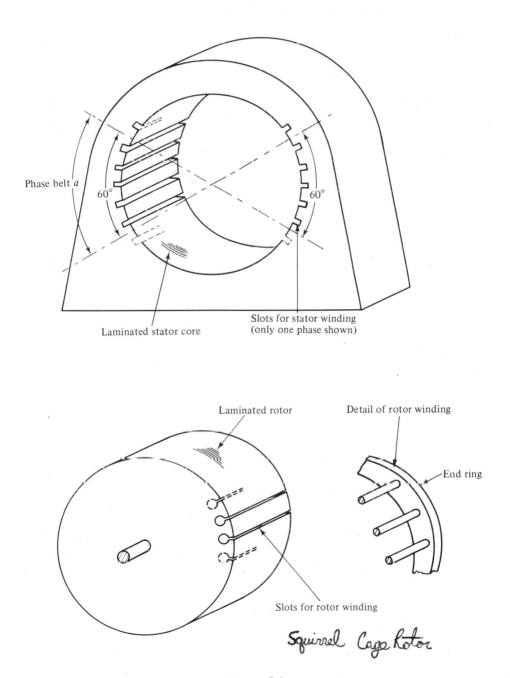

Phase belt a

$60°$ $60°$

Laminated stator core

Slots for stator winding
(only one phase shown)

Laminated rotor   Detail of rotor winding

End ring

Slots for rotor winding

Squirrel Cage Rotor

Figure 8.1

8.3.1 Time–Space Dependency of Stator Flux

Consider the magnetic flux Φ_a created by the current in phase a of the distributed two-pole stator winding in Fig. 8.1. The flux will clearly have a vertical orientation or "magnetic axis" as shown in Fig. 8.2. (Disregard for the moment the two flux components Φ_{a+} and Φ_{a-} also shown in the figure.) Φ_a will *pulsate* with the same frequency as the voltage which energizes the winding. Its density, $B_a(x, t)$, will vary both with space x and time t, the former being a tangential coordinate (Fig. 8.2) *fixed with respect to the stator.* For $B_a(x, t)$ we can write

$$B_a(x, t) = \hat{B} \cos \omega t \cos \left(\frac{p}{D} x\right) \qquad \text{T.} \qquad (8.1)$$

Here $\hat{B}$ represents the peak value of the flux density. It is directly proportional to the rms value of the exciting voltage (compare Eq. 5.8).

The factor $\cos \omega t$ tells us that the flux will pulsate in time with the same radian frequency, ω, as the voltage. The factor $\cos (px/D)$ informs us that the

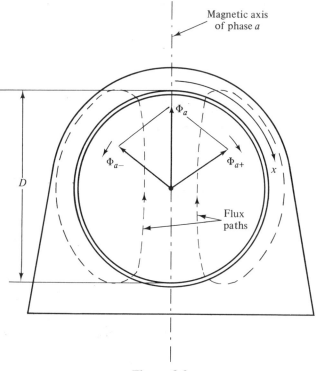

Figure 8.2

flux density varies harmonically around the stator periphery peaking at the center of the magnetic north pole at $x = 0$, reaching a negative peak at the center of the magnetic south pole at $x = \pi D/p$ and completing a full cycle (or wavelength) for $x = 2\pi D/p$ (compare with Fig. 3.32).

In Fig. 8.3 we have plotted $B_a(x, t)$ (dashed line) versus x for four different

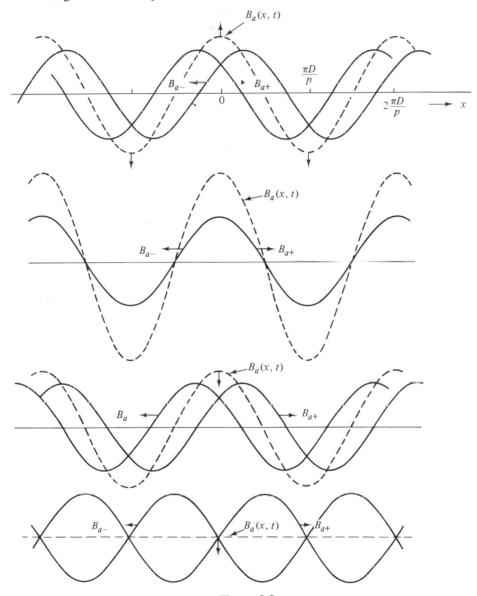

Figure 8.3

time instants. Its pulsating character is clearly shown. (Again, disregard for the moment the two components B_{a+} and B_{a-}.)

8.3.2 Equivalence Between Pulsating and Revolving Fluxes

By making use of the trigonometric identity

$$\cos \alpha \cos \beta = \tfrac{1}{2}\cos (\alpha - \beta) + \tfrac{1}{2}\cos (\alpha + \beta), \tag{8.2}$$

we can write Eq. (8.1) in the following alternate form:

$$B_a(x, t) = \underbrace{\frac{\hat{B}}{2} \cos \left(\frac{px}{D} - \omega t\right)}_{B_{a+}} + \underbrace{\frac{\hat{B}}{2} \cos \left(\frac{px}{D} + \omega t\right)}_{B_{a-}} \qquad \text{T.} \tag{8.3}$$

The flux caused by the current in the a-phase has thus been resolved into two components, B_{a+} and B_{a-}. The first of these components, B_{a+}, was identified in Chapter 4 (Eq. 4.44) as a wave traveling in the positive x-direction. The wave amplitude is $\tfrac{1}{2}\hat{B}$ and the wave velocity s_+ is

$$s_+ = \frac{\omega D}{p} = \frac{2\pi f D}{p} \qquad \text{m/s.} \tag{8.4}$$

This is readily understood if we note that the angle $((px/D) - \omega t)$ will be constant for all x and t values that satisfy the ratio

$$\frac{x}{t} = \frac{\omega D}{p} \qquad \text{m/s.}$$

The angle $((px/D) + \omega t)$ will be constant for x and t values satisfying the ratio

$$\frac{x}{t} = -\frac{\omega D}{p} \qquad \text{m/s.}$$

We thus conclude that the second component, B_{a-}, of Eq. (8.3) represents a wave traveling in the negative x-direction with the velocity given by

$$s_- = -\frac{\omega D}{p} = -\frac{2\pi f D}{p} \qquad \text{m/s.} \tag{8.5}$$

In summary, a pulsating flux wave can be considered equivalent to two oppositely revolving flux waves of constant amplitude and equal speed.† The

---

† If we express the velocity in rpm, it can readily be shown that the flux revolves at the *synchronous* speed

$$n_s = 120\frac{f}{p} \qquad \text{rpm.}$$

situation is clarified graphically in Fig. 8.3 where the pulsating flux $B_a(x, t)$ (dashed) is clearly shown to equal the sum of the two oppositely traveling component waves, B_{a+} and B_{a-}.

(Figure 8.2 depicts an alternate and sometimes more useful way of representing the above equivalence. The flux Φ_a pulsating along the vertical magnetic axis between the peak values $\pm\hat{\Phi}_a$ is resolved into the two oppositely revolving fluxes, Φ_{a+} and Φ_{a-}, respectively, each having a magnitude of $\frac{1}{2}\hat{\Phi}_a$.)

Example 8.1   The air gap in the 2-pole machine in Fig. 8.1 has a diameter of $D = 1$ m. Find the velocity of the flux wave! Stator frequency is $f = 60$ cps.

SOLUTION: Equation 8.4 gives directly

$$s_+ = \frac{\omega D}{p} = \frac{2\pi f D}{p} = \pi \cdot 60 \cdot 1 = 188.5 \text{ m/s}.$$

If we express the velocity in rpm we get

$$n_s = 120 \cdot \frac{60}{2} = 3600 \text{ rpm}.$$

....

8.3.3   Composite Three-Phase Flux Picture

We are now ready to tackle the task of finding the total composite flux picture as created by *all three* of the phase windings of the three-phase motor.

First we write expressions for $B_b(x, t)$ and $B_c(x, t)$ analogous to $B_a(x, t)$ of Eq. (8.1). The magnetic axes of phases b and c are displaced, respectively, $2\pi/3$ and $4\pi/3$ radians from that of phase a. The voltages of these two phases are lagging that of phase a by $2\pi/3$ and $4\pi/3$ radians, respectively. We thus have:

$$\left.\begin{aligned} B_b(x, t) &= \hat{B} \cos\left(\omega t - \frac{2\pi}{3}\right)\cos\left(\frac{px}{D} - \frac{2\pi}{3}\right), \\ B_c(x, t) &= \hat{B} \cos\left(\omega t - \frac{4\pi}{3}\right)\cos\left(\frac{px}{D} - \frac{4\pi}{3}\right). \end{aligned}\right\} \tag{8.6}$$

By adding (8.1) and (8.6) we obtain the *total* flux density

$$B_{tot}(x, t) = B_a(x, t) + B_b(x, t) + B_c(x, t). \tag{8.7}$$

If we first apply the trigonometric identity (8.2) to each term and then use the identity (4.77) we can readily prove that the expression (8.7) reduces to

$$B_{tot}(x, t) = \tfrac{3}{2}\hat{B} \cos\left(\frac{px}{D} - \omega t\right) \quad \text{T.} \tag{8.8}$$

which is the equation for a *single clockwise revolving flux-wave* of amplitude

$$B_{\max} = \tfrac{3}{2}\hat{B}.$$

This analysis result is so crucially important that we give it the alternate geometrical interpretation presented in Fig. 8.4.

The pulsating flux Φ_a is thus resolved into the clockwise (CW) rotating component Φ_{a+} and the counterclockwise (CCW) revolving component Φ_{a-} (as was already done in Fig. 8.2). Because the voltage in b-phase is lagging that of phase a by 120 electrical degrees, the flux Φ_b will be lagging Φ_a by 120°. Consequently, the revolving components Φ_{b+} and Φ_{b-} will lag their counterpart components Φ_{a+} and Φ_{a-} by 120°. (As the magnetic axis of the b-phase is by design shifted 120° from that of the a-phase, the position of each revolving component Φ_{b+} and Φ_{b-} must of course be measured relative to its own magnetic axis.)

Finally, the resolution into rotating component fluxes Φ_{c+} and Φ_{c-} of the flux Φ_c pulsating along the magnetic axis of the c-winding can be performed in a manner identical to that of Φ_b. All angles are now, of course, 240°.

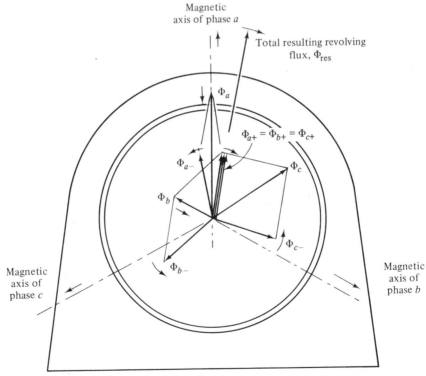

Figure 8.4

We have thus obtained six different rotating component fluxes. From Fig. 8.4 we summarize the following features:

1) The three CCW revolving components Φ_{a-}, Φ_{b-}, and Φ_{c-} are lagging each other by 120° and 240°, respectively. They thus add up to zero. *No negatively rotating flux therefore exists in a symmetrical three-phase motor.*

2) The three CW revolving components Φ_{a+}, Φ_{b+}, and Φ_{c+} are all of equal space phase. They thus add up to a *total* CW rotating resulting flux Φ_{res} of *density amplitude*

$$B_{max} = 3\frac{\hat{B}}{2}.$$

This observation evidently agrees with Eq. 8.8.

3) Were we to reverse the phase sequence of the stator voltages from abca . . . to acba . . . , the resulting flux Φ_{res} would rotate in CCW direction.

We shall find in the next section that the rotor will actually rotate in the direction of Φ_{res}. *Reversing the phase sequence* (which is simply accomplished by shifting *any* two of the three phases) *therefore represents a simple means of reversing the direction of rotation.*

8.3.4 Flux Harmonics

We have assumed in the above analysis that the flux wave is purely sinusoidal. This is not quite correct. Like an ocean wave which is surrounded by ripples of all sizes, our fundamental flux wave is accompanied by a large number of small-amplitude ripples of many different frequencies.

These ripples will cause generally undesirable phenomena in the form of "cogging," "crawling," and also magnetic noise and vibrations. These second-order side effects are of practical significance and cannot be ignored. However, by various design tricks they can be minimized to tolerable levels. The scope of our basic treatment does not permit us to enter into a discussion of these matters. The interested reader finds a good treatment in Alger, 1970.

8.4 THE TORQUE-CREATING MECHANISM

As the revolving flux Φ_{res} sweeps across the rotor *and* the stator windings, it will induce emf's in both. In the short-circuited rotor cage winding these emfs will cause currents. The currents will interact with the stator flux to create an electromechanical torque. Under the influence of the torque the motor will accelerate to its full operating speed.

The following questions now are important:

■ What type of torque is obtained—constant or pulsating?
■ What will be the magnitude of the starting torque?
■ As the motor speeds up, how will the torque change?
■ What will be the motor operating speed?
■ What power will the motor absorb from the network from which it is being run?

We can best address these questions by first investigating the nature of the induced rotor currents.

8.4.1   The Rotor Currents at Standstill

Consider the induction motor at standstill, with the stator winding energized. This condition occurs at start or stall of the motor. The speed of the flux wave relative to the stator and rotor windings is now the same for both, that is, n_s rpm. The frequencies of the induced emf's will therefore be the same in both windings, that is, the frequency, f Hz, of the energizing voltage.

The instantaneous value of the emf induced in a particular conductor will be proportional to the instantaneous flux density at the conductor in question (compare Eq. 3.58). The stator and rotor windings will thus experience "emf waves" that will accompany and be in phase with the flux wave.

Consider for a moment the rotor emf wave. In each rotor copper bar the emf's will excite currents. If the impedance of the rotor bars were purely resistive the instantaneous magnitude of the currents would be directly proportional to the instantaneous magnitude of the emf's and the result would be a current wave in phase with the emf wave. However, the impedance of the rotor bars contains a considerable reactive component† and the *current wave* will thus be *lagging* the emf (and flux) wave by a certain angle, γ.

The total wave picture is depicted in Fig. 8.6, where one full cycle of the wave system is shown. We make the following observations in regards to this picture:

1) As the flux wave is essentially of sinusoidal form the same will apply to both the emf and current waves.
2) The emf and current waves will have as many maxima and minima (poles) as does the stator flux wave. This number is solely determined by the number of poles of the stator winding (two in Fig. 8.1).

---

† The reactance is due to the self-inductance of the rotor winding. Each rotor bar surrounds itself with a magnetic field (Fig. 8.5) which is induced by its own current.

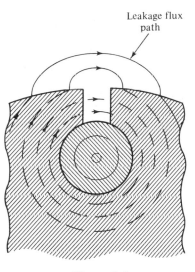

Leakage flux
path

Figure 8.5

3) The rotor end-rings serve as short-circuit paths for the copper bars. The currents in the rings vary sinusoidally around the periphery as shown in Fig. 8.6, having their instantaneous maximum values opposite to those bars which experience the instantaneous current zeros. The current waves in the rings and bars travel, or course, with equal speed.

8.4.2 Torque Formula

Returning for a moment to the synchronous machine we remind ourselves that its electromechanical air gap torque resulted from the interaction between the following:

1) A *rotor bound* flux wave of amplitude B_{max} revolving with the synchronous speed n_s rpm.

2) A *stator bound* current wave revolving with the same speed but lagging the flux wave by the angle γ. The current wave although existing only in the discrete stator slots could be considered in a macro sense as "smeared out" over the total stator surface to form a surface or sheet current wave of amplitude A_{max}.

We derived a formula (Eq. 4.87) for the torque and restate it here for easy reference:

$$T_{em} = V_{rot}B_{max}A_{max}\cos\gamma \qquad N\cdot m \qquad (8.9)$$

where V_{rot} is the rotor volume.

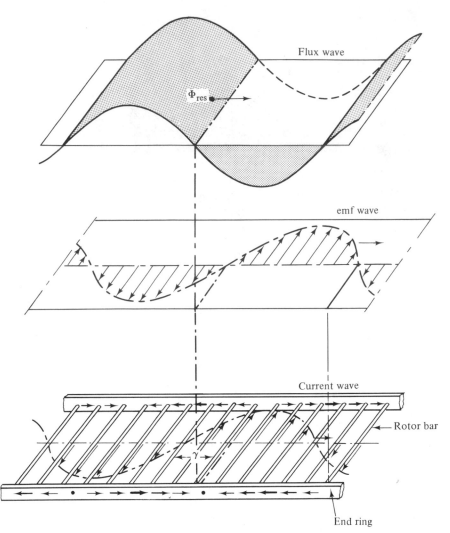

Φ_{res}

Flux wave

emf wave

Current wave

Rotor bar

End ring

Figure 8.6

The three-phase induction motor has a *stator bound* revolving flux wave of amplitude B_{max}. The current wave is *rotor bound*, lagging the flux wave by γ degrees and existing in the discrete rotor slots. If, as before, we "smear out" the current over the rotor surface† we obtain a continuous surface or sheet

---

† The cage winding shown in Fig. 8.1 contains a discrete *finite* number of rotor bars. By letting the number of bars approach infinity the cage approaches a cylindrical shell. In fact, some induction motors have this type of rotor winding design ("drag cup" motors). When the rotor winding consists of a continuous cylindrical shell the rotor currents do in fact appear as a continuous sheet current.

current wave of amplitude A_{max} A/m revolving at the speed n_s rpm relative to the stator and lagging the B-wave by γ degrees.

The only difference between the synchronous and induction machines is the reversed "home bases" of the two waves. Clearly this fact will not affect† the torque and we conclude therefore that the formula (8.9) applies to induction motors as well.

In particular, we are reminded that the torque is *constant*.

8.4.3 Current Wave In Running Rotor—Concept of Slip

Under the influence of the above torque the rotor will (unless constrained) accelerate and reach an operating speed n rpm. How will the torque change as the rotor speeds up?

The relative speed n_{rel} between the rotor and the flux wave is

$$n_{rel} = n_s - n \qquad \text{rpm.} \tag{8.10}$$

As the *rotor frequency* f_r, of the rotor emf's and currents will be proportional to this relative speed we can write

$$\frac{f_r}{f} = \frac{n_{rel}}{n_s} = \frac{n_s - n}{n_s} = 1 - \frac{n}{n_s}, \tag{8.11}$$

where f as before represents the stator or source frequency. The rotor frequency thus can be computed from the formula

$$f_r = \frac{n_s - n}{n_s} f = sf \qquad \text{Hz.} \tag{8.12}$$

We have here introduced the *slip*, s, defined by

$$s = \frac{n_s - n}{n_s} \qquad \text{pu.} \tag{8.13}$$

The slip is a measure of the relative speed between the rotor and the stator flux wave. At standstill, n is zero and the slip then equals 1 pu (or 100 percent). Should the rotor be running at synchronous speed the slip would be zero. The graph in Fig. 8.7 shows the relation between rotor speed and slip. Note in particular that the slip is negative for speeds in excess of the synchronous. Note also that negative speeds correspond to slips in excess of 100 percent.

As the motor speeds up, the relative speed decreases thus reducing both the rotor frequency and the magnitude of the rotor emf's. As a result, the magnitude of the rotor currents will also decrease.

---

† You can pull a boat by standing either in the boat or on the shore.

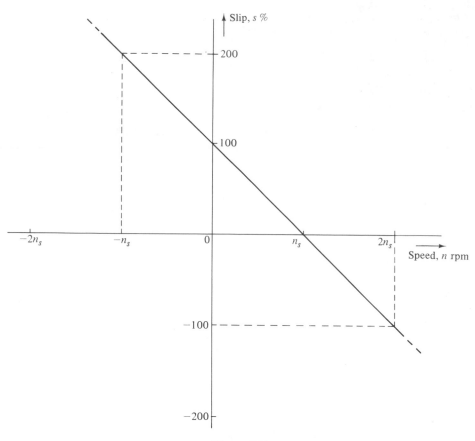

Figure 8.7

The rotor emf and current waves, in spite of their lower speed relative to the rotor, *still have the same synchronous speed, n_s, relative to the stator.* The flux, emf, and current waves thus, as before, constitute a wave system that revolves with the speed n_s rpm relative to the stator, *independent of the rotor speed, n.*

Example 8.2 The two-pole motor depicted in Fig. 8.1 is energized from a 60-Hz ac source. The rotor is running at 3520 rpm. Find slip and rotor frequency!

SOLUTION: The stator flux will travel at the synchronous speed

$$n_s = 120\frac{f}{p} = 120 \cdot \frac{60}{2} = 3600 \text{ rpm.}$$

From Eq. (8.13) we have for the slip

$$s = \frac{3600 - 3520}{3600} = 0.0222 \text{ pu.}$$

Equation (8.12) yields the rotor frequency

$$f_r = 0.0222 \cdot 60 = 1.333 \text{ Hz.}$$

Example 8.3 What would be the rotor frequency in the previous example if the rotor were forced to run with synchronous speed *against* the flux wave?

SOLUTION: As the rotor speed is now $n = -3600$ rpm, we obtain

$$s = \frac{3600 - (-3600)}{3600} = 2 \text{ pu.}$$

Thus $f_r = 2 \cdot 60 = 120$ Hz.

8.4.4   Torque Dependency on Slip—A Qualitative Analysis

As the motor speeds up and approaches the speed n_s of the flux wave, both the frequency and magnitude of the rotor currents will diminish. Our first guess would be that the torque would also decrease. Decreased rotor current magnitude undoubtedly means a decrease in A_{max}, the current wave amplitude. According to the torque formula (8.9) this tends to *decrease* the torque.

However, the reduced rotor frequency f_r will result in a lower rotor reactance and a corresponding decrease in the phase angle γ. Cos γ will thus increase which thus tends to *increase* the torque.

As the torque (Eq. 8.9) depends upon the *product* $A_{max} \cdot \cos \gamma$, and as the first of these factors decreases whereas the second increases with rotor speed, it is not immediately clear whether the result will be a torque increase or decrease.

For a typical motor, as the rotor speed first increases from standstill, the increase in cos γ will dominate over the decrease in A_{max}. The torque, which at zero speed has a *standstill* or *starting* value, T_{st}, will thus first increase with speed as shown in Fig. 8.8. When the rotor speed has reached a certain value, n_{max}, the incremental increase in cos γ exactly equals the incremental decrease in A_{max}. The torque growth thus will be zero—the torque has reached its highest value T_{max}. As the speed grows beyond n_{max}, the decrease in A_{max} dominates over the increase in cos γ, resulting in a dropping torque. Finally, when the rotor speed reaches the value n_s there is no longer any relative speed

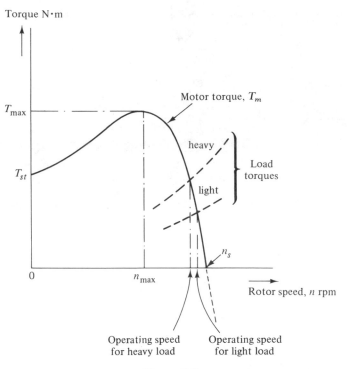

Figure 8.8

between the flux wave and rotor. No rotor emf's and currents are thus induced and the torque will equal zero.

It is obvious, therefore, that an induction motor can never pull a load at synchronous speed n_s. *Its speed, if run as a motor, will always be less than n_s.* For this reason the induction motor is sometimes referred to as "asynchronous."

8.4.5 Determination of Motor Operating Speed

Assume that we have found either by measurement or by analysis (see later sections in this chapter) the motor torque curve shown in Fig. 8.8. It is then very easy to determine the operating speed n of the motor. As the speed obviously will depend upon the load being pulled it is necessary to have a knowledge of the torque–speed characteristics of the load in question. Remember that the same was true in the case of dc motors.

In Fig. 8.8 are plotted two different load curves, labeled "light" and "heavy." The intersections between these curves and the motor torque curve indicate torque balance. These points thus give directly the operating speeds of the motor under the loads in question.

As the load torque increases, the induction motor responds by a speed reduction. We can actually plot the operating speed versus load torque and obtain the drooping speed–torque curve depicted in Fig. 8.9. In this respect the induction motor resembles the dc shunt motor (the torque–speed curve of which has been included in Fig. 8.9 for comparison).

Figure 8.9 also demonstrates a distinct difference in the behavior of the two motor types. Obviously, the induction motor cannot deliver a torque in excess of T_{max} in Fig. 8.8. Were the load to be continuously increased the speed would decrease to the value n_{max}. A further increase in the load torque would result in a speed collapse (or "stall"). The dc shunt motor would not stall (but the high armature current would overheat the motor).

Typically, an induction motor will normally operate at speeds of 95 to 99 percent of synchronous speed, corresponding to a slip range of 5 to 1 percent.

8.4.6 The Induction Generator

If the induction motor load were to be dropped its rotor would accelerate to slightly less than synchronous speed. The always present losses will require some small torque and to deliver this torque the motor must run at a slight slip.

Assume now that by means of a prime mover we were to accelerate the motor *beyond* synchronous speed. How will the machine behave?

As the rotor now runs *faster* than the stator flux the relative speed n_{rel} (Eq. 8.10) becomes negative. Consequently the induced rotor emf's reverse

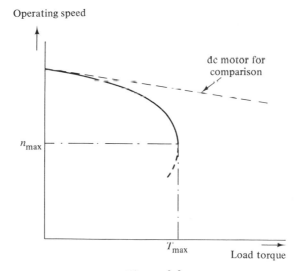

Figure 8.9

direction, and so do the rotor currents. As a result, the electromechanical torque will reverse direction—indicating, as in the case of dc motors, *generator action*. (See the dashed motor torque curve portion in Fig. 8.8.)

The induction machine now acts like a generator receiving mechanical energy from the prime mover and transforming it into electrical energy which, via the stator terminals, is supplied to the electrical grid. For example, were we to power the mine elevator discussed in the previous chapter by an induction motor the motor could serve as a "dynamic brake" during the descent phase of the work cycle. In this respect it would thus match the dc motor.

Note that a crucial assumption in the above discussion was the presence of a flux wave. This means that the induction motor can turn into a generator *only if its stator winding is connected to a source which can uphold its voltage*. An induction machine cannot, for example, turn into a generator and feed energy into a set of resistors.

8.4.7 "Wound-Rotor" Induction Motors

The above qualitative discussion indicates clearly that the magnitude of the motor torque depends to a great extent upon the magnitude of the induced rotor currents. As the magnitudes of these currents in turn depend strongly upon the impedance of the rotor bars and end rings, it follows that this impedance will to a great extent affect the magnitude of the torque. (Specifically *how* the torque will depend upon the rotor impedance will be shown in Section 8.5.9.)

Varying the rotor impedance seems thus an obvious way of varying the torque. However, a squirrel-cage rotor winding has a built-in impedance and therefore does not provide this option of torque control. In cases where torque control is important, "wound-rotor" winding design is preferred to the cage winding.

In a wound-rotor induction motor the rotor winding consists of a symmetrical three-phase winding, of the same type found in the stator. However, the rotor winding does not need to be identical to the stator winding. *The only important restriction is that the two must be wound with an equal number of poles.*

Of the six rotor winding terminals, three are connected together to form an "internal" neutral. The remaining three terminals are connected to three slip rings. Three external variable resistors, via carbon brushes, are connected to the three slip rings. The total rotor circuit is closed by means of an "external" neutral, usually grounded. The arrangement is depicted schematically in Fig. 8.10. It is important to note that in order to obtain complete symmetry between the three rotor phases the three variable external resistors must be mechanically interlocked so as to be of equal magnitude in any position.

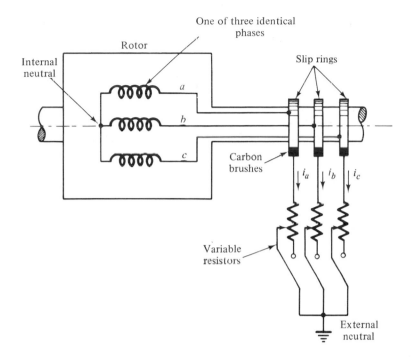

Figure 8.10

A brief explanation of the workings of this motor type follows:

As the stator flux wave sweeps by the rotor winding, a *symmetrical* three-phase set of emf's will be induced in the latter (for the same reason that a three-phase set of emf's will be induced in the stator of a synchronous machine). Because the three rotor circuits are identical, the three-phase emf's will give rise to a *symmetrical* three-phase set of rotor currents. Both the emf's and currents are, of course, of slip frequency, *sf*.

We learned in chapter 4 how a symmetrical three-phase set of currents flowing in a three-phase winding gives rise to a traveling current wave. We appreciate therefore the fact that the symmetrical three-phase set of rotor currents when flowing through the three-phase rotor winding will result in a rotor current wave. The wave will travel with the speed sn_s rpm relative to the *rotor* and thus with the speed n_s rpm relative to the *stator*.

Thus, with a stator-bound flux wave and a rotor-bound current wave, both traveling at the same speed, the conditions are again met for creation of a constant motor torque. We shall later analyze (Sect. 8.5.9) how the magnitude of the variable resistor affects the magnitude of the torque.

8.5   THREE-PHASE INDUCTION MOTOR PERFORMANCE ANALYSIS

The torque formula (8.9) proved extremely valuable in conveying a *qualitative* picture of the three-phase induction motor performance. It is not very practical, however, for the purpose of obtaining *quantitative* performance data, the reason being that the variables B_{max}, A_{max}, and $\cos \gamma$ which enter into the formula are very difficult to measure. We need to develop a formula that is based upon more easily measurable variables.†

Preparatory to finding such a formula it will prove useful to develop an *equivalent electric circuit* for the induction motor. This circuit will permit us to compute conveniently the more relevant electric variables that will determine the motor performance.

8.5.1   Transformer Analog of Induction Motor

Before we attempt to find a circuit equivalent for the induction motor we shall point out some far-reaching similarities between this type of motor and a transformer.

When the primary winding of a transformer is energized from an ac source a core flux will result that will induce emf's in both the primary and secondary windings. If the secondary winding is connected to a "load" a secondary current will flow, the magnitude and phase of which will depend upon the load impedance. In response, a primary current will arise which will be of opposite polarity and having a magnitude proportional to the secondary current so as to restore magnetic mmf balance in the core. The primary current drawn from the ac source accounts for the energy which, upon transformation, is supplied to the secondary load.

When the "primary" (the stator) winding of an induction motor is energized from an ac source a magnetic flux will result that will induce emf's in both the primary and the "secondary" (the rotor) winding. Secondary currents will flow, the magnitude and phase of which will depend, as we have seen, upon the speed of the rotor. A primary current will arise which will be of opposite polarity and have a magnitude proportional to the secondary current so as to preserve magnetic mmf balance in the iron core. This primary current drain from the ac source accounts for the energy supplied which, upon transformation, is used to run the rotor and its load.

We take advantage of these obvious similarities between an induction motor and a transformer as we proceed to develop an equivalent circuit for the former.

---

† The reader will remember that in a discussion of synchronous machines we also found it necessary to develop a more practical formula (4.97) for obtaining quantitative data.

8.5.2   The Concept of "Ideal Motor" (IM)

In deriving the equivalent transformer circuit in Chapter 5, the analysis proceeded from very idealized assumptions. We introduced first the concept of "ideal transformer" (IT), a device that was characterized by (1) zero winding resistances (and thus zero copper losses), (2) zero core reluctance (and as a consequence also zero leakage fluxes and reactances), and (3) zero core losses.

After developing the very simple mathematical model for the IT (Fig. 5.5) we removed these nonphysical assumptions one by one, and arrived eventually at a model that represented the physical transformer (Fig. 5.10).

We shall follow the same procedure here but we must exercise special care in doing so. For example, we *cannot* use the above assumptions in defining an "ideal motor" (IM). Were we to make the assumptions of zero winding resistances and reactances then, in effect, we would say that the rotor impedance of the squirrel-cage motor is zero. This would be an absurd assumption as it would lead to infinite currents.† As we have already suggested, and as we shall further confirm, the rotor resistance of an induction motor plays a *dominant* role in its theory and we therefore must retain it in our equations.

However, without introducing much inaccuracy into our motor models but adding considerable simplicity and clarity to our analysis, we shall define an induction motor "ideal" when characterized by the following features:

1) a magnetic iron having zero reluctance (that is, infinite permeability), and zero core losses;

2) an air-gap width which approaches zero;

3) zero friction and windage losses.

The air gap of a physical motor is made as narrow as practically possible. As shown in Fig. 8.11 the magnetic flux path crosses the air gap twice. By reducing the air gap we thus increase the air-gap flux and the torque of the motor. Ideally, the air-gap width should approach zero, which means that the air-gap reluctance will vanish. If in addition the iron permeability is infinite that part of the magnetic path will also have zero reluctance.

We remember that zero reluctance of the IT core resulted in the requirement that its magnetic path must encircle zero total current (Fig. 8.12)—we were led to the equation for "mmf balance":

$$N_1 I_1 = N_2 I_2 \qquad A \cdot t. \tag{8.14}$$

For the ideal motor the need for mmf balance requires that the magnetic path must encircle zero total current. We have already concluded that the rotor

---

† In the case of a transformer, this caused little difficulties as the load impedance will always be in series with the secondary.

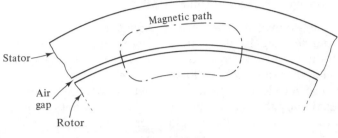

Figure 8.11

surface contains a sinusoidally varying surface current sheet. *We now conclude that the stator surface must contain a matching surface current sheet but of opposite polarity if mmf balance shall be maintained for an arbitrary position of the magnetic path.* The current situation in an ideal motor is depicted in Fig. 8.13 *and is not much different from the actual situation in a physical motor.* Of course, in reality this stator surface current will be found in the discrete stator slots.

The rotor current sheet revolves with the speed of n_s rpm relative to the stator and so must the stator current sheet. In order to generate a stator current wave traveling at synchronous speed *the stator currents must constitute a symmetrical three-phase set.* As these currents are drawn from the ac supply source *we thus conclude that the three-phase induction motor constitutes a balanced three-phase load on the network.*

8.5.3   Rotor Quantities Referred to the Stator

Consider the ideal transformer depicted in Fig. 8.14. Its 100-turn primary P is energized from a 100-volt source. Energy can be tapped from either of two

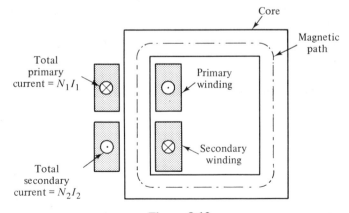

Figure 8.12

Figure 8.13

secondaries, S^{I} and S^{II} having 10 and 5 turns, respectively. The voltages measured across S^{I} and S^{II} are 10 and 5 V, respectively.

If S^{I} delivers energy to a 1-Ω load, the secondary current I_2^I will equal 10 A and the secondary power will amount to $10^2 \cdot 1 = 100$ W. The secondary current will give rise to a core mmf of $10 \cdot 10 = 100$ A $\cdot$ t.

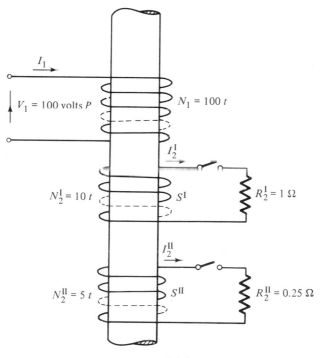

Figure 8.14

If S^{II} supplies energy to a 0.25-Ω load, the secondary current I_2^{II} will equal 20 A and the secondary power will amount to $20^2 \cdot 0.25 = 100$ W. I_2^{II} will cause a core mmf of $20 \cdot 5 = 100$ A $\cdot$ t.

In both cases, the secondary (and thus the primary) power equals 100 W, and in both cases the secondary mmf equals 100 A $\cdot$ t.

The primary winding will respond with a current I_1 to nullify the secondary core mmf. According to Eq. (8.14) the primary current can be computed from

$$100 \cdot I_1 = N_2 I_2 = 100 \text{ A} \cdot \text{t} \qquad (8.15)$$

which equation yields $I_1 = 1$ A in *both cases.*

Clearly, from the primary side the two load cases look identical. No measurements performed from the primary side can tell the cases apart.

The reason why these loads behave in an *equivalent* manner is that their respective load impedances *when referred to the primary* have identical values. According to formula (5.20) we have in the two cases

$$R_2' = 1.00 \cdot \left(\frac{100}{10}\right)^2 = 100 \ \Omega,$$

$$R_2' = 0.25 \cdot \left(\frac{100}{5}\right)^2 = 100 \ \Omega.$$

The primary "sees" the secondary mmf of 100 A $\cdot$ t and reacts by creating an equal and opposite mmf of its own.

A similar situation exists in an induction motor. When the motor is pulling a certain mechanical load at a certain slip, a certain rotor current wave is being created by the mechanism earlier discussed. The stator "sees" this current wave and reacts by creating an equal and opposite one of its own. The stator has no way of telling whether the rotor current wave is created in a cage winding or the three-phase winding of a wound rotor. From an electrical point of view the two are equivalent. Neither does the stator know whether the rotor current wave is created by a relatively small current flowing in a rotor winding of many conductors per slot or by a large rotor current in a winding of few conductors.

In conclusion, the behavior of the motor *as felt by the network* is determined by the rotor impedances "referred to the stator side."

8.5.4   Circuit Equations in the IM Case

The analysis in this section is based upon the following assumptions:

1) The motor has IM magnetic features, that is, the stator and rotor current waves are equal and of opposite sign (see Fig. 8.13).

2) The stator and rotor both have three-phase type, Y-connected windings.†
    There are an equal number of slots on the stator and rotor.‡ The number
    of conductors per stator slot is a times the number of conductors per rotor
    slot.

3) The stator winding is fed from a three-phase source measuring $|V|$ volts
    rms between each phase and ground. Phase a of the stator is chosen as
    reference phase.

4) In the analysis to follow the subscript "1" refers to "primary" (the stator);
    the subscript "2" refers to the "secondary" (the rotor).

5) All reactances used in the analysis are assumed computed at 60 Hz. Any
    reactance is of the form $X = \omega L$. As we have different frequencies in
    rotor and stator it is important to agree upon the frequency at which the
    reactance is computed.

Voltage equilibrium for the *stator* circuit (Fig. 8.15) reads

$$V = E_1 + I_1(R_1 + jX_1) \qquad \text{V} \tag{8.16}$$

where

E_1 = induced emf per phase of stator winding;

R_1 = stator resistance, ohms per phase;

X_1 = stator reactance, ohms per phase;

I_1 = stator current, amps per phase.

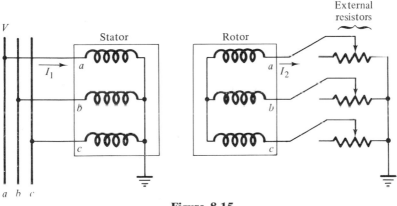

Figure 8.15

---

† We said earlier that the stator cannot tell whether the rotor is of "cage" or
"wound" type. The analysis is simplified by assuming that rotor and stator both have
similar winding types.

‡ This assumption is really not necessary. We make it to assure, for simplicity, that
stator and rotor windings have equal distribution factors.

Writing the voltage equilibrium equation for the *rotor* circuit yields

$$E_2 = I_2(R_2 + jsX_2) \qquad \text{V} \qquad (8.17)$$

where

E_2 = induced emf in rotor winding, volts per phase;

R_2 = rotor resistance, ohms per phase (including
the external resistance if any);

X_2 = rotor reactance, ohms per phase;

I_2 = rotor current, amps per phase.

Note that the reactance term in (8.17) is multiplied by s because the equation is referring to slip frequency sf.

In addition to the above two voltage equations we have these relations between primary and secondary emfs and currents:

$$E_2 = s\frac{E_1}{a} \qquad \text{V,} \qquad (8.18)$$

$$I_2 = aI_1 \qquad \text{A,} \qquad (8.19)$$

where

$$a \equiv \frac{\text{number of conductors per stator slot}}{\text{number of conductors per rotor slot}}.$$

Equation (8.18) tells us that the rotor emf is reduced both due to slip (s) and to winding ratio (a).

Equation (8.19) is identical to Eq. (8.14) and expressses mathematically what Fig. 8.11 depicts pictorially. We may write the equation as follows:

$$I_1 = \frac{I_2}{a} \equiv I_2' \qquad (8.20)$$

The current I_2' thus defined is "the rotor current referred to the stator." (Compare Eq. 5.22.)

By substitution of the expressions for E_2 and I_2 into Eq. (8.17) and by subsequent elimination of E_1 between equations (8.16) and (8.17) we obtain the following equation:

$$V = I_1\left[R_1 + \frac{a^2 R_2}{s} + j(X_1 + a^2 X_2)\right] \qquad \text{V.} \qquad (8.21)$$

In analogy with Eq. (5.20) we now define

$$\left.\begin{array}{l} R_2' \equiv a^2 R_2 \qquad \Omega, \\ X_2' \equiv a^2 X_2 \qquad \Omega. \end{array}\right\} \qquad (8.22)$$

These are the "rotor impedances referred to the stator side."

In terms of these impedances Eq. (8.21) is finally put into the form

$$V = I_1\left[R_1 + \frac{R_2'}{s} + j(X_1 + X_2')\right] = I_2'\left[R_1 + \frac{R_2'}{s} + j(X_1 + X_2')\right]. \quad (8.23)$$

8.5.5 Equivalent Circuit in the IM Case

Equation (8.23) provides the basis for the very simple circuit shown in Fig. 8.16(a). This is the "equivalent circuit" for the induction motor modeled as an ideal motor. In a practical case all of the four impedance elements can be easily found either by tests or by computation (the latter being the case at the design stage). The circuit can thus give us the value for the current I_1 for any speed (expressed in slip s) of the motor. From Eq. (8.19) we can then obtain I_2. As we shall see in later sections, torque and power can then easily be computed.

The equivalent circuit thus provides a very practical means for obtaining *quantitative* performance data in a convenient way.

8.5.6 The Circle Diagram in the IM Case

Using the circuit in Fig. 8.16(a), we develop some interesting features characteristic of the induction motor. In particular let us study how the motor current

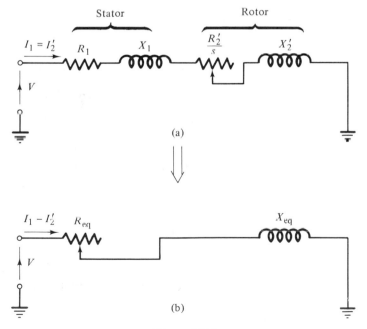

Figure 8.16

depends upon the speed. In Fig. 8.16(b) we have somewhat simplified the circuit by lumping the resistive elements into an equivalent resistance:

$$R_{eq} \equiv R_1 + \frac{R_2'}{s} \quad \Omega. \tag{8.24}$$

Similarly, we combine the reactive elements into an equivalent reactance:

$$X_{eq} \equiv X_1 + X_2' \quad \Omega. \tag{8.25}$$

We make the important observation that the speed will affect R_{eq} but not X_{eq}.

From Fig. 8.16(b) we note that the current equals

$$I_2' = I_1 = \frac{V}{R_{eq} + jX_{eq}} \quad A. \tag{8.26}$$

The rms value of current is

$$|I_2'| = |I_1| = \frac{|V|}{\sqrt{R_{eq}^2 + X_{eq}^2}} = \frac{|V|}{X_{eq}\sqrt{1 + \left(\dfrac{R_{eq}}{X_{eq}}\right)^2}} \quad A. \tag{8.27}$$

Its phase angle φ (relative to V) can be computed from

$$\varphi = \cos^{-1}\left[\frac{R_{eq}}{\sqrt{R_{eq}^2 + X_{eq}^2}}\right] = \cos^{-1}\left[\frac{1}{\sqrt{1 + \left(\dfrac{X_{eq}}{R_{eq}}\right)^2}}\right]. \tag{8.28}$$

Equations (8.27) and (8.28) confirm the following features:

Increasing speed (that is, decreasing slip) results in decreasing $|I_1|$ and φ.

We obtain a very interesting interpretation of the current variation by the following simple geometric consideration:

Decompose the current I_1 into the "in-phase" component I_p and the "out-of-phase" component I_q shown in Fig. 8.17. For these components we have

$$I_p = |I_1| \cos \varphi = \frac{|V|}{R_{eq}^2 + X_{eq}^2} \cdot R_{eq} \quad A, \tag{8.29}$$

$$I_q = |I_1| \sin \varphi = \frac{|V|}{R_{eq}^2 + X_{eq}^2} \cdot X_{eq} \quad A. \tag{8.30}$$

By squaring and adding these two last equations we obtain

$$I_p^2 + I_q^2 = \frac{|V|^2}{R_{eq}^2 + X_{eq}^2} = \frac{|V|}{X_{eq}} \cdot I_q. \tag{8.31}$$

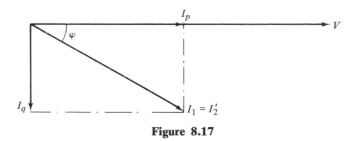

Figure 8.17

The last step follows directly from (8.30).

Equation (8.31) can be rewritten to read

$$I_p^2 + \left(I_q - \frac{|V|}{2X_{eq}}\right)^2 = \left(\frac{|V|}{2X_{eq}}\right)^2. \tag{8.32}$$

This is the equation for a circle in the $I_p - I_q$ plane! The circle has the radius $|V|/2X_{eq}$. Its center is located in the point $[0, |V|/2X_{eq}]$. *As the slip, s, changes thus causing a change in R_{eq} the tip of the current I_1 will move along a circular contour* (Fig. 8.18). This is called an "induction motor circle diagram."

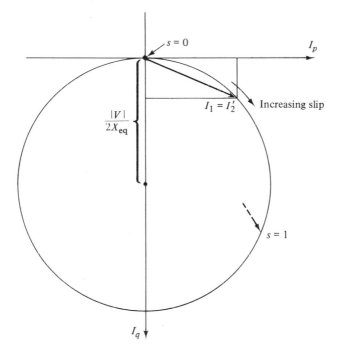

Figure 8.18

Example 8.4   A 3-phase, 6-pole induction motor rated at 10 hp, 220 volts, 60 Hz has the following impedance parameters referred to the stator:

$$R_1 = 0.295 \ \Omega/\text{phase},$$
$$R_2' = 0.150 \ \Omega/\text{phase},$$
$$X_1 = 0.510 \ \Omega/\text{phase},$$
$$X_2' = 0.210 \ \Omega/\text{phase}.$$

a) Find the circle diagram for this motor if it is being operated from a 220-volt, 3-phase network. Model the motor as an IM.

b) Find stator current if the motor runs at synchronous speed.

c) Find stator current at standstill. Also compute the heat losses and the power drained from the network.

SOLUTION: With the numerical data given we compute

$$|V| = \frac{220}{\sqrt{3}} = 127.0 \text{ volts/phase}$$

$$X_{eq} = 0.510 + 0.210 = 0.720 \ \Omega/\text{phase}$$

a) The radius of the circle will be

$$\frac{|V|}{2X_{eq}} = \frac{127.0}{2 \cdot 0.720} = 88.2 \qquad \text{A/phase.}$$

b) At synchronous speed we have $s = 0$. Thus

$$R_{eq} = 0.295 + \frac{0.150}{0} = \infty.$$

From Eqs. (8.27) and (8.28) we thus obtain

$$|I_1| = 0 \qquad \cos \varphi = 1.$$

c) At standstill we have $s = 1$. Thus

$$R_{eq} = 0.295 + \frac{0.150}{1} = 0.445 \ \Omega/\text{phase.}$$

From Eqs. (8.27) and (8.28) we obtain

$$|I_1| = \frac{127.0}{\sqrt{0.445^2 + 0.720^2}} = 150.0 \text{ A/phase,}$$

$$\varphi = \cos^{-1}\left(\frac{0.445}{\sqrt{0.445^2 + 0.720^2}}\right) = 58.28°.$$

We have marked the position of the current vector I_1 for these two cases in Fig. 8.18.

According to the equivalent circuit the real power drained from the ac network will be

$$P_1 = R_{eq} |I_1|^2 = \left(R_1 + \frac{R_2'}{s}\right) |I_1|^2 \qquad \text{W/phase.} \qquad (8.33)$$

In this case we have at standstill

$$P_1 = \left(0.295 + \frac{0.150}{1}\right) \cdot 150.0^2 = 10.01 \text{ kW/phase}$$

or

$$3 \cdot 10.01 = 30.03 \text{ kW total three-phase}$$

The power dissipated in the stator and rotor windings is

$$P_\Omega = R_1 |I_1|^2 + R_2 |I_2|^2 = R_1 |I_1|^2 + R_2 |aI_1|^2$$
$$= (R_1 + a^2 R_2) |I_1|^2 = (R_1 + R_2') |I_1|^2 \text{ W/phase.} \qquad (8.34)$$

Thus

$$P_\Omega = P_1.$$

In other words, *all* the power drained from the network goes into *losses*. This is, however, only true at standstill, because $R_{eq} = R_1 + R_2'$ only for $s = 1$. (Compare also next example.)

Caution: With a total loss power of 30 kW this motor would rapidly overheat. A stalled motor must be speedily disconnected from the source.

···

Example 8.5

a) Compute the stator current and ohmic losses if the above motor is running at 97% of synchronous speed.

b) Compute also the power drained from the ac source.

SOLUTION: The slip would now be 3% or 0.03 pu. Equation (8.24) yields

$$R_{eq} = 0.295 + \frac{0.150}{0.03} = 5.295 \text{ }\Omega\text{/phase.}$$

a)   Equations (8.27) and (8.28) give

$$|I_2'| = |I_1| = \frac{127.0}{\sqrt{5.295^2 + 0.720^2}} = 23.77 \text{ A/phase}$$

and

$$\varphi = \cos^{-1}\left(\frac{5.295}{\sqrt{5.295^2 + 0.720^2}}\right) = 7.74°.$$

Where will this current phasor be positioned in the circle diagram in Fig. 8.18?

The ohmic losses according to Eq. (8.34) will be

$$P_\Omega = 0.445 \cdot 23.77^2 = 251 \text{ W/phase}.$$

Thus, total loss in all three phases = 753 W. Compare this value with the 30 kW loss at standstill! Note in addition that the running motor has better cooling capability.

b) The power drained from the ac source is computed from formula (8.33):

$$P_1 = R_{eq} \cdot |I_1|^2 = 5.295 \cdot 23.77^2 = 2.992 \text{ kW/phase},$$

or,

$$3 \cdot 2.992 = 8.975 \text{ kW (total three-phase)}.$$

8.5.7 Motor Power and Torque in the IM Case

In the previous example the IM drained 8.975 kW from the ac source. Only 0.753 kW (or 8.39%) was actually dissipated in the resistors as ohmic heat. Where did the remaining 8.222 kW (or 91.61%) go? (Note in the stalled motor in Example 8.4 that 100% of the received power went into losses.)

As the IM has no other "power sinks" to account for these 8.222 kW *we can only conclude that they pass through the motor to the load being pulled. In other words they constitute the motor power P_m which corresponds to the electromechanical torque T_m of the motor.*†

Generally, we can write

$$P_m = P_1 - P_\Omega \qquad \text{W/phase}. \tag{8.35}$$

By making use of formulas (8.33) and (8.34) we thus get

$$P_m = \frac{1-s}{s} R_2' |I_1|^2 \qquad \text{W/phase}, \tag{8.36}$$

or

$$P_m = 3\frac{1-s}{s} R_2' |I_1|^2 \qquad \text{W (total three-phase)}. \tag{8.37}$$

---

† Note that 8.222 kW corresponds to an output of 11.0 hp. This motor is rated at 10 hp and we conclude therefore that when running at 3% slip this motor is slightly overloaded.

Between motor torque and motor power we have the relation

$$P_m = \omega_m T_m \quad \text{W,} \tag{8.38}$$

where

$$\omega_m = 2\pi \cdot \frac{n}{60} = \frac{\pi}{30}(1 - s)n_s \quad \text{rad/s.} \tag{8.39}$$

By combining (8.37), (8.38), and (8.39) we thus obtain

$$T_m = \frac{90}{\pi n_s} \cdot \frac{R_2'}{s} \cdot |I_1|^2 \quad \text{N} \cdot \text{m.} \tag{8.40}$$

Equations (8.37) and (8.40) are our most important performance equations. Formula (8.40) yields, of course, the same result as formula (8.9). The important difference, however, is that formula (8.40) expresses the motor torque in easily measurable quantities.

Example 8.6   Consider again the motor and the impedance data described in Example 8.4. As the motor is of 6-pole design the synchronous speed $n_s = 1200$ rpm.

Compute the torque (a) at standstill, and (b) at a speed corresponding to $s = 0.03$.

SOLUTION:

a) From Example 8.4 we have for $s = 1$ pu,

$$|I_1| = 150.0 \text{ A.}$$

Formula (8.40) yields directly for the *starting* torque,

$$T_m = \frac{90}{\pi \cdot 1200} \cdot \frac{0.150}{1} \cdot (150.0)^2 = 80.6 \text{ N} \cdot \text{m.}$$

b) From Example 8.5 we have for $s = 0.03$ pu,

$$|I_1| = 23.77 \text{ A.}$$

The torque formula now gives the *running* torque:

$$T_m = \frac{90}{\pi \cdot 1200} \cdot \frac{0.150}{0.03} \cdot (23.77)^2 = 67.4 \text{ N} \cdot \text{m.}$$

8.5.8   Maximum Torque (The IM Case)

The formula (8.40) clearly indicates that the motor torque varies strongly with speed (or slip). We had earlier (see Fig. 8.8) alluded to the fact that somewhere in the speed range 0–n_s rpm, (or slip range 1–0 pu) one can expect to find a maximum value, T_{max}, for the torque. It is important to know the value of this torque as it informs of the maximum load torque that can be accommodated without danger of stall.

Before we proceed to find T_{max} we substitute the value for $|I_1|$ (formula 8.27) into (8.40) which then yields

$$T_m = \frac{90}{\pi n_s} \cdot \frac{R_2'}{s} \cdot \frac{|V|^2}{\left(R_1 + \dfrac{R_2'}{s}\right)^2 + (X_1 + X_2')^2} \qquad \text{N} \cdot \text{m.} \qquad (8.41)$$

The torque is now expressed as a function of slip only.

Several observations can be made in regards to this formula:

1) Although the slip range $0 < s < 1$ is of particular importance in normal motor operation the formula gives the torque in the *total* slip range $-\infty > s > +\infty$.

2) The torque is positive for *all s* > 0. This is the speed region for *motor operation.*

3) The torque is negative for *all s* < 0. This is the speed region for *generator* operation.

4) The torque magnitude approaches zero for $s = 0$ and $s = \pm\infty$.

5) The negative torque magnitudes are numerically larger than the positive ones. (This follows because the denominator of 8.41 is numerically smaller for negative *s*.)

If we put all the above pieces together we obtain a torque curve like that depicted in Fig. 8.19. We have particularly identified the important speed† region, $0 < s < 1$.

We can expect to find a positive torque maximum T_{max} occurring for $s = s_{max}$. We can also expect to find a negative maximum. The positive maximum located in the important speed region shall be the object for our search.

The optimum search is straightforward. One seeks the value, s_{max}, that satisfies the equation‡

$$\frac{dT_m}{ds} = 0. \qquad (8.42)$$

---

† Note that the torque curve in Fig. 8.8 appears "inverted" as compared to Fig. 8.19 because it is plotted versus *n* rather than *s*.

‡ If one realizes that the torque (8.41) is a function of R_2'/s one simplifies the analysis by solving the simpler equation

$$dT_m/d\left(\frac{R_2'}{s}\right) = 0.$$

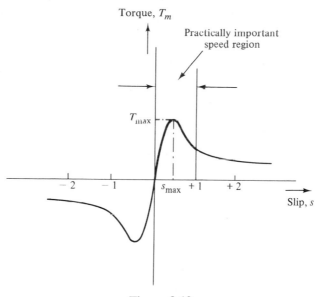

Figure 8.19

A relatively simple analysis reveals that

$$s_{max} = \pm \frac{R_2'}{\sqrt{R_1^2 + (X_1 + X_2')^2}} \text{ pu.} \tag{8.43}$$

If we limit attention to the positive torque maximum we substitute the positive s_{max} value into (8.41) and obtain

$$T_{max} = \frac{45}{\pi n_s} \cdot \frac{|V|^2}{R_1 + \sqrt{R_1^2 + (X_1 + X_2')^2}} \quad \text{N} \cdot \text{m.} \tag{8.44}$$

Example 8.7   Find s_{max} and T_{max} for the motor discussed in Examples 8.4, 8.5, and 8.6.

SOLUTION: Equation (8.43) first gives

$$s_{max} = \frac{0.150}{\sqrt{0.295^2 + 0.720^2}} = 0.193.$$

Equation (8.44) then yields

$$T_{max} = \frac{45}{\pi \cdot 1200} \cdot \frac{127.0^2}{0.295 + \sqrt{0.295^2 + 0.720^2}} = 179.4 \, \text{N} \cdot \text{m}.$$

Compare this value with those computed in Example 8.6!

8.5.9 Resistive Torque Control

We had earlier surmised that by inserting external resistances into the rotor circuit we should be able to control, within limits, the magnitude of the motor torque. The theory developed above permits us to get a good picture of the effect of these resistors.

Consider for a moment the two formulas (8.43) and (8.44). The first of these formulas informs us that the value of s_{max} is directly proportional to the rotor resistance. *By adding rotor resistance we can thus move the torque maximum toward higher s-values, that is, in the direction of lower speeds.*

The second equation informs us that T_{max} *is independent of the rotor resistance.*

By combining these two observations we realize that insertion of rotor resistance shifts the torque curves in a manner indicated by the graphs in Fig. 8.20.

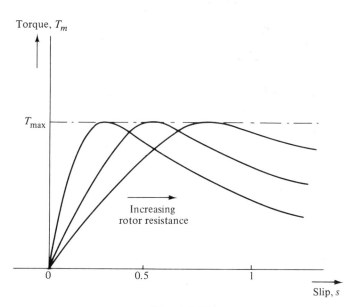

Figure 8.20

Example 8.8 By shifting the torque curve for the machine in the previous example so that the maximum torque moves from $s_{max} = 0.193$ to $s_{max} = 1.00$ we in effect arrange so that maximum torque occurs at $n = 0$, that is, at start. This will result in a fast starting motor.† How much resistance must be inserted in the rotor circuit?

SOLUTION: From (8.43) we get

$$s_{max} = 1 = \frac{R_2'}{\sqrt{0.295^2 + 0.720^2}}.$$

This equation yields

$$R_2' = 0.778 \ \Omega/\text{phase}.$$

If we deduct the rotor winding resistance ($0.150 \ \Omega$) we obtain the required *external* resistance:

$$R_{2\,\text{ext}}' = 0.778 - 0.150 = 0.628 \ \Omega/\text{phase}.$$

It should be remembered that this resistance value *is referred to the stator side*. To obtain the actual ohmic value we must divide by a^2. ~

8.6 MODEL CORRECTION FOR NONIDEAL MOTOR CHARACTERISTICS

All previous analysis was based upon the "ideal motor" assumption. It is appropriate at this time to make corrections in our various mathematical models to account for the "nonideal" behavior of a physical motor.

8.6.1 Inaccuracy of IM Models at Low Load Level

Should the mechanical load be disconnected from the shaft of an ideal motor, its speed would rise to n_s rpm. The slip would thus drop to zero and the impedance element R_2'/s in the equivalent circuit would approach infinity. The IM equivalent circuit thus would predict zero current and power at this no-load type of motor operation.

A physical motor would behave quite differently. Following the load drop its slip would *not* drop to zero but to a value usually well below 1 percent. This small but nonzero slip will cause sufficient rotor current and torque to overcome windage and friction losses. (See also discussion on page 373.) The

---

† As the motor accelerates one can shift s_{max} toward lower s-values (by disconnecting rotor resistance) thus seeking to match s_{max} with the actual s. In this manner the motor torque is at its peak value during the total startup time.

impedance element R_2'/s would be large but not infinite. Should we use the IM equivalent circuit to model the actual physical motor at such low load level, it would produce current and power values which are in poor agreement with reality. We exemplify.

Example 8.9   Consider the 6-pole motor discussed in Example 8.4. The motor is made subject to a no-load test (it carries no mechanical load but its own friction and windage) at rated voltage. Speed, stator current, and power are measured. The test results are

> speed = 1198.6 rpm,
>
> stator current = 7.51 A/phase,
>
> power drain from network = 0.503 kW (total three-phase).

Compare these test results with those power and current values you can compute from the IM equivalent circuit!

SOLUTION: The measured speed corresponds to a slip of

$$s = \frac{1200 - 1198.6}{1200} = 0.001167 \text{ pu.}$$

From the IM equivalent diagram we obtain

$$R_{eq} = 0.295 + \frac{0.150}{0.001167} = 128.9 \ \Omega/\text{phase},$$

$$X_{eq} = 0.510 + 0.210 = 0.720 \ \Omega/\text{phase}.$$

For the current we thus have

$$I_1 = I_2' = \frac{V}{R_{eq} + jX_{eq}} = \frac{127.0}{128.9 + j0.72} = 0.985 \ \underline{/-0.32°}.$$

For the real power drawn from the source we get

$$P_1 = 127.0 \cdot 0.985 \cdot \cos 0.32° = 125 \text{ W/phase}$$

or 375 W (total three-phase).
....

We note that the computed current value, 0.985 A, compares poorly with the actual measured value, 7.51 A. There is likewise a considerable spread between the computed (375 W) and measured (503 W) power values. An even more noticeable disagreement occurs for the phase angle. The computed value is $\varphi = 0.32°$ lagging. The actual power factor, according to measurements, is

$$\cos \varphi = \frac{503/3}{127.0 \cdot 7.51} = 0.176,$$

which corresponds to a phase angle between voltage and current of $\varphi = 79.9°$.

Conclusion: The IM model, in this case, does not very well represent the physical device. We shall later identify the reasons for these disagreements (see Example 8.10).

8.6.2   Existence of Exciting or Magnetizing Current (I_m)

In developing the theory that led to the "ideal" model for the induction motor we neglected the reluctance of both the iron core and the air gap. This assumption, in effect, leads to the conclusion that it requires zero mmf and thus zero resulting current for maintaining the core flux. In reality, the iron core does *not* have infinite permeability and the air gap is *not* of zero width.† Thus, the physical core requires a *nonzero* mmf for maintaining the flux. Differently expressed, we can state that the physical core does not represent an infinite reactance *as viewed from the source* but a *finite* reactance of value X_m ohms/phase.

In addition, the physical core absorbs a finite amount of *real* power, in the form of hysteresis and eddy current losses. These losses *as viewed from the source* can be represented by a finite resistance of value R_m ohms/phase. Together, R_m and X_m constitute a *magnetizing impedance* which absorbs a magnetizing or exciting current, I_m, *which is essentially independent of the mechanical loading of the motor.*

8.6.3   Corrected IM Circuit Model

The logical way‡ of accounting for the presence of I_m is shown in Fig. 8.21. The magnetizing current component I_m is drawn via the shunt elements R_m and X_m. As before, the rotor slip results in the torque-creating rotor current I_2

---

† Example 3.23 shows how even a *very* small air gap in a magnetic circuit drastically increases the mmf requirement (that is, the current need) for maintaining the flux. Contemplate what factors determine the minimum air-gap width for a real motor!

‡ Many will disagree and point out that a "better" model of the physical motor can be had by placing the shunt elements *after* the stator impedance, $R_1 + jX_1$, as shown dashed in Fig. 8.21. This is debatable for several reasons. For one thing, there are *additional* losses beyond windage, friction, core, and copper losses. They are sometimes conveniently lumped into a group labeled "stray" losses. These losses are due to harmonic effects (that we have neglected), rotor hysteresis, and other causes. Generally, they are hard to model. Sometimes one simply lumps them into the core-loss group, sometimes one divides them between core and copper losses.

The point being made is that the model corrections we are discussing are representing second-order effects. The question of whether the shunt impedance should be connected before or after the stator impedance thus corresponds to a second-order effect of a second-order effect.

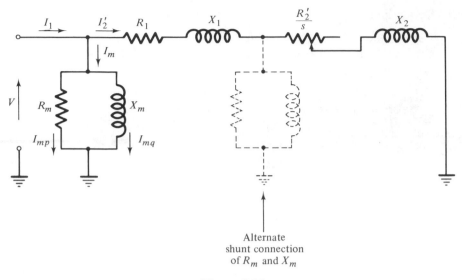

Figure 8.21

which, reflected to the stator side, shows up in our circuit as I_2'. The primary current is the vectorial sum of the two:

$$I_1 = I_m + I_2' \quad \text{A.} \tag{8.45}$$

As long as the source voltage remains constant, the I_m-component is unchanged. As before, the mechanical load and thus the slip will greatly affect I_2'.

8.6.4 Effect on Circle Diagram

As before, the tip of the current vector I_2' will follow a circle, according to Fig. 8.18. As the primary current I_1 is the vectorial sum of I_2' and the *constant* vector I_m, we obtain a corrected circle diagram of the type shown in Fig. 8.22.

This circle diagram reveals the following important facts:

1) For a heavily loaded motor, that is, when the slip is relatively large, the currents I_2' and I_1 tend to become of equal magnitude and phase *as measured in relative terms.* For this mode of operation Eq. (8.20) is thus best satisfied and is the IM model of greatest validity.

2) For low load, that is, when the slip approaches zero, the disagreement between I_1 and I_2' becomes greatest. In this region, as shown in Example 8.9, the IM model becomes invalid.

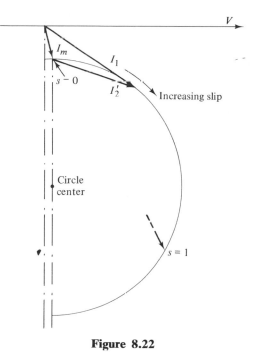

Figure 8.22

Example 8.10 Clear up the model disagreements we noted in Example 8.9. Also, from the test data compute the magnetizing impedance and core losses.

SOLUTION: In view of the above theory we now have a better understanding of the reasons for the disagreements that we encountered in Example 8.9.

I. The current we computed (0.985 A) is not I_1 but I_2'. Figure 8.23 shows an enlarged view of the circle diagram in the region of zero slip and also the relationship between I_1, I_2' and I_m.

II. The power we computed (375 W) is the sum of ohmic losses in rotor and stator windings plus the motor power, that is, the power needed to overcome windage and friction losses.

As the ohmic losses at this test amount only to a very low 17 W (how do you get this figure?), we can for all practical purposes say that windage and friction constitute the total computed power or 375 W.

III. From the diagram in Fig. 8.23 we can compute the magnetizing current I_m as follows:

Step 1. I_m is first resolved into the in-phase component I_{mp} and the out-of-phase component I_{mq}.

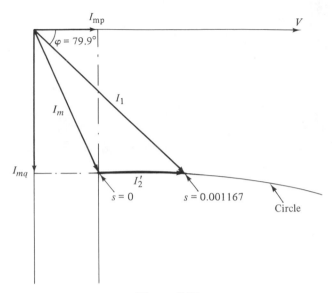

Figure 8.23

Step 2. Because I_2' for all practical purposes is parallel with V we obtain

$$\left.\begin{array}{l} I_{mp} \approx |I_1| \cos \varphi - |I_2'| = 7.51 \cos 79.9° - 0.985 = 0.332 \text{ A,} \\ I_{mq} \approx |I_1| \sin \varphi = 7.51 \sin 79.9° = 7.39 \text{ A.} \end{array}\right\} \quad (8.46)$$

Step 3. $|I_m| = \sqrt{I_{mp}^2 + I_{mq}^2} = \sqrt{0.332^2 + 7.39^2} = 7.40$ A.

Step 4. $\underline{/I_m} = -\tan^{-1}\left(\dfrac{I_{mq}}{I_{mp}}\right) = -87.4°.$

IV. The magnetizing impedance elements in Fig. 8.21 are obtained from

$$\left.\begin{array}{l} R_m = \dfrac{|V|}{I_{mp}} = \dfrac{127.0}{0.332} = 383 \text{ } \Omega/\text{phase,} \\[2mm] X_m = \dfrac{|V|}{I_{mq}} = \dfrac{127.0}{7.39} = 17.2 \text{ } \Omega/\text{phase.} \end{array}\right\} \quad (8.47)$$

V. Total core loss equals $503 - 375 = 128$ W. (Note that you can also compute it as $3V^2/R_m$.)

....

Example 8.11   Use the corrected model thus derived for our sample motor to "polish" the current and power data computed in Example 8.5 based upon the IM model. We make the very reasonable assumption that the windage and friction power stays about the same at the 3% lower speed.

SOLUTION: In Example 8.5 we computed $I_2' = 23.77\,\underline{/-7.74°}$. In Example 8.10 we found $I_m = 7.40\,\underline{/-87.4°}$. From Eq (8.45) we thus have

$$I_1 = 23.77\,\underline{/-7.74°} + 7.40\,\underline{/-87.4°} = 26.13\,\underline{/-23.92°}.$$

The power drained from the ac source is

$$P_1 = 3\,|V|\,|I_1|\cos\varphi = 3 \cdot 127.0 \cdot 26.13 \cdot \cos 23.92°$$
$$= 9.100\,\text{kW (total three-phase)}$$

(as compared to 8.975 kW based upon the IM model).

The output power P_{out} equals the input power minus losses:

$$P_{out} = P_1 - (P_{core} + P_{WF} + P_\Omega)$$

where

$$P_{core} + P_{WF} = 0.503 \quad \text{(tested)};$$
$$P_\Omega = 3R_1\,|I_1|^2 + 3R_2'\,|I_2'|^2$$
$$= 3 \cdot 0.295 \cdot 26.13^2 + 3 \cdot 0.150 \cdot 23.77^2$$
$$= 0.857\,\text{kW (total three-phase)};$$
$$\therefore P_{out} = 9.100 - (0.503 + 0.857) = 7.740\,\text{kW}.$$

(This value should be compared with $P_m = 8.222$ kW computed based upon IM model.)

For the motor efficiency we have

$$\eta = \frac{P_{out}}{P_1} = \frac{7.740}{9.100} \cdot 100 = 85.1\%.$$

(The IM model yielded the better value 91.6%.)

8.7 OPERATIONAL CONSIDERATIONS

The greatest advantages of the induction motor are its simplicity in design, ruggedness, and reliability and its ability to run directly off the ac network. Offsetting these advantages are the following negative aspects:

1) high starting current,
2) limited means for control of speed and torque,
3) low efficiency when operating at high slips.

8.7.1 The Problem of High Starting Current at Direct Start

By "direct start" we mean that the motor is switched directly to the network. As seen from the equivalent circuit the motor at standstill ($s = 1$) offers a very low† impedance to the network. The equivalent reactance X_{eq} amounts to the small leakage value. The equivalent resistance R_{eq} equals the sum of the rotor and stator winding resistances. This means that as the motor is connected to the network a high starting current will follow. As the motor speeds up, the equivalent resistance R_2'/s will increase with a resulting decrease in current.

The high starting current will cause undesired voltage drops in the feeding transformer and/or feeder lines. These voltage drops may be bothersome to other load objects. In practice, the problem becomes particularly acute if the load inertia is large resulting in long starting times, due to the low starting torque. Direct start can also be damaging to the motor load which way not tolerate the sudden torque shock.

Example 8.12   Consider the situation shown in Fig. 8.24(a). A 25-kVA transformer feeds power to a 220-V bus. The transformer impedance is $Z_t = 0.010 + j0.050$ pu based upon its own rating.

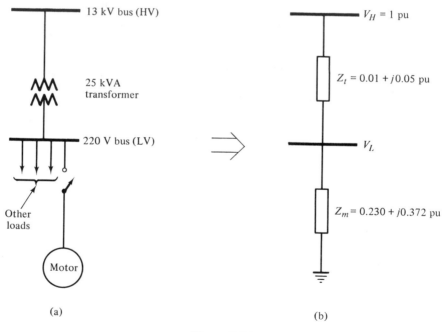

(a)   (b)

Figure 8.24

If our previous 10-hp sample motor is started directly off the 220 bus what will be

a) the magnitude of the starting current,

b) the voltage drop of the 220-V bus?

We assume that the voltage of the HV bus remains fixed.

SOLUTION: The impedance of the motor for $s = 1$ is

$$Z_m = R_{eq} + jX_{eq} = 0.445 + j0.720 \; \Omega/\text{phase}.$$

Using the formula (5.62), this impedance is expressed in terms of the same kVA base (25/3) and voltage base ($220/\sqrt{3}$) as the transformer impedance:

$$Z_m = (0.445 + j0.720) \cdot \frac{\dfrac{25000}{3}}{\left(\dfrac{220}{\sqrt{3}}\right)^2} = 0.230 + j0.372 \; \text{pu/phase}.$$

a) Since the HV bus voltage V_H is assumed to remain 1 pu, the motor current will thus be

$$I_1 = \frac{V_H}{Z_t + Z_m} = \frac{1.00}{0.240 + j0.422};$$

$$\therefore \; |I_1| = 2.06 \; \text{pu/phase}.$$

Note that although the motor is rated at only 10 hp (or 7.46 kW) its starting current amounts to more than twice the rated current of the 25-kVA transformer.

b) The voltage V_L of the 220 bus is obtained by simple "voltage division":

$$V_L = \frac{Z_m}{Z_t + Z_m} V_H = \frac{0.230 + j0.372}{0.240 + j0.422} \cdot 1.00;$$

$$\therefore \; |V_L| = 0.901 \; \text{pu}.$$

In other words, the LV bus will experience a momentary voltage drop of about 10%. Such a drop would be highly bothersome on lighting loads, for example.

We discuss now a few means for alleviating this voltage drop problem.

A. Insertion of Secondary Resistance. If the motor is of the wound-rotor type (which rarely would be the case with a 10-hp motor because a squirrel-cage motor is considerably cheaper) secondary resistance would normally be inserted to shift the maximum torque to $s = 1$. (Compare Example 8.8.) This added resistance, will of course, also somewhat reduce the starting current and the voltage dip. The effect is, however, not too pronounced because the major portion of the voltage drop across the transformer impedance (which is predominantly reactive) is caused by the *out-of-phase* component of current. Addition of secondary resistance will reduce the *in-phase* current component.

For example, if the rotor resistance R_2' were to be quintupled from $0.150\,\Omega$ to $0.750\,\Omega$ in the previous example, the voltage drop would be reduced only from 10% to about 6%, as the reader can readily check.

B. Using Starting Compensator. The only possible means of softening the shock of the starting current in a cage-rotor motor is to reduce the starting voltage. A common way of doing this is to use a starting autotransformer sometimes referred to as a "compensator" (Fig. 8.25). The motor is started with the circuit breaker in position S. When the motor attains running speed the breaker is thrown to position R. The starting current and thus the voltage drop is reduced in the same ratio as the transformer tap.

The price one pays for this type of starting current control is *greatly reduced starting torque* and thus prolonged starting periods. Note from formula (8.41) that the motor torque varies as the *square* of the voltage.† For example, if the transformer tap is set at 50% in order to reduce the voltage drop to half value the torque will be reduced in ratio 1:4.

Note that the motor is disconnected from the network during the breaker transition period S→R. If the transition is slow then the flux may decay and/or fall back one pole. A very strong *but short* reclosing current surge will then occur. To prevent this one sometimes adds in the S lead a transition impedance which will preserve current continuity during changeover.

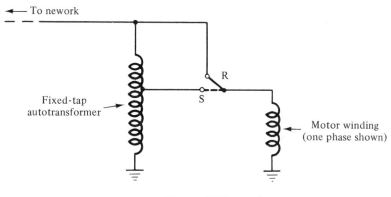

Figure 8.25

---

† The physical reason is the following: A reduction in stator voltage in ratio r results in a reduction of *both* the stator flux and rotor current in the same ratio. According to formula (8.9) this means a torque reduction in ratio r^2.

C. The Y–D Starting Method. This starting method requires that the stator winding has all six terminals accessible. By means of three double-throw switches (not detailed in Fig. 8.26) the stator winding at start is connected in Y. After the motor has accelerated the winding is reconnected in Δ.

In normal (run) operation each phase winding thus is connected to the *line* voltage. During start the same winding is experiencing only phase voltage, that is, a voltage reduction in ratio $1:\sqrt{3}$. The starting torque is thus reduced in ratio 1:3.

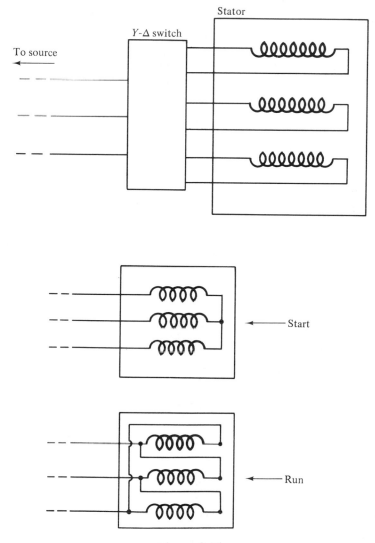

Figure 8.26

8.7.2 Efficiency Reduction at Low Speed Operation

A distinct disadvantage of the induction motor is the reduced efficiency at low operating speeds. The major reason is the high ohmic losses in rotor and stator resistances at high slips.

From formula (8.34) we note that these losses, based upon IM model, are

$$P_\Omega = |I_1|^2 (R_1 + R_2')   \text{W/phase.}$$

The output or motor power is

$$P_m = |I_1|^2 \frac{1-s}{s} R_2'   \text{W/phase.}$$

If we neglect the relatively small core and friction and windage losses, the efficiency will be

$$\eta = \frac{P_m}{P_m + P_\Omega} = \frac{\dfrac{1-s}{s} R_2'}{R_1 + \dfrac{1}{s} R_2'} = \frac{(1-s)R_2'}{sR_1 + R_2'}.$$

If, for simplicity, we set $R_1 \approx R_2'$, we get the simple efficiency formula

$$\eta = \frac{1-s}{1+s}. \tag{8.48}$$

We note that the efficiency will be highest at low slip, that is, at high speed. At half synchronous speed, corresponding to $s = 0.5$, the efficiency will drop to 33%.

If the motor is of squirrel-cage type, *all* of the ohmic losses are dissipated within the machine. If the motor is of wound-rotor type with external resistors, at least part of the secondary losses occur outside the machine. These facts have important consequences:

1) The motor can be subject to heavy heat stresses when starting high inertia loads. (In fact, the energy dissipated in heat during start equals the kinetic energy of the rotating masses. Compare Chap. 7.)

2) High sustained load torques will cause sustained low speeds. Careful attention must be given to heat balance within the motor in such situations. (The danger is increased by the fact that a slow rotor has decreased cooling capacity.)

8.7.3 Torque–Speed Control

One of the outstanding features of the separately excited dc machine is its ability to deliver a torque of *any* magnitude and direction at *any* speed.† Such a degree of freedom does not characterize the induction motor. In fact, we have seen that one of the very basic features of the induction motor is the strong dependency of its torque upon speed.

Furthermore, as the induction motor speed is *closely* related to its synchronous speed, it becomes clear that we have very narrow control margins in an actual situation. Our options are summarized in Fig. 8.27. The figure shows the torque–speed curve T_m of an induction motor. It also depicts the torque–speed curve T_L of a certain load that is being pulled. The motor is running at the subsynchronous speed n_0 rpm. Assume now that it is desired to lower the operating speed.

Two possibilities exist:

A. Speed Control by Voltage Manipulation.   The stator voltage V of the motor is lowered by some available means (autotransformer, for example). As the torque amplitude is proportional to $|V|^2$ the reduced voltage results in the reduced torque (curve I). Torque equilibrium will now occur at the lower speed, n_1, which will thus be the new operating speed.

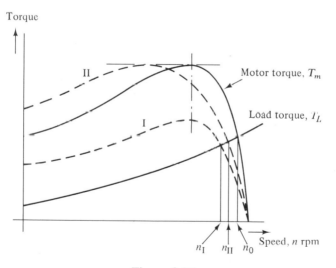

Figure 8.27

---

† Within the limitations set by the ratings, of course.

B. Speed Control by Secondary Rheostat Manipulation. By adding external resistance the motor torque curve is shifted towards left (curve II). Again torque equilibrium occurs at a lower speed, n_{II}. The three speeds, n_0, n_1, and n_{II}, all lie normally within a narrow subsynchronous range.

The speed of an induction motor may be varied over a wider range only if provisions are taken to change its synchronous speed, n_s. The formula $n_s = 120f/p$ tells us that n_s can be varied by changing either p or f. Both alternatives are practically possible. For example, it is not difficult to reconnect a four-pole stator winding to an eight-pole stator winding by making the proper coil connection. This would then change n_s from 1800 rpm to 900 rpm.

Neither is it difficult by means of present-day solid-state circuitry to vary the frequency f within wide ranges. Since the primary voltage V is related to the frequency f and the flux Φ through the relation

$$|V| \sim f \cdot \Phi, \tag{8.49}$$

it is necessary to vary $|V|$ in proportion to f in order to keep Φ, and thus the maximum torque, constant.

Example 8.13 Consider once more the 10-hp motor with the impedance data of Example 8.4. The motor is connected to a 220-V network and driving a load the torque of which varies quadratically with speed n according to the formula

$$T_L = 60 \left(\frac{n}{n_s}\right)^2 \qquad \text{N} \cdot \text{m}. \tag{8.50}$$

a) Determine its speed.

b) Determine the speed change resulting from a terminal voltage reduction of 15%.

SOLUTION: In terms of slip s, the load torque can be written

$$T_L = 60(1 - s)^2 \qquad \text{N} \cdot \text{m}. \tag{8.51}$$

If all the known parameters are substituted into formula (8.41), the electromechanical motor torque can be written

$$T_m = \frac{57.773}{s} \cdot \frac{1}{\left(0.295 + \dfrac{0.150}{s}\right)^2 + 0.720^2} \qquad \text{N} \cdot \text{m} \tag{8.52}$$

a) By requiring torque balance we thus obtain the equation

$$60(1 - s)^2 = \frac{57.773}{s} \cdot \frac{1}{\left(0.295 + \dfrac{0.150}{s}\right)^2 + 0.720^2}. \tag{8.53}$$

This is a fourth-order equation in s and the most practical solution method is therefore by "trial-and-error." First we make a reasonable guess[†] for s, and by back substitution into (8.53) we converge readily on the solution

$$s = 0.02477 \text{ pu.}$$

This slip corresponds to the speed

$$n = 1170 \text{ rpm.}$$

b) If the terminal voltage is reduced in the ratio 0.85 the motor torque will decrease in the ratio 0.85^2, that is, to the new value

$$T_m = \frac{41.741}{s} \cdot \frac{1}{\left(0.295 + \dfrac{0.150}{s}\right)^2 + 0.720^2} \cdot \tag{8.57}$$

Our new torque balance equation thus reads

$$60(1 - s)^2 = \frac{41.741}{s} \cdot \frac{1}{\left(0.295 + \dfrac{0.150}{s}\right)^2 + 0.720^2} \cdot \tag{8.58}$$

Trial-and-error solution of this equation yields

$$s = 0.03529,$$

which corresponds to the speed

$$n = 1158 \text{ rpm.}$$

Result: A voltage reduction of 15% resulted in a speed drop of only 12 rpm or about 1%.

···

[†] We know that the equation (8.53) will render a solution, s, that satisfies the inequality

$$s \ll 1.$$

The torque expressions (8.51) and (8.52) can thus be approximated:

$$T_L = 60(1 + s^2 - 2s) \approx 60(1 - 2s), \tag{8.54}$$

$$T_m \approx \frac{57.773}{s} \cdot \frac{1}{\left(\dfrac{0.150}{s}\right)^2} = 2568s. \tag{8.55}$$

Thus, we have *linearized* the torque curves around the operating slip. By solving the approximate equation

$$60(1 - 2s) \approx 2568s \tag{8.56}$$

we obtain $s = 0.022$ which is a good first trial value.

8.8   SINGLE-PHASE INDUCTION MOTORS

For reasons of cost and simplicity one cannot always count on three-phase ac power supply. In a multitude of domestic, commercial, and sometimes even industrial cases where only single-phase ac power is available and where the power need is limited (usually below 1 hp), single-phase induction motors, sometimes collectively referred to as "fractional-horsepower motors," fill an important need. Together with the earlier mentioned commutator motor (the *universal* motor) single-phase induction motors account for practically 100 percent of the small electrical motor market.

8.8.1   Flux and Current Waves in Single-Phase Motor

A single-phase induction motor is obtained if only one phase of a three-phase motor is being energized. We can thus discuss the basic features of this motor by returning for a moment to Figs. 8.1 and 8.2. With only the *a*-phase energized the motor flux is pulsating. Figure 8.2 depicts the equivalence between this pulsating flux and two oppositely revolving flux waves of equal magnitude. Assuming that the rotor is at standstill it will thus experience two fluxes of equal size rotating in each direction. Each flux wave will induce a corresponding rotor current wave. Thus there will exist in the motor a traveling wave system, shown in Fig. 8.28, consisting of:

1) the stator flux B_+ revolving with speed $+n_s$ rpm,

2) the stator flux B_- revolving with speed $-n_s$ rpm,

3) the rotor current A_+ revolving with speed $+n_s$ rpm,

4) the rotor current A_- revolving with speed $-n_s$ rpm.

The current waves will be lagging their respective flux (and emf) waves by angles γ_+ and γ_-.

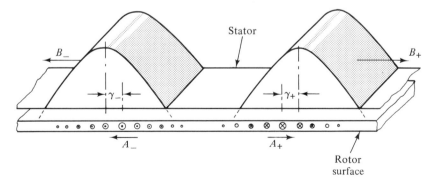

Figure 8.28

In addition, but not shown in the figure, there will be stator current waves accompanying the rotor current waves. These waves will be of opposite polarities so as to achieve mmf balance, in the same manner as in a three-phase motor.

8.8.2 Torque Situation—A Qualitative Assessment

Our first observation is that the flux and current picture in a single-phase machine is considerably more complex than that in a three-phase machine. *As a result the mathematical analysis of single-phase induction motors is more complex than that of three-phase motors.*

Our second observation is that *the single-phase motor lacks a starting torque.*

This important fact is realized immediately from a look at Fig. 8.28 which—we emphasize—was drawn for a stationary rotor. The flux wave B_+ will interact with the current wave A_+ to give a *constant* torque component T_+ that will attempt to accelerate the rotor to the right. The flux wave B_- and current wave A_- will similarly create the *constant* torque component T_- acting to the left. Because, by reasons of symmetry, T_+ and T_- are of equal magnitude the *resulting constant* torque is zero.

Our third and equally important observation is that *the single-phase induction motor will contain a pulsating torque component of twice the network frequency.*

The flux wave B_+ and current wave A_- glide past each other with a relative speed of $2n_s$ rpm. The resulting torque between them will thus be sinusoidal of the frequency $2f$ Hz. (This can best be seen from formula (8.9) by setting the angle γ equal to $2\omega t$ where $\omega = 2\pi f$.) The same applies to the torque created between the flux wave B_- and current wave A_+. The total 120-Hz vibratory torque component resulting from the interactions between B_+ and A_- and B_- and A_+ will always be present in a single-phase motor. We must live with it as best we can. The resulting noise and vibrations, should they become bothersome, can be effectively minimized by means of elastic motor supports.

8.8.3 A Three-Phase Analog of the One-Phase Motor

As was pointed out, the single-phase induction motor has zero driving torque at rotor standstill, that is, if its slip relative to both flux waves is unity. However, should we *help* the rotor gain speed in the direction of *either flux wave*, the motor will immediately develop its own torque in the chosen direction and take off on its own.

We can obtain a very useful analogy† of the torque situation in the running single-phase motor by considering the three-phase dual-motor drive system in Fig. 8.29(a).

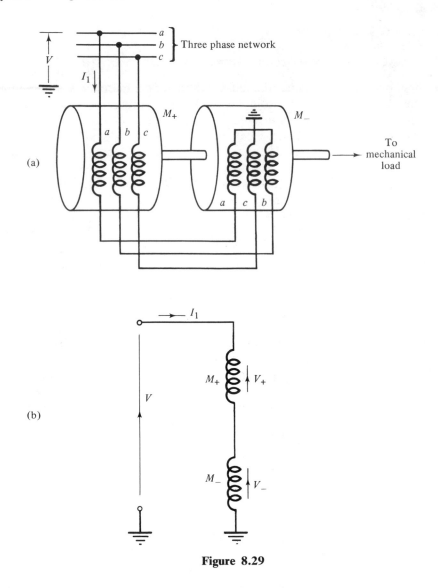

(a)

(b)

Figure 8.29

---

† The analog gives a truthful picture of the *constant* torque component in a running one-phase motor. Because three-phase induction motors lack pulsating torque components, it will *not* simulate the 120-Hz torque component in the one-phase motor.

Two *identical* three-phase motors M_+ and M_- are mechanically coupled together to drive a common load. Electrically they have their stator windings coupled in *series* onto the ac network. *However, their phase sequences are opposite because two phase leads have been shifted on M_-.* The motor M_+ thus delivers a torque T_+ in direction $(+)$ whereas the torque T_- of M_- acts in direction $(-)$. By realizing that the stator currents, I_1, are equal in both motors we obtain the following expressions for the torques by use of formula (8.40):

$$T_+ = \frac{90}{\pi n_s} |I_1|^2 \frac{R_2'}{s_+} \qquad \text{N} \cdot \text{m}, \tag{8.59}$$

$$T_- = \frac{90}{\pi n_s} |I_1|^2 \frac{R_2'}{s_-} \qquad \text{N} \cdot \text{m}. \tag{8.60}$$

The synchronous speeds of the two motors are $+n_s$ and $-n_s$ respectively. Thus we obtain for their slips

$$\left.\begin{array}{l} s_+ = \dfrac{n_s - n}{n_s} = 1 - \dfrac{n}{n_s}, \\[3mm] s_- = \dfrac{-n_s - n}{-n_s} = 1 + \dfrac{n}{n_s}. \end{array}\right\} \tag{8.61}$$

These equations yield upon addition the following relationship between s_+ and s_-:

$$s_- = 2 - s_+. \tag{8.62}$$

By making use of this relationship we obtain the following ratio between the two torques:

$$\frac{T_+}{T_-} = \frac{2 - s_+}{s_+}. \tag{8.63}$$

Note that for rotor standstill we have $s_+ = 1$ and thus

$$\frac{T_+}{T_-} = 1;$$

or, in words:

The dual drive like the one-phase motor has no starting torque.

Suppose now the system is helped on its way by giving it a small "speed push" in either direction. For example, if we start it off in the direction of the torque T_+ with a speed of 10 percent of n_s, then we have $s_+ = 0.9$ and formula (8.54) yields

$$\frac{T_+}{T_-} = \frac{1.1}{0.9} = 1.22.$$

At this speed M_+ is obviously stronger than M_- *and the motor will now take off on its own in the* $(+)$ *direction* (assuming a non-excessive load torque).

Had we instead given the initial speed push in the opposite direction then we would have had $s_+ = 1.1$ and formula (8.54) would yield

$$\frac{T_+}{T_-} = \frac{0.9}{1.1} = 0.82.$$

Now T_- exceeds T_+ and the motor keeps running in the $(-)$ *direction.*

8.8.4   Physical Explanation of Dual-Drive Torque Behavior

The ability of the motor system in Fig. 8.29 to develop driving torque in either direction has its identical counterpart in the one-phase motor. We need to explain this important phenomenon from a physical point of view. This feature directly relates to the relationship between the magnitudes of the rotating fluxes of the two motors. *As we spin the dual-drive rotor in the direction of one of the two counter-revolving fluxes the magnitude of this flux will increase whereas the magnitude of the other flux will decrease.* The reason why this will happen is as follows:

The *total* source voltage V will divide between the two motors in the following proportions (see Fig. 8.29b);

$$\left. \begin{array}{l} V_+ = \dfrac{Z_+}{Z_+ + Z_-} V \quad V \\[4mm] V_- = \dfrac{Z_-}{Z_+ + Z_-} V \quad V \end{array} \right\} \tag{8.64}$$

where Z_+ and Z_- represent the impedances as viewed into each phase of the stator windings. If we for simplicity use the IM model of each motor, these impedances are

$$\left. \begin{array}{l} Z_+ = R_1 + \dfrac{R_2'}{s_+} + j(X_1 + X_2') \quad \Omega/\text{phase}, \\[4mm] Z_- = R_1 + \dfrac{R_2'}{s_-} + j(X_1 + X_2') \quad \Omega/\text{phase}. \end{array} \right\} \tag{8.65}$$

The flux magnitude in a motor is directly proportional to the rms value of its stator voltage. This means that the flux in M_+ is proportional to $|V_+|$, and the flux in M_- to $|V_-|$.

At standstill, $s_+ = s_- = 1$ and consequently $Z_+ = Z_-$. The voltage V thus divides in equal ratio between the two motors. Their fluxes are equal and so are their torques.

As we speed up the motor in the (+)-direction the slip s_+ will decrease and s_- will increase. Consequently, Z_+ will increase and Z_- will decrease. The motor M_+ will receive a greater share of the source voltage than M_- receives and the magnitude of its flux will grow as that of M_- will decrease. Thus, the torque T_+ will dominate over T_-.

Example 8.14   Assume that the speed of the dual-drive rotor shown in Fig. 8.29 equals 97% of $+n_s$.

a) Find the ratio between $|V_+|$ and $|V_-|$.

b) Find the ratio between T_+ and T_-.

c) Find the total torque in N · m.

d) Compare this total torque with the torque obtainable if M_+ *only* were connected to the source.

Use the IM model for each machine and the parameter data given in Example 8.4.

SOLUTION: We first find the impedance per phase as viewed into each stator. From Eq. (8.65) we have

$$Z_+ = 0.295 + \frac{0.150}{0.03} + j(0.510 + 0.210) = 5.295 + j0.720 \quad \Omega/\text{phase},$$

$$Z_- = 0.295 + \frac{0.150}{1.97} + j(0.510 + 0.210) = 0.371 + j0.720 \quad \Omega/\text{phase}.$$

a) From Eq. (8.64) we obtain

$$\frac{V_+}{V_-} = \frac{Z_+}{Z_-};$$

thus

$$\frac{|V_+|}{|V_-|} = \frac{|5.295 + j0.720|}{|0.371 + j0.720|} = 6.598.$$

b) From Eq. (8.63) the torque ratio is found to be

$$\frac{T_+}{T_-} = \frac{2 - 0.03}{0.03} = 65.7.$$

c) The stator current is

$$|I_1| = \frac{|V|}{|Z_+ + Z_-|} = \frac{127.0}{5.846} = 21.72 \quad \text{A/phase}.$$

From Eq. (8.59) we thus obtain

$$T_+ = \frac{90}{\pi \cdot 1200} \cdot (21.72)^2 \frac{0.150}{0.03} = 56.31 \quad \text{N} \cdot \text{m},$$

$$T_- = \frac{1}{65.7} \cdot 56.31 = 0.86 \quad \text{N} \cdot \text{m},$$

$$T_{\text{tot}} = T_+ - T_- = 55.45 \quad \text{N} \cdot \text{m},$$

d) If only M_+ were connected to the source and running at slip 0.03, the stator current would be

$$|I_1| = \frac{|V|}{|Z_+|} = \frac{127.0}{|5.295 + j0.720|} = 23.77 \quad \text{A/phase}.$$

Its torque would be, according to Eq. (8.59),

$$T_m = \frac{90}{\pi \cdot 1200} \cdot (23.77)^2 \frac{0.150}{0.03} = 67.42 \quad \text{N} \cdot \text{m}.$$

This example teaches several things:

1) If one of the motors operates close to synchronous speed, its stator winding absorbs almost all of the available source voltage.

2) Consequently the flux and the torque of the other machine are of almost negligible importance.

3) The total torque is approaching the torque obtainable from a single machine operating at the same speed.

If we were to perform the computations in the above example for several slip-values, then we would obtain a torque-*speed curve* of the type shown in Fig. 8.30. We note several points of interest:

1) Whereas the torque of a single three-phase motor passes through its zero value at exactly the synchronous speed, the zero torque for the dual drive occurs *below* n_s.

2) For speeds close to n_s the dual-drive torque is of nearly the same magnitude as that of a single-motor drive.

3) The torque for positive speeds equals the torque for negative speeds but with opposite sign.

8.8.5 The Dual Three-Phase Drive Versus the Single-Phase Motor

The dual-drive system just described contains the two counter-revolving fluxes in two *separate* motors. A single-phase motor contains the two counter-revolving fluxes in the *same* motor. For this reason the two systems are

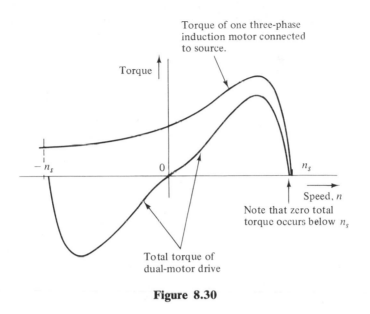

Figure 8.30

identical in some respects, nonidentical in others. For example, the total torque of the dual drive is *constant*, whereas the total torque of the single-phase motor contains a 120-Hz vibratory component. This we pointed out earlier.

If we limit our attention to the total *constant* torque of the single-phase motor we can, in fact, make full use of the knowledge which we possess from the dual drive. Let us translate this knowledge to the single-phase motor. We have depicted, schematically, a single-phase (two-pole) motor in Fig. 8.31. If the rotor is at standstill the flux situation will be identical to the one shown in Fig. 8.2. The two counter-revolving flux waves are of equal magnitudes and the resulting flux pulsates along the magnetic axis of the machine.

Assume then that we rotate the rotor by external means in CW (+) direction as shown in Fig. 8.32. The revolving flux Φ_+ will now dominate over Φ_- and the two unequal revolving fluxes will combine into the resulting flux Φ_{res} *which will no longer pulsate along the magnetic axis. Instead, Φ_{res} will take on the character of a revolving flux.* Unlike a "true" revolving flux, however, it will not have a constant wave amplitude. It constitutes, in fact, a hybrid between a "true" pulsating flux (like Φ_a in Fig. 8.2) and a "true" revolving flux (like Φ_{res} in Fig. 8.4). If symbolized by a vector, its tip will trace out an elliptical path as shown in Fig. 8.32. We may resolve Φ_{res} into a "direct" (or *d*-axis) component, Φ_d, and a "quadrature" (or *q*-axis) component Φ_q. The latter two components are of "true" pulsating character. Timewise they are 90° out of phase, Φ_d leading Φ_q.

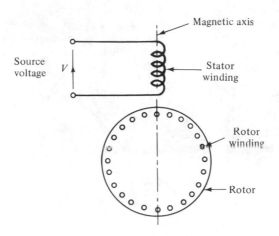

Figure 8.31

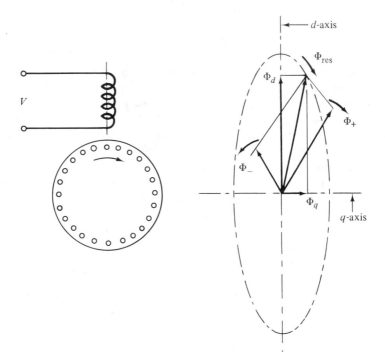

Figure 8.32

The spinning of the rotor has thus resulted in the birth of a flux component pulsating along the quadrature magnetic axis.† The faster the rotor spins, the more dominant will Φ_+ become over Φ_- and the larger will be the magnitude of Φ_q. Were we to spin the rotor in CCW $(-)$ direction, Φ_- would dominate over Φ_+. The tip of the Φ_{res} vector would now move along the elliptic contour in the opposite direction. Again a pulsating quadrature flux is created—this time its time phase will be 90° *leading* relative to Φ_d.

All the formulas (8.59 through 8.65) relating to the torque–speed performance of the three-phase dual-drive (and which resulted in the torque curve of Fig. 8.30) are valid for the single-phase motor as well. In applying those formulas to the single-phase motor the impedance parameters R_1, R_2', X_1, and X_2' shall be replaced by *half* the corresponding values for the single-phase machine.‡

8.8.6 Starting Techniques

Figure 8.30 convincingly shows that the single-phase motor (except for the vibratory torque component) produces good torque, particularly at high speed. But no motor is practically acceptable that does not start. Just as we are forced to provide a car engine with a starter arrangement, so we must devise some starting scheme for the single-phase motor. In fact, the various single-phase induction motors that are in use can be grouped on the basis of their starting methods. The names by which the different motor types are known are often descriptive of the starting techniques employed.

All starting methods in use are based upon the following principle:

If it is true that two oppositely revolving fluxes of unequal magnitude give rise to two pulsating fluxes of unequal time phase in d- and q-directions (Fig. 8.32), then the opposite is also true—two pulsating fluxes in d- and q-directions of unequal time phase will give rise to two oppositely revolving fluxes of unequal magnitude

---

† We can actually *measure* the existence of this flux component by placing a test winding on the stator having a magnetic axis coinciding with the q-axis and measuring the emf induced in this winding. As Φ_q increases with the rotor speed, the magnitude of this induced emf will vary with speed. (AC tachometers are based upon this phenomenon.)

‡ A strict proof of this will make use of Kron's generalized machine transformation theory (Kron, 1930; White–Woodson, 1959) and would go beyond our scope. The reader will not have any difficulties in accepting the *heuristic* explanation that the single-phase motor in essence consists of two identical series connected machines (this explains the "halfing" of the impedance), each associated with one of the counter-revolving flux components.

The validity of this principle is simply demonstrated in Fig. 8.33. In addition to the *main* winding having its magnetic axis coinciding with the d-axis, there has been added the *auxiliary* or *starter* winding with its magnetic axis along the q-direction. Assume that somehow we have arranged so that the magnetizing currents in the windings are of *different time phase*. In Fig. 8.33, the q-current is leading the d-current by $\alpha°$. Current I_d gives rise to a pulsating flux in d-direction which can be resolved into the two *equal* and counter-revolving components Φ_{d+} and Φ_{d-}.

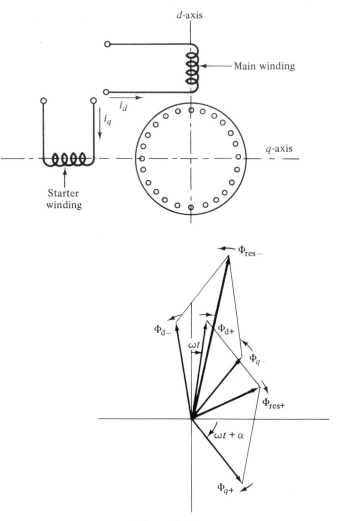

Figure 8.33

Similarly, current I_q causes the two equal and counter-revolving flux components Φ_{q+} and Φ_{q-}. The position of these flux components relative to the q-axis is $\alpha°$ ahead of Φ_{d+} and Φ_{d-}, respectively. By separate vectorial addition of the two CW revolving fluxes and the two CCW fluxes we obtain the two resulting revolving fluxes $\Psi_{\text{res}+}$ and $\Psi_{\text{res}-}$. Obviously the latter dominates over the former. In this case then the motor would obviously possess a resulting starting torque in the CCW (or $-$) direction. *A general rule is that the motor will start in the direction of the winding carrying the lagging current.*

The greater the dominance of one revolving flux over the other, the better the starting torque. The dominance is increased with increasing α-angle. (Note that if α equals 90° and the pulsating fluxes in d- and q-directions are equal the resultant flux component $\Phi_{\text{res}+}$ disappears completely.)

A. Resistance Split-Phase Motor.   The simplest way of obtaining the required time phase differential α between the two winding currents is to insert a resistance† in the starter winding (Fig. 8.34). The added resistance makes the

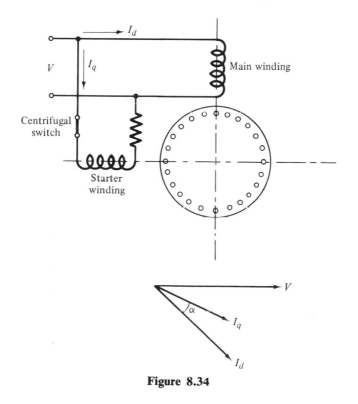

Figure 8.34

---

† (or by designing the starter winding with high resistance-to-reactance ratio)

q-winding less inductive than the d-winding. Consequently its current I_q will lead I_d. This method results in fairly small phase angle α and thus relatively weak starting torque. The method is used for motors of less than 1/3 hp rating. Washing machines and dishwashers are typical applications.

Once the motor has started the main winding produces the necessary torque. The starter winding is then unnecessary and may actually reduce the overall torque of the motor. For this reason, it is usually disconnected by means of a centrifugal switch which operates, for example, at 70 percent of full speed.

B. Capacitor Motor. The most effective way of obtaining a phase angle α of 90° (and even larger) is by inserting a capacitor in series with the starter winding. The capacitor motor (Fig. 8.35) thus has the best possible starting torque and is the most common type of single-phase motors. It comes in sizes as large as 5 hp. As the capacitor must have a capacitance, C, of 250 μF or larger, the capacitor usually is of electrolytic type.

Example 8.14 As measured at 60 Hz, the two windings of a single-phase motor have the following impedances:

$$\text{main winding: } Z_m = 3.1 + j2.9,$$

$$\text{starter winding: } Z_s = 7.0 + j3.1.$$

Find the capacitor size that will produce the phase angle $\alpha = 90°$!

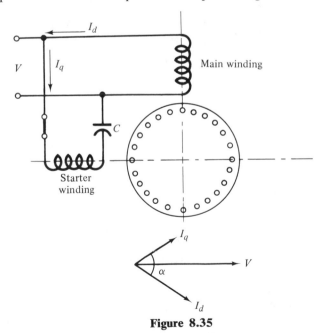

Figure 8.35

SOLUTION: From Fig. 8.35 we have

$$\underline{/I_d} = \underline{/V} - \underline{/Z_m} = \underline{/V} - \underline{/3.1 + j2.9},$$

$$\underline{/I_q} = \underline{/V} - \underline{\Big/ Z_s + \frac{1}{j\omega c}} = \underline{/V} - \underline{\Big/ 7.0 + j3.1 - j\frac{1}{\omega c}}.$$

For the angle α we thus have

$$\alpha = 90° = \underline{/I_q} - \underline{/I_d} = \underline{/3.1 + j2.9} - \underline{\Big/ 7.0 + j\Big(3.1 - \frac{1}{\omega c}\Big)}.$$

This equation can be written

$$90° - \tan^{-1}\Big(\frac{2.9}{3.1}\Big) = \tan^{-1}\left(\frac{\frac{1}{\omega c} - 3.1}{7.0}\right)$$

Solving for the unknown C yields

$$C = 250.7 \ \mu F.$$

C. Shaded-Pole Motor. Very small single-phase induction motors obtain their starting torque by means of "magnetic shading," the principle of which is depicted in Fig. 8.36. The flux created by the main winding will split up into two parts, Φ_d and Φ_q. The latter flux portion will pass through the parallel magnetic path encircled by the "shading coils," which consist of one short-circuited copper

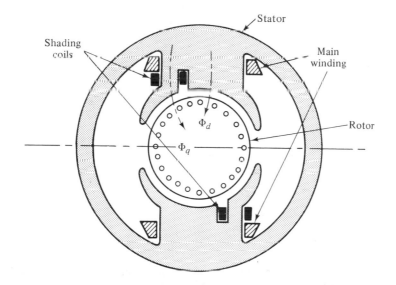

Figure 8.36

turn per pole. Currents will be induced in these coils and, according to Lenz'
Law, they will have such a direction as to prevent changes in the flux from
taking place. The end effect is that Φ_q will lag Φ_d by a certain angle α. The
conditions are therefore met for creation of a starting torque. The motor will
run in the direction of the q-axis, that is, in CCW direction.

8.8.7 Induction-Start Synchronously Running Motors

In many motor applications it is very important that the speed be constant. For
example, the accuracy requirement of an electric clock certainly necessitates a
synchronously running motor. The need for frequency reproducibility in a
high-quality recording system may also exclude an asynchronously running
induction motor. In such and similar cases, one may choose a motor which
starts as an induction motor but runs as a synchronous motor. Figure 8.37
depicts a detail of a rotor† with these particular features.

The basic difference from a normal induction motor rotor is the *salient*
design. The number of salient rotor poles must match the number of poles of
the stator winding. The induced currents in the squirrel-cage winding will give
the torque necessary for starting. As the rotor reaches speeds close to synchronous speed *and if the slip speed is below a critical value* the rotor locks into
synchronism with the stator flux. The torque necessary for this "lock-in" is of

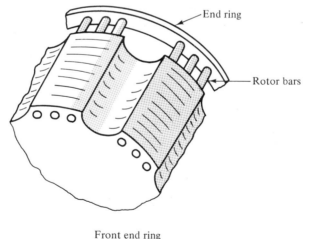

Front end ring
not shown

Figure 8.37

---

† The stator may be either three-phase or single-phase (with starter winding).
Three-phase units are built in sizes up to 150 hp.

the *reluctance type* discussed in Example 3.27. This type of motor lacks the brushes and slip rings of a normal synchronous machine. Its rotor winding has the simplicity and ruggedness of an induction motor cage. The motor is widely used in constant speed applications where the requirements for high torque are not too severe.

Synchronous machines, as we learned in Chapter 4, lack starting torque and require a prime mover or motor to bring them up to synchronous speed before being synchronized. For some applications when the synchronous machine† must be *selfstarting*, this feature is added by means of a rotor-cage winding. Upon being energized the rotor starts as an induction motor. When its speed nears the synchronous value it snaps into synchronism due to the synchronous torque emanating from the dc current in the field winding. The damper winding (see p. 125) can sometimes serve the purpose of "starter winding."

8.9 SUMMARY

Ac induction motors are the most common electrical motors in use. Larger sizes (in excess of about 5 hp) are invariably of three-phase design. Smaller units, usually of "fractional hp," are typically made single phase.

Three-phase induction motors deliver a constant torque and load the supply network symmetrically. Single-phase motors deliver a torque containing a 120-Hz pulsating torque component. In addition, of course, they load the supply network unsymmetrically. Single-phase motors must be equipped with special starter windings

Compared to dc motors, three-phase induction motors have low starting torque. It is possible, however, by secondary resistor control to make the motor deliver its maximum torque at start.

The attraction of ac induction motors lies in their simplicity of design, ruggedness, low price, and ability to run directly off the ac network.

EXERCISES

8.1 It was explained in the text that the induction motor cannot reach synchronous speed because the rotor currents are zero for $s = 0$. If a dc source (via sliprings) were to inject a dc current into the rotor winding, explain how the motor would behave! Compare the synchronous motor!

8.2 A 6-pole, 3-phase, 60-Hz induction motor is running at a speed of 1162 rpm. Two stator phases are being suddenly reversed.

    a) What is the motor slip s before phase reversal?

    b) What is the slip immediately after phase reversal?

---

† Large machines in the megawatt range rarely can stand this treatment.

c) Use the motor data given in Example 8.4 and compute the current immediately following phase reversal. (Neglect any transients.) Use IM model.

8.3 Consider the 10-hp induction motor and the data given in Example 8.4. Compute the motor torque, motor power, power drawn from the network, and ohmic loss power when running at the following speeds: (a) 1150 rpm, (b) 1250 rpm, (c) -1200 rpm, (d) ∞ rpm.

Use IM model and identify the generator cases.

8.4 Consider the circle diagram in Fig. 8.18. At what slip s will the stator current reach its maximum rms value?

a) At this slip, prove that the real power supplied to or from the supply network equals zero.

b) What will be the ohmic loss power?

c) From where will the latter be supplied?

Use IM model in your analysis!

8.5 A 3-phase, 8-pole, squirrel-cage induction motor is designed to run off a 440-V, 60-Hz network. The motor has the following equivalent circuit parameters, expressed in Ω/phase, referred to the stator:

$$R_1 = 0.087, \quad R_2' = 0.070, \quad X_1 = 0.200, \quad X_2' = 0.168.$$

a) Construct a circle diagram for the machine. Use IM model.

b) Compute the horsepower output, stator current, powerfactor, efficiency, and torque for a rotor slip of 3.1%! Identify the current position in a circle diagram!

8.6 What will be the starting current drawn by the motor in exercise 8.5? Identify this current in the circle diagram.

8.7 In this exercise we account for motor losses. Tests show that the motor in exercise 8.5 has a strayload loss of 520 W. The motor draws 2.72 kW at a power factor of 0.31 in a no-load test.

Correct the motor model and the circle diagram for the above nonideal features and then rework exercises 8.5 and 8.6. Assume that half of the no-load losses are made up of core losses.

8.8 Find the maximum torque of the motor in exercise 8.5.

8.9 The induction machine in exercise 8.5 is being run as an "induction generator" on a 440-V network at a speed of 930 rpm, and is driven by a gas turbine. What power does it deliver to the network? What is the efficiency of this generator? Use IM model in your analysis!

8.10 The 3-phase induction motor with the data given in exercise 8.5 is connected to an 11-kV bus via a 100-kVA, 3-phase transformer of voltage rating 11/0.44 kV.

a) Compute the voltage on the LV side at direct start. The transformer impedance is $0.75 + j5.1$ % based on its ratings.

b) What is the LV voltage when the motor is running at a slip of 3.1%?

Assumption: The 11-kV bus experiences no voltage drops.

8.11 The formula (8.44) for maximum torque was derived on the assumption that the motor terminal voltage was fixed.

Compute the maximum torque for the motor in exercise 8.10 if the transformer impedance is taken into account!

8.12 Often it is found practical to express the induction motor torque, T_m, in relation to its maximum torque, T_{max}. Use formulas (8.41) and (8.44) and prove that the torque-ratio T_m/T_{max} is

$$\frac{T_m}{T_{max}} = \frac{1 + \sqrt{R^2 + 1}}{1 + \frac{1}{2}\sqrt{R^2 + 1}\left(\frac{s}{s_{max}} + \frac{s_{max}}{s}\right)}, \tag{8.66}$$

where

$$R = \frac{X_1 + X_2'}{R_1} \tag{8.67}$$

and s_{max} is determined by Eq. (8.43).

8.13 Plot T_m/T_{max} for the slip range

$$0.1 > \frac{s}{s_{max}} > 10.$$

In computing the torque ratio, use the following widely different R-values: (a) $R = 0$, (b) $R = 3$, (c) $R = \infty$.

Your three plots will not show a significant difference which would indicate that the parameter R does not have a great influence on the torque ratio. How would you interpret this fact?

8.14 A 3-phase, 100-hp induction motor develops its rated power at a rotor slip of 1.8%. The maximum torque is 250% of rated torque (that is, the torque developed at rated power). The motor has a R-ratio of $R = 8$.

Use IM modeling and find

a) slip, s_{max}, at maximum torque,

b) stator current at maximum torque,

c) starting torque,

d) starting circuit.

Express your answers in parts (b), (c), and (d) in terms of current and torque at rated speed.

REFERENCES

Alger, Philip L. *Induction Machines*. New York: Gordon and Breach, Science Publishers, Inc., 1970.

Kron, G. "Generalized Theory of Electrical Machinery." *AIEE Translation* Vol. 49 (1930): pp. 666–683.

Veinott, C. G. *Fractional- and Subfractional Horsepower Electric Motors.* New York: McGraw-Hill, 1970.

White, D. C., and H. H. Woodson. *Electromechanical Energy Conversion.* New York: Wiley, 1959.

THE ELECTRIC
ENERGY FUTURE

chapter 9

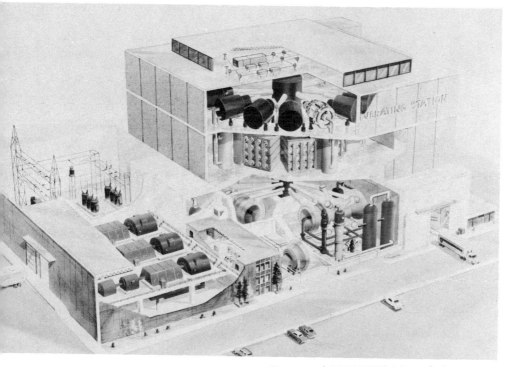

*Conceptual 1170 MW(e) laser fusion reactor
(LFR) generating station. (Courtesy University
of California, Los Alamos Scientific Laboratory)*

9.1   GENERAL

In the previous chapters of this book we have presented a picture of present electric energy technology. The presentation has been made by means of simple models and is subject to a considerable selection process. In this final chapter we shall do some gazing in the crystal ball and attempt to make some predictions of the probable future energy developments in United States.

As this is being written (1976) the whole energy field is undergoing a most extraordinary and historical development on a worldwide scale. In a few short years energy has become a household word. The cohesive policies of the OPEC nations in restricting oil production and increasing petroleum prices have brought home to the energy-consuming public the extent to which our society is dependent upon an unrestricted flow of energy.

A plethora of proposals is being offered for the solution of the "energy crisis." For example, of all the bills introduced in the 94th U.S. Congress, more than 20 percent are directly or indirectly related to energy. Experts in the energy area hold widely differing views as to where major efforts should be made. Most agree that the United States should strive for "energy independence," that is, our total future energy needs should be met from domestic sources.

Before we attempt to make predictions in regard to the future electric energy picture, let us consider the probable developments in the general energy field.

9.2   ENERGY-DEMAND GROWTH

We pointed out in Chapter 1 that the energy demand during the last few decades grew at an exponential rate, a fact that has served to accelerate the arrival of many of the energy supply problems that we face today. All available evidence points to a somewhat slower energy demand growth in the next decades. There are several reasons for this decelerating tempo in our energy-use patterns.

First, there is the sharp increase in the energy prices that occurred following the oil embargo of 1973. Oil prices which had held steady for decades suddenly quadrupled, sending shock waves throughout the economic systems of the world. The higher prices immediately resulted in sharp reductions in energy consumption.

Second, a whole spectrum of energy-conservation actions are being adopted worldwide. Better-insulated homes, smaller cars which consume less gas, and similar measures will result in better utilization of our energy resources over the long haul.

The slowdown in energy demand due to price increases and conservation measures will put a temporary dent in the energy-growth curve. However, the increase in population, the continued upward shift in prosperity, the never-ending process of adding new "energy slaves" for our use will mean an inexorable increase in our national long-term energy demand. However, there are no sure methods of predicting the future rate magnitudes. It used to be common practice to base energy-demand predictions upon "linear" extrapolations of the geometric demand curves of the type shown in Fig. 1.2. This method is clearly not applicable in today's world.

The energy consumption of a nation is closely related to the lifestyle of its people. There is evidence that our national outlook on environment, resources, etc., is undergoing basic changes which will eventually be reflected in our energy-use rates.

9.3   THE UNITED STATES ENERGY BASE—AN INVENTORY

Although we are unable to predict with any certainty the exact magnitudes of our future energy-growth rates, we can easily make conclusions about some obvious trends in our future energy-use patterns. These will, by necessity, be dictated by the domestic energy resource availability. Figure 9.1 shows† a recent inventory of the U.S. energy resources. All such inventories are by necessity subject to margins of considerable uncertainty, but they give "ballpark" estimates‡ of our resources. Let us discuss each resource type separately.

9.3.1   Natural Gas

The *proven* U.S. reserves of natural gas amount to about 240 Q. As the *total* U.S. annual energy demand in 1976 (compare Fig. 1.1) equals 75 Q, our proven gas reserves would last us only a few years should we base our *total* energy need upon gas.

These figures obviously demonstrate the scarcity of this fuel. Natural gas is easy to transport and use and, since it also is the least polluting of our fossil fuels, it is in fact our premium energy source. For this reason and because of its scarcity our national policy is to save it for only the most high-grade purposes

---

† The energy unit used in Fig. 9.1 is quadrillion Btu or "quad," that is, 10^{15} Btu. This unit is often found in estimates of this type. The unit is given the symbol Q. Since 1 Btu = 1055 joules, we have the relationship: 1 Q = 1.055 · 10^{18} J. (Some references use other definitions for the Q-unit; for example, one often finds $Q \equiv 20^{18}$ Btu.)

‡ The following is a good example of the inherent difficulties in predicting resource availability. In the early nineteen seventies the major U.S. oil companies considered the eastern Gulf, off the western coast of Florida, one of the best oil prospecting regions. They paid hundreds of millions of dollars to the United States government for the offshore drilling rights. Subsequent drillings have all proved negative and at this writing (1976) the drilling projects have been abandoned.

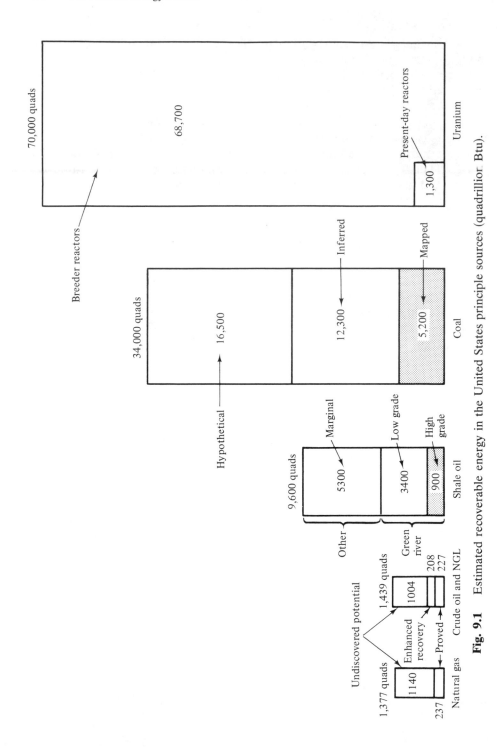

Fig. 9.1   Estimated recoverable energy in the United States principle sources (quadrillior Btu).

like domestic and commercial uses. We cannot any longer count on natural gas for bulk production of electric energy.

9.3.2 Oil

Our proven domestic reserves of crude petroleum and natural gas liquids (NGL) total the energy equivalent of about 230 Q. By application of so-called enhanced† extraction techniques, another 200 Q of oil may be recovered from existing oil fields. Our proven oil resources (like those of our gas reserves) are thus comparatively modest when we put them in relation to our annual energy demands.

It should be noted that existing oil and gas fields represent the most easily accessible fossil fuel resources. Future exploration must be centered in offshore and arctic regions and exploration costs will therefore drastically increase.

9.3.3 Shale Oil

Our shale-oil deposits, estimated at the energy equivalent of 9600 Q, are often considered by many as our great hidden future energy reserve. Measured in energy equivalent, this is more than the entire known oil deposits around the Persian Gulf.

However, most of the deposits are marginal, diffuse, and difficult to exploit. Our most high-grade shale deposits ("Green River Shale") contain about 25 gallons per ton. Shale-oil exploitation is associated with massive environmental problems, few of which have been solved. We must also remember that shale-oil extraction methods have not yet passed the experimental stage; a proven technology does not yet exist. This is an important consideration which often is not fully appreciated in discussion of energy matters. Long, technological time lags are the real barriers for a fast changeover to a new energy source.

9.3.4 Coal

As shown in Fig. 9.1, coal constitutes one of our nation's underutilized energy resources. As long as oil and natural gas were available at cheap prices, there was no economic incentive for development of our coal resources. This situation has drastically reversed since 1973. As this is being written a massive shift from oil to coal is taking place, particularly in electric power generation.

---

† One method employs steam pumped into the oil wells. The latest scheme consists of pumping detergents into the wells and actually "washing" the oil out of the ground.

As Fig. 9.1 seems to indicate that our coal deposits could sustain us for centuries at present energy demand rates, it would appear that our present energy troubles are only temporary.

However, coal comes at a very high environmental price. We have the obvious environmental impacts resulting from mining. Open-pit or strip-mining can have devastating effects upon entire regions as exemplified in Appalachia.

We also have the air-pollution hazards associated with all coal burning. The particulates in the stack exhausts can be fairly well controlled by means of electrostatic precipitators. However, the gaseous pollutants are the real villains. As this is being written the state of pollution-control technology is very rudimentary; the most effective remedy consists of building higher stacks. This, of course, simply spreads the pollution over larger areas.

Our nation is fortunate to possess huge regions of unsurpassed air quality. We should not underestimate the resource value of our "Big Sky Country." We should therefore be highly critical of any uncontrolled coal development that is not coupled with a most sincere and unprecedented effort of stack clean-up.

9.3.5   Uranium

The burning of fossil fuels to obtain their energy content results in their being totally and irreversibly destroyed. This presents us with an uncomfortable moral issue. What right do we have to deplete—in a few generations—a finite resource that in the long term would really have its greatest value as a recyclable raw material for a myriad of products vitally needed by this and future generations? Looked at in a wider time perspective, our fossil-fuel burning thus takes on the dimension of a giant waste.

No such moral hangups need to bother us in the case of uranium. This mineral, used as fuel in nuclear fission reactors, cannot be used for any other purposes. It is therefore solely and truly an energy resource and a very clean one. Another great feature of uranium is its incredible *energy density*. Although in a typical nuclear reactor only a small fraction of its nuclear mass is transformed into energy, those energy quantities are still enormous. For example, it takes an almost continuous string of coal cars to feed the ravenous appetite of a 1000-megawatt, coal-fired power station. The annual fuel charge of an equal-sized nuclear plant amounts to only a few truckloads.

Nuclear fission power constitutes the most interesting energy development in our time. It also demonstrates the long, technological time lags associated with the introduction of a new energy resource. Although the controlled fission process has been known for more than a third of a century and despite massive development efforts, atomic power, in 1976, has captured only about two percent of the total U.S. energy market.

Figure 9.1 indicates that nuclear fuels constitute our most plentiful energy resource, sufficient to tide us over many centuries. However, the development and application of nuclear fission power is burdened by technical, regulatory, environmental, and financial concerns—all interlocked in a complex Gordian knot.

The greatest impediment against an orderly nuclear development has been a lack of public acceptance of the radiation risks associated with a possible nuclear accident.

No major new technology should be expected to reach adulthood without having had its share of childhood diseases. Our young nuclear technology is no exception. We have had some scares but the fact remains that no other branch of technology has accumulated even a remotely comparable safety record. As of this writing (1976) we may actually claim zero fatalities connected with the peaceful use of the atom. This must be seen against the fact that by 1977 the nuclear share of the U.S. electric energy production will amount to almost ten percent.

Our society accepts presently the risks associated with fossil fuel mining; for example, hundreds of fatalities annually plus the tens of thousands of cases of "black lung" disease.

Nobody would come up with the absurd idea of grounding our commercial air fleet because (and this is a certainty) this year hundreds of persons will die in air crashes.

And yet—several states are seriously discussing the possibilities of prohibiting new nuclear plant construction and closing down presently existing units.

Nuclear fission power will take its deserved position as our leading energy source only when we will accept the fact that there exist no zero-risk energy alternatives.

9.3.6   The Exotics

As could have been expected, the energy crisis has spawned an explosion of ideas for more or less realistic energy alternatives that have become known as "exotics." Geothermal energy, solar power, wind and tide, ocean currents and thermal gradients, burning of garbage and animal waste, fuel cells, fusion power—these are but part of the growing energy catalog. As we scrutinize these alternatives and make realistic assessments, the exotics do not offer a promising energy solution for this century.

Some, like geothermal and tidal power, offer only limited regional hope. Others, like ocean currents and thermal gradient schemes, are far too diluted† to represent an economically feasible solution. Solar power, likewise, is a very diluted resource which in addition is characterized by random availability. It was pointed out in Chapter 2 that the combination of low-intensity solar

---

† An energy resource is called "diluted" if its utilization requires vast and unwieldy recovery apparatus.

collectors and a heat-storage facility (water or rocks) represents a very attractive heat–energy complement for the domestic sector and conceivably for the commercial sector, also.

The outstanding environmental features and the almost limitless availability of solar energy makes it a very attractive energy resource. Consequently, vast federal funding can be expected for solar energy research.

It is of interest to look at some of the quantitative aspects of solar energy. It is estimated that over a period of one year, one square kilometer of the continental United States receives a solar-energy radiation insolation amounting to about $0.006Q$. This corresponds to an average power input of about 190 MW per square kilometer. As the total 1976 U.S. annual energy need is 75 Q, a land area of about 130,000 square kilometers would produce the total U.S. energy need if a ten-percent conversion efficiency is assumed. This is about two percent of the total land area of the 48 contiguous states, or in excess of the total area of all U.S. cities. This tells us something about the degree of diluteness of this energy resource.

A. Solar-Electric Conversion.   If solar energy ever shall be useful beyond the limited area of domestic low-grade heat, it must be transformed into an energy form that can be easily transported and that is versatile in its use. Electric power immediately comes to mind. What are the prospects of solar-electric bulk power conversion?

The first requirement is the use of *high-intensity* solar collectors (that is, focusing mirrors) so as to obtain the high-pressure, high-temperature steam needed for the turbines. A major problem presents itself in the fact that such collectors function only in direct sunlight. A single cloud will thus render a high-intensity collector useless.

The second requirement is an energy storage facility which will extend utilization over a 24-hour period.

Consider the conceptual design for an electric plant of 500 MW average power output. In view of the previously given insolation figures we need obviously about 26 square kilometers of collectors. The focusing of this vast mirror array toward the central steam generators will clearly not be a trivial technological problem.

The whole generating facility must be designed as a pumped hydro storage in order to "average" the power. It is not difficult to appreciate the magnitude of the engineering obstacles to be expected in the design of this type of power center. It may in fact prove impossible to work with such a vast unit. We may be forced to utilize an alternate solution consisting of, say, ten 50-MW plants.

Such a solar-electric pumped storage facility must, of course, be located in one of those relatively few regions of our country where abundant sunshine is available. The electricity must then be transported to regions where the energy is needed. Extremely long transmission lines will thus be required.

As one contemplates the technological obstacles associated with such a 500-MW solar-electric power center (including the required transmission lines) it is not hard to become a believer in nuclear fission power.

B. Controlled Nuclear Fusion Processes. Of all the known exotic energy processes, none is of greater future importance and potential than controlled nuclear fusion. This is the energy process that is ongoing in the interior of the sun and stars. When we master this process, we in effect will have "brought down" the sun. We will then also have achieved the "final solution" to our energy problems. The primary fuel in this process will be deuterium (heavy hydrogen) which exists in sufficient quantities in the oceans to satisfy any conceivable energy demands for all the future of the human race. The cost of obtaining heavy hydrogen by isotope separation is so low that it would be less than one percent of the cost of coal on a per-unit-of-energy basis.

Another great feature of the fusion process is the absence of dangerous radiating by-products like those present in the fission processes. Uncontrolled nuclear fusion has been achieved on earth in the hydrogen bomb. Controlled nuclear fusion must be carried out in the following three steps:

1) Heat a small quantity of fusion fuel above its ignition point—about 100 million degrees kinetic temperature.

2) Maintain the fuel in heated condition until the release of fusion energy exceeds the energy input.

3) Convert the released energy, which is in the form of high-grade heat, into a useful form. As in the case of nuclear fission, we still will need the present-day steam-electric cycle for the future fusion process.

Fusion research is advancing at this time along two alternate paths: (1) nuclear fusion by magnetic confinement; (2) laser-initiated fusion.

Confinement of the hot fuel gas has always been the central problem in fusion work. No materials can obviously survive the high temperatures required for the process. The idea was therefore suggested early to confine the fuel gases within a strong magnetic field—a "magnetic bottle." At fusion temperature, the fuel gas exists only in the plasma state, that is, ions and electrons. Those charged particles immersed in a magnetic field execute helical coil-spring-like orbits. If the magnetic field has the right geometry, the end result will be that the plasma remains in a certain finite region.

Laser-initiated fusion is based upon the idea that if enough energy is concentrated upon a small enough spherical pellet, the pellet will implode resulting in extremely high compression densities. If those densities become high enough, conditions would be created which are similar to those inside the sun, and the fusion process would get started. (This is in fact the way a hydrogen bomb is "ignited.")

It is envisioned that those high-energy concentrations can be achieved by means of large lasers. The concept calls for rapid injection of small pellets into the focus of a synchronously firing pulsed laser beam.

9.4 FUTURE ROLE OF ELECTRIC POWER

A. *The Near Future* (*Present to Year 2000*). Having thus made an inventory of the available prime-energy resources we can now draw certain conclusions about the United States energy picture in the next few decades. It is probably unwise to suggest that we can say with some degree of certainty what lies beyond the year 2000, and therefore we limit our quantitative predictions to the remaining quarter of the present century. Figure 9.2 summarizes our guess work.

The fossil fuels which in 1976 carry 94 percent of the U.S. energy load will continue to serve as our energy "workhorse." However, the load will shift from oil and natural gas to the solid fuels, coal and uranium. By the year 2000, nuclear fission energy should be our most important prime source, assuming we will have overcome our present hangups about this energy resource. Hydropower and some of the exotics (solar) will play a relatively insignificant role in our total energy picture. Long before the end of the century our nation should be able to achieve energy independence. Actually, we should be able to get by on less oil by the year 2000 than we presently do. The assumption is made, of course, that we will have developed coherent national energy policies.

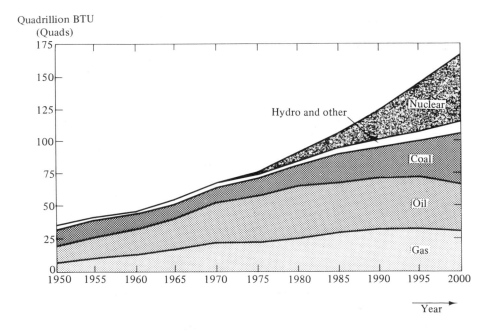

Fig. 9.2 United States energy supply by type of fuel, 1950–2000.

As energy of uranium and coal can be utilized only in the electric form, by the year 2000 we will be well on our way to the "all-electric society." The reduced need for oil imports, in all probability, will have stabilized the world energy prices. In fact, with some degree of luck in the continuing development of nuclear fission technology, we may even see energy price reductions by the turn of the century.

B. *The Distant Future* (*Beyond the Year 2000*). The nuclear "take-over" of the energy market should continue well into the twenty-first century. Maybe sometime in the period from 2000 to 2050 in some laboratory, history's first controlled-fusion process will be conclusively demonstrated. This will be a momentous event in the history of mankind. In all probability this will mark the beginning of an era of both cheap and abundant energy. With an unlimited work force of energy slaves, man will be able, for better or worse, to fully control his world. It is to be hoped that his wisdom will match his power.

C. *Is an "All-Electric Society" Possible or Desirable?* As coal and uranium become our basic prime-energy resources, later to be replaced by heavy hydrogen, the majority of our energy will be made available in the electric form. Is this an acceptable development from the users point of view? Will he be able to switch over to electric power those devices which presently run directly on fuel?

As we study the present situation it is clear that direct-energy use dominates the consumer scene. Figure 9.3 is a somewhat more detailed version of the flow chart in Fig. 1.1. The figure indicates that the principal consumers of energy in the United States fall into three areas: (1) industrial, (2) transportation, (3) domestic and commercial.

In Fig. 9.4(a) we have shown the relative importance of these three areas of use. We have also indicated the portion which presently (1976) is made up of electric power. It is particularly significant that practically no use is made of electric power in the transportation sector. This is not true in other parts of the world. In Europe and Japan essentially all railroads are electrified—one of the major reasons why their railroad systems are of superior quality.

As electric power gradually becomes the dominant energy form, what portions of the three sectors can be electrified? Figure 9.4(b) shows our predictions. The industrial, domestic, and commercial sectors could easily be totally electrified. For example, there is not a single device in a modern home or commercial establishment that could not be run on electricity.

The transportation sector, however, contains certain devices that could not be run electrically in the foreseeable future. Whereas it is highly probable that our total fleet of automobiles will be "electrics" forty to fifty years from now, it is equally probable that airplanes and trucks will still be directly powered by liquid fuels as they are today.

Would an all-electric society be a desirable development? What would life be like in that future day?

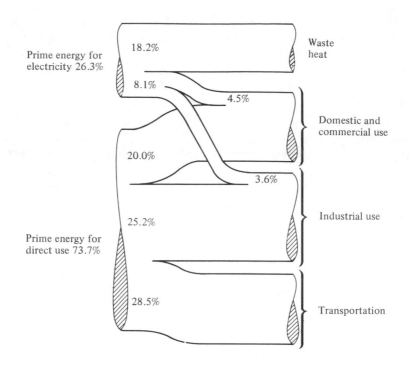

Figure 9.3

First and foremost, the air pollution scourge would be eliminated. Very few exhaust pipes and smokestacks would remain. Quiet and nonpolluting electric cars would handle the private transportation needs. Gas stations would have been replaced by battery recharging centers. Urban life would be greatly improved as a result.

An abundance of energy would permit us to desalinate seawater and recycle our waste products with great efficiency. Our freshwater resources would thus be preserved and the water-pollution problems would be essentially resolved.

In the final analysis, cheap and abundant energy resources are the primary requirement for economic well-being. An energy-affluent society should therefore have a better chance of providing a high standard of living for most of its population.

9.5   FURTHER TECHNOLOGICAL TRENDS

The inexorable long-term shift toward an all-electric economy will bring with it many changes in electric power technology. Let us briefly comment on these.

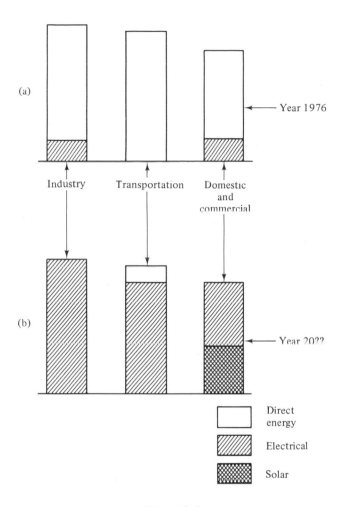

Figure 9.1

9.5.1 Electric Generation

It is highly probable that the steam–electric generation method in present use will continue to completely dominate the scene for the next century. No other known method of transforming high-grade heat into electric energy does a better job.

It should be mentioned that fluids other than water can be used for the steam cycle. However, the easy availability, nontoxicity, and other advantages make water the natural choice.

Future material improvements will undoubtedly permit higher steam temperatures and pressures, thus pushing the thermal efficiency maybe as high as fifty percent. This increased efficiency will reduce the problem of thermal pollution. But since the sizes of the electric plants will increase, the problem will remain with us. Many future power plants therefore will be located along our coastlines where the thermal problem becomes only a minor localized nuisance.

For example, consider the cooling-water requirements of a giant 10,000-MW nuclear center of the future. With an assumed thermal efficiency of 33 percent, the cooling requirements would be 20,000 MW. By use of the conversion table in Fig. 2.14, we find that this corresponds to 480 m^3/second of condenser-cooling water if the exit temperature elevation is set at 10° C ambient. In a reasonably deep coastal zone this waste heat would be diffused within a radius of about 1 km. (Of course, we can even question whether an elevation of the water temperature by 10° C in fact constitutes a real problem in the first place.)

The period between 1950 and 1975 saw a spectacular development in the design of synchronous generators. The "power-per-pound" increased by a factor of four due to the greatly improved, forced-cooling methods and to better insulating materials. Both gas coolants (hydrogen) and liquids (water) are being used to remove the ohmic heat losses in both rotor and stator windings, thus making it possible to increase their current-carrying capability and power ratings.

It appears that we have reached the practical power limits of generators of this type (with ratings in the 1000–1500 MW range). Larger power ratings would result in machine sizes that would not meet shipping constraints and/or would exceed mechanical stress limits in the rotor.

Super Conductivity.   In the electric-generating field, as well as in the electric-transmission and storage-technology areas, considerable interest is presently focused on the use of superconducting conductors. By cooling an electric conductor to temperatures close to the absolute zero (–273°C) the conductor loses its resistance to electric current. Thus its ohmic losses drop drastically. Current densities of tens-of-millions amps per cm^2 can be tolerated in such conductors without ill effects.

As cryogenic technology is now being mastered, one can envision the possibility of supercooling not only generators but also underground transmission lines and storage coils (see below). In fact, research is ongoing in all these areas. It has been estimated that by employing these techniques the present maximum generator unit power rating of 1000–1500 MW could be increased to about 10,000 MW without an increase in machine size.

9.5.2 Electric Transmission

A. Present Choices.   Great advances are anticipated in future electric transmission technology. Technologically speaking, transmission lines constitute in many ways the weakest links of the electric power systems.

Overhead lines are bulky, vulnerable in storms, and unsightly. They produce "noise pollution," and are the cause of many accidents! Neither is their energy-transmission capability very impressive in comparison with some other methods of transporting energy. Consider, for example, a 400-kV overhead line carrying at best 2000 MW of power at a cost of about $1.00 per MWh per 100 miles. Compare this line with a typical 36-inch natural gas pipeline moving energy at the rate of about 1 billion Btu per hour (corresponding to 12,000 MW) at a cost of about $0.1 per MWh per 100 miles. Even an electric power engineer must agree that an underground gas pipeline is preferable to an overhead power line defiling our great Western vistas—if a choice would exist.

Both existing and proposed overhead lines are being challenged with increasing opposition, particularly in urban areas. As we move toward ultra-high voltage (UHV)† there are fears that the exposed electric field between the conductors and ground will negatively affect plants, small animals, and possibly humans.

Realistically assessed, we will have to live with overhead power lines for many years yet. Presently the only alternative is the underground (UG) cable. Typically these cables are of high-pressure, oil-filled cellulose paper insulated design. Not only are UG cables vastly more expensive than overhead lines but they can be used (for ac transmission) over very short distances only—at the most, a few kilometers.

This latter serious limitation is due to the very large shunt capacitance of the cable. In an overhead line the spacings between the phases are very large, typically many meters. This is necessary because of the poor insulating characteristics of air. The large spacings are the cause of the bulkiness of a power line, but at the same time they are the cause of a great reduction in the shunt capacitance between the phases and ground. The capacitive reactance, $1/\omega C$, thus has a relatively large magnitude. This limits the capacitive leakage current and permits us to transmit electric power over distances measured in many hundreds of kilometers.

In a UG cable, the shunt capacitance will be very large for two reasons. First, the distances between the conductors and the grounded outer sheath are very small, typically only a few centimeters. Second, the relative dielectric constant, ϵ, is much larger than in air.

---

† UHV is defined as in excess of 800 kV.

The large shunt capacitance is, of course, of no concern in HVDC transmission.†

B. Compressed Gas-Insulated Transmission.   Transmission lines in which compressed gas is used as an insulating medium offer near-term opportunities in the 1000-MW-plus range. These transmission systems are attractive because of their simplicity and low capital investment need. Each phase conductor is placed inside a tube and centered by means of circular spacers. The tube space is filled with compressed gas, usually sulphur hexafluoride (SF_6). Each tube in a 345-kV line will have a diameter of about 50 cm.

Gas-insulated lines have some very attractive features that make them contenders for replacement of overhead lines in critical areas.

1) Their losses are considerably smaller than those of cables.
2) They can match the load-carrying capability of overhead lines of equal voltage.
3) Unlike cables, they can be designed for UHV.
4) No external electric fields exist.

Because the dielectric constant of compressed gas is approximately unity, the shunt capacitance (despite the small distances) will be considerably less for a gas-insulated line than for a cable. A gas-insulated line can thus transmit power over much larger distances than cable lines.

Several problems are presently being encountered with these systems. The systems are susceptible to metallic particles, even dust, which might get into the system and would cause a dielectric breakdown. A method must be found for eliminating these particles or nullifying their adverse effect on the system. In addition, less expensive spacers for holding the conductor in the center of the outer tube must be found which are not susceptible to flashovers of the insulator surfaces.

C. Superconductors.   No doubt the ultimate electric transmission line will be superconductive. Furthermore, it will be operated on HVDC. It should be noted that a zero-resistance ac line will represent lossless transmission, but its power capacity is still determined by its reactance.‡

A zero-resistance dc line will not only have zero losses but will also be capable of limitless power transmission (because the line will have zero voltage

---

† As this is being written a 130-km submarine HVDC power cable, the world's longest, is being laid between Norway and Denmark. The line will transmit 250 MW at an operating voltage of ±250 kV. In this case overhead conductors or ac cables would, of course, be unthinkable.

‡ Remember that $P_{max} = \dfrac{|V_1| \cdot |V_2|}{X}$.

drop). These two features clearly represent the best that we can ever hope to achieve. Of course, wireless bulk power transmission would be still better but is unfortunately beyond the realm of the possible.

9.5.3 Electric Energy Storage

Figure 1.4 shows how the electric power demand undergoes hourly variations throughout a typical day or week. Ideally, the electric energy should be generated at the constant *average* rate. This would require less installed generating capacity and would result in better economy and longer operating life of the equipment.

This mode of electric power system operation would require electric energy storage facilities. A natural gas system operates approximately in this manner. In spite of a highly fluctuating demand the production of gas at the wellheads takes place at a constant rate. The necessary storage takes place in storage containers (including caverns) and in the pipelines themselves.

An electric storage facility that could deliver, for example, 1000 MW during two hours must have a storage capacity of 2000 MWh, or $7.2 \cdot 10^{12}$ joules. Such facilities do not presently exist. For example, the energy stored in an electrical "pipeline" in the form of electrostatic energy ($\frac{1}{2}Cv^2$) in the shunt capacitance represents a vastly smaller energy amount. Were we to extract it at the rate of 2000 MW it would last only for about 1 millisecond.

As was demonstrated in Example 6.1, some energy is stored in kinetic form in the rotating generator masses. However, were we to draw from this storage source, the frequency would drop at a rapid rate. (Compare Example 6.1.)

This means that electric energy must be generated at the instant it is being demanded. This means also that we must have enough generating capacity to be able to handle the peak load. This, of course, greatly adds to the equipment cost.

Pumped-hydropower storage and compressed-gas storage, both described in Chapter 2, represent hybrid electric–mechanical storage facilities. They have become increasingly popular as supplies for "peaking power" (see Fig. 1.4).

Energy storage in electric and magnetic fields, both discussed in Chapter 3, are the only "true" electrical storage methods known. The electric field storage, quantitatively described by the formula

$$w_e = \tfrac{1}{2}Cv^2,$$

requires either a very large capacitance, C, or high voltage, v. The magnetic field storage, following the formula

$$w_m = \tfrac{1}{2}Li^2,$$

calls for a large inductance or/and high current. This storage method seems presently to be the only one that could handle energy amounts in the 1000-MWh range. By supercooling the magnetic coil, one could conceivably obtain current values (and magnetic field densities) of required magnitudes.

Large superconducting magnets have been built for laboratory purposes in the elementary-particle field of physics. The largest magnet in existence has a storage capacity of 800 MJ. This is still only 0.22 MWh of electric energy.

Within limits imposed by existing technology, this figure can be raised and recent studies (Peterson, 1975) have mentioned energy storage figures of 10^7 MJ.

9.6 SUMMARY

We have speculated in this final chapter over the future developments in the U.S. energy field with particular focus on the electric sector. Due to the very dramatic events that have occurred in the world energy markets since 1973, one is wise not to make sweeping claims of questionable accuracy in any such predictions. From a 1976 vantage point, it seems reasonable to make a few safe assumptions, however.

Whether we like it or not, it seems almost sure that we must shift our energy base away from oil and natural gas to the domestically plentiful solid fuels, coal and uranium. As a corollary fact, electric energy will play an increasingly important role in the U.S. energy supply picture.

Exotic new energy alternatives will not capture any significant portion of the U.S. energy market before the year 2000. Two factors are important in making this assessment:

1) No exotic new energy source (except for the limited-use, low-grade solar heat) can claim the existence of a proven technology.

2) Even though such a technology is developed, the timelags for changeover are very long—of the order of magnitude of one human generation.

The changes that will come about in electric power technology will be of an evolutionary type rather than a revolutionary type. Gradual improvements in electric transmission system technology will tend to move electric power underground. We will undoubtedly also see increased use in HVDC.

The advantages to the economy, to the environment, and to overall security, that could ensue from long distance UG HVDC transmission and large-scale magnetic field energy storage, will accelerate the development of the cryogenic technology needed for a full utilization of superconductivity.

REFERENCES

Boom, R. W., et al. "Superconductive Energy Storage for Large Systems," paper presented at 1974 Applied Superconductivity Conference (S-MAG), September 30–

October 2. Argonne National Laboratory and Fermilab. Copies may be obtained from University of Wisconsin, Department of Electrical Engineering.

"Energy's Hazy Future: Two Scenarios," *IEEE Spectrum*, Vol. 12, no 5 (May 1975): pp. 32–40.

Gulf Oil. "The US Energy Base—An Inventory," *The Orange Disc*, Vol. 22, no. 1 (September–October 1975). Copies may be obtained from Gulf Oil Publications, P.O. Box 1166, Pittsburgh, Pennsylvania.

Heer, John E., and D. J. Hagerty. "Refuse Turns Resource," *IEEE Spectrum*, Vol. 11, no. 9 (September 1974): pp. 83–87.

Kirtley, J. L., and M. Furuyama. "A Design Concept for Large Super Conducting Alternators," *IEEE Transaction*, Vol. PAS-94, no. 4 (July–August 1975): pp. 1264–1269.

Klein, M., et al. "HVDC to Illuminate Darkest Africa," *IEEE Spectrum*, Vol. 11, no. 10 (October 1974): pp. 51–58.

Nagel, T. J. "Electric Power's Role in the US Energy Crisis," *IEEE Spectrum*, Vol. 11, no. 7 (July 1974): pp. 69–72.

Peterson, H. A., et al. "Superconductive Energy Storage Inductor–Convertor Units for Power Systems," *IEEE Transactions*, Vol. PAS-94, no. 4 (July–August 1975): pp. 1337–1348.

Ramakumar, R., et al. "Solar Energy Conversions and Storage Systems for the Future," *IEEE Transactions*, Vol. PAS-94, no. 5 (November–December 1975): pp. 1926–1934.

Spencer, D. F. "The Spectrum of Future Electric Generation Alternatives." Available from EPRI, 3412 Hillview Avenue, Palo Alto, California.

PHASOR ANALYSIS

appendix A

A.1   VECTOR REPRESENTATION OF SINUSOIDS. CONCEPT OF PHASOR.

Consider the *harmonic*, or sinusoidal, time functions

$$u(t) = u_{max} \sin \omega t,$$
$$v(t) = v_{max} \sin (\omega t - \alpha).$$

(A.1)

These functions may represent steady-state voltages, currents, velocities, or any other physical variables. Graphically, the functions may be depicted (Fig. A.1) as the vertical projections of the rotating vectors U and V. These vectors have the lengths u_{max} and v_{max}, respectively, and rotate with the angular velocity, ω, rad/s.

The algebraic sum

$$w(t) = u(t) + v(t)$$

(A.2)

is also sinusoidal and can thus be represented by the projection of a third rotating vector, W. This vector is obtained as the *vectorial sum* of the vectors U and V as shown in Fig. A.1.

Analysis of sinusoidally varying quantities can thus be performed by a *geometric* study of vectors. The studies can be further simplified by "freezing" the rotating vectors in fixed positions. Such "time-frozen" vectors are often referred to as *phasors*. One may thus think of a phasor as a snapshot of a rotating vector.

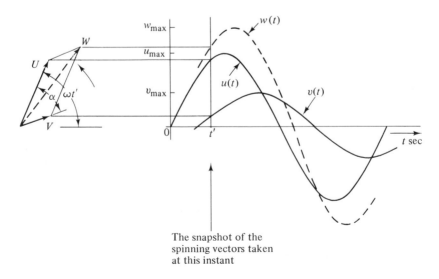

The snapshot of the
spinning vectors taken
at this instant

Figure A.1

Example A.1   Find the sum w of the two sinusoids

$$\left.\begin{array}{l} u(t) = 6\sin\omega t \\ v(t) = 5\sin(\omega t - 37°). \end{array}\right\} \tag{A.3}$$

SOLUTION: The two vectors U and V are shown in an x-y coordinate system (Fig. A.2). For simplicity they have been "frozen" at the instant when the U vector coincides with the x-axis. According to our definition, the rotating vectors have thus turned into phasors. The U-phasor coinciding with the x-axis, is referred to as the *reference* phasor.

The sum phasor W is found by first determining its components, W_x and W_y along the x and y axes, respectively. We obtain

$$W_x = 6 + 5\cos 37° = 9.993,$$
$$W_y = 0 - 5\sin 37° = -3.009.$$

The magnitude $|W|$ of the sum phasor W thus equals

$$|W| = \sqrt{9.993^2 + 3.009^2} = 10.436.$$

The angle $\underline{/W}$ of phasor W relative to the reference phasor equals

$$\underline{/W} = \tan^{-1}\left(\frac{W_y}{W_x}\right) = \tan^{-1}\left(\frac{-3.009}{9.993}\right) = -16.76°.$$

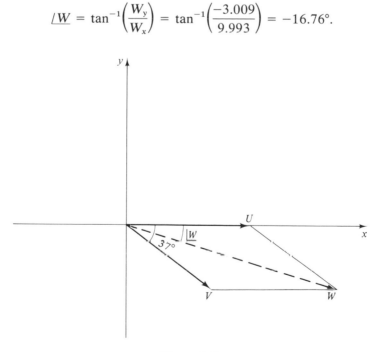

Figure A.2

For the time function $w(t)$, we can thus write

$$w(t) = 10.436 \sin (\omega t - 16.76°).$$

....

Example A.2   A sinusoidal current

$$i = i_{max} \sin \omega t \qquad A \tag{A.4}$$

is injected into a circuit consisting of a resistor R in series with an inductor L and a capacitor C. Describe the steady-state voltages v_R, v_L, v_C defined in Fig. A.3!

SOLUTION: The voltage across the resistor according to Eq. (3.29) is

$$v_R = Ri = Ri_{max} \sin \omega t \qquad V. \tag{A.5}$$

The voltage across the inductor is obtained from Eq. (3.84):

$$v_L = L\frac{di}{dt} = \omega L i_{max} \cos \omega t = \omega L i_{max} \sin (\omega t + 90°). \tag{A.6}$$

Finally, by use of Eq. (3.12) we have for the capacitor voltage

$$v_C = \frac{1}{C} Q = \frac{1}{C} \int i \, dt$$

$$= \frac{1}{C} \int i_{max} \sin \omega t \, dt = -\frac{i_{max}}{\omega C} \cos \omega t$$

$$= \frac{i_{max}}{\omega C} \sin (\omega t - 90°) \qquad V. \tag{A.7}$$

In summary: The resistor voltage v_R is sinusoidal and in phase with the current i. The inductor voltage v_L and capacitor voltage v_C are likewise sinusoidal but respectively leading and lagging the current by 90°. If we represent the above time variables with the phasors V_R, V_L, V_C, and I, respectively, we obtain the *phasor diagram* shown in Fig. A.4.

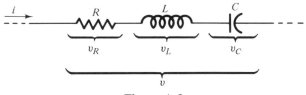

Figure A.3

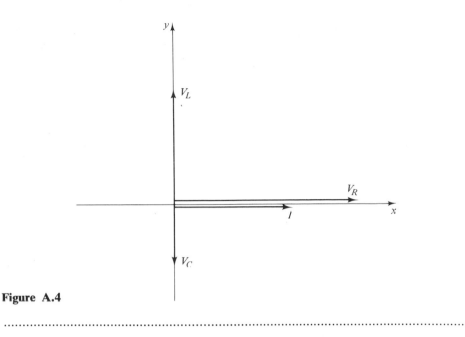

Figure A.4

Example A.3   For the circuit in the previous example we have the following numerical values:

$$i_{max} = 10 \text{ A,}$$
$$R = 20 \, \Omega,$$
$$\omega - 377 \text{ rad/s} \quad (60 \text{ Hz}),$$
$$L = 0.040 \text{ H,}$$
$$C = 300 \, \mu F.$$

Find the amplitudes of the voltages across R, L, and C, and also the total voltage across the circuit!

SOLUTION: From Eq. (A.5) we get

$$v_{R \text{ max}} - Ri_{max} = 20 \cdot 10 = 200 \text{ V.}$$

From (A.6):

$$v_{L \text{ max}} = \omega L i_{max} = 377 \cdot 0.040 \cdot 10 = 150.8 \text{ V.}$$

From (A.7):

$$v_{C \text{ max}} = \frac{i_{max}}{\omega C} = \frac{10}{377 \cdot 300 \cdot 10^{-6}} = 88.4 \text{ V.}$$

The total voltage v across the circuit in Fig. A.3 is the sum of the individual voltages. By representing v by its phasor V, we obtain the latter simply by vectorial addition (Fig. A.5) of the phasors V_R, V_L, and V_C. We get directly

$$|V| = \sqrt{200^2 + (150.8 - 88.4)^2} = 209.5 \text{ V},$$

$$\underline{/V} = \tan^{-1}\left(\frac{150.8 - 88.4}{200}\right) = 17.33°.$$

For the total voltage v we can thus write

$$v = 209.5 \sin(\omega t + 17.33°). \tag{A.8}$$

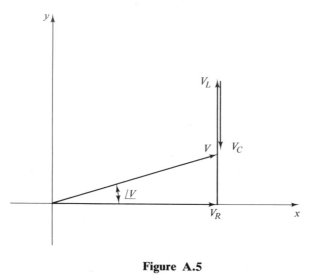

Figure A.5

A.2   PHASOR REPRESENTATION BY MEANS OF COMPLEX NUMBERS

Complex algebra is proven a most valuable analytical tool for the study of phasors. We give first a very brief exposé of complex algebra followed by some examples on its use in working with phasors.

Complex Numbers–Definition

Figure A.6 shows a *complex number plane.* The complex number z is defined by its coordinates, x and jy, along the *real* and *imaginary* axes, respectively. We write the complex number x in its *cartesian* form:

$$z = x + jy. \tag{A.9}$$

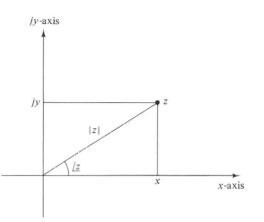

Figure A.6

The factor "j" is defined by the equation

$$j^2 = -1. \tag{A.10}$$

We say that x is the "real part of z," and y the "imaginary part."

The distance $|z|$ from the origin to the coordinate point z is referred to as the *modulus* or *magnitude* of z.

We clearly have

$$|z| = \sqrt{x^2 + y^2}. \tag{A.11}$$

The angular orientation $\underline{/z}$ of z in reference to the real axis is obtained from

$$\underline{/z} = \tan^{-1}\left(\frac{y}{x}\right). \tag{A.12}$$

From Fig. A.6 we obviously have

$$\left.\begin{array}{l} x = |z| \cos \underline{/z}, \\ y = |z| \sin \underline{/z}. \end{array}\right\} \tag{A.13}$$

Euler's formula† states that

$$e^{j\underline{/z}} = \cos \underline{/z} + j \sin \underline{/z} \tag{A.14}$$

---

† This formula can be proven as follows: From the series

$$e^x = 1 + x + \frac{x^2}{2!} + \frac{x^3}{3!} + \frac{x^4}{4!} + \cdots,$$

we obtain

$$e^{j\varphi} = 1 + j\varphi - \frac{\varphi^2}{2!} - j\frac{\varphi^3}{3!} + \frac{\varphi^4}{4!} - + \cdots$$

$$= 1 - \frac{\varphi^2}{2!} + \frac{\varphi^4}{4!} - + \cdots + j\left(\varphi - \frac{\varphi^3}{3!} + - \cdots\right)$$

$$= \cos \varphi + j \sin \varphi.$$

The complex number z can thus be written in the alternate *polar* form:

$$z = x + jy$$
$$= |z| \cos \underline{/z} + j |z| \sin \underline{/z}$$
$$= |z| (\cos \underline{/z} + j \sin \underline{/z}) = |z| e^{j \underline{/z}}. \qquad (A.15)$$

Complex Algebra

Addition, subtraction, multiplication, and division of complex numbers are defined in terms of the same operations valid for real numbers. We demonstrate by working one example on each operation.

Example A.4   Given the two complex numbers:

$$z_1 = 3 + j1, \qquad z_2 = 4 - j2.$$

Find $z_1 + z_2$, $z_1 - z_2$, $z_1 z_2$ and z_1/z_2!

SOLUTION: The complex number *sum* is obtained as follows:

$$z_1 + z_2 = (3 + j1) + (4 - j2)$$
$$= (3 + 4) + j(1 - 2) = 7 - j1.$$

Note that the real and imaginary parts are added separately.

Note also (Fig. A.7) *that if each complex number is associated with a phasor, then the complex sum of the numbers corresponds to a vectorial addition of the phasors.*

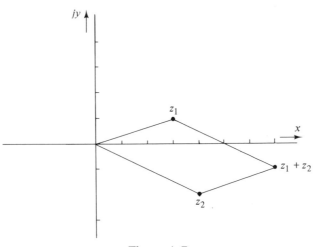

Figure A.7

Next we compute the complex *difference*:

$$z_1 - z_2 = (3 + j1) - (4 - j2)$$
$$= (3 - 4) + j(1 + 2) = -1 + j3.$$

The *product* and *quotient* can be obtained from either the cartesian or the polar forms. We start with the former:

$$z_1 z_2 = (3 + j1)(4 - j2)$$
$$= 3 \cdot 4 + (j1)(-j2) + j(1 \cdot 4 - 3 \cdot 2)$$
$$= 14 - j2;$$

and

$$\frac{z_1}{z_2} = \frac{3 + j1}{4 - j2} = \frac{(3 + j1)(4 + j2)}{(4 - j2)(4 + j2)}$$
$$= \frac{3 \cdot 4 + (j1)(j2) + j(1 \cdot 4 + 2 \cdot 3)}{16 + 4}$$
$$= \frac{10}{20} + j\frac{10}{20} = 0.5 + j0.5.$$

By using the polar number representation, we obtain

$$z_1 z_2 = (3 + j1)(4 - j2) = 3.162 e^{j18.435°} \cdot 4.472 e^{-j26.565°}$$
$$= 14.140 e^{-j8.130°} = 14.140(0.990 - j0.141)$$
$$= 14.00 - j2.00 \quad \text{(as before)}.$$

And

$$\frac{z_1}{z_2} = \frac{3.162 e^{j18.435°}}{4.472 e^{-j26.565°}} = 0.707 e^{j45°}$$
$$= 0.5 + j0.5 \quad \text{(as before)}.$$

Example A.5 Use complex algebra to solve Example A.3.

SOLUTION: If we express the voltage phasors V_R, V_L, and V_C as complex numbers we get

$$V_R = 200,$$
$$V_L = j\,150.8,$$
$$V_C = -j\,88.4.$$

The total voltage phasor V is then obtained by addition:

$$V = V_R + V_L + V_C = 200 + j150.8 - j88.4$$
$$= 200 + j62.4 = 209.5 e^{j17.33°}$$

Expressed as a time function the total voltage is:

$$v(t) = 209.5 \sin(\omega t + 17.33°). \qquad (A.16)$$

Compare Eq. (A.8).

....

A.3 IMPEDANCES

Consider the phasor P and the complex number z. The product zP can be written

$$zP = |z| e^{j\angle z} \cdot |P| e^{j\angle P} = |z||P| e^{j(\angle z + \angle P)}.$$

We make the following observation:

Multiplication by z results in a magnitude amplification in ratio $|z|$ and a phase advancement by $\angle z$ degrees.

In particular, when a phasor is multiplied by the factor j, it is rotated counter-clockwise through the angle $90°$ with unchanged magnitude. Multiplication by the factor $-j$ results in a clockwise rotation through $90°$. Consider now the four phasors depicted in Fig. A.4. In view of the formulas (A.5), (A.6), and (A.7), and in consideration of the above multiplication rules we can obviously write

$$\left. \begin{aligned} V_R &= RI, \\ V_L &= j\omega LI, \\ V_C &= -j\frac{1}{\omega c} I = \frac{I}{j\omega C}. \end{aligned} \right\} \qquad (A.17)$$

The voltage phasor V representing the total voltage across the circuit is the sum (Fig. A.5) of the individual component voltage phasors, that is,

$$V = RI + j\omega LI + \frac{1}{j\omega C} I$$

$$= \left(R + j\omega L + \frac{1}{j\omega C}\right)I. \qquad (A.18)$$

The multiplication factors R, $j\omega L$, and $1/j\omega C$, which when multiplied by the current phasor I give us the voltage phasors, are referred to as the *impedances* of the individual elements. We symbolize them by the symbol Z and have for the resistor, inductor, and capacitor, respectively,

$$\left. \begin{aligned} Z_R &\equiv R, \\ Z_L &\equiv j\omega L, \\ Z_C &\equiv \frac{1}{j\omega C}. \end{aligned} \right\} \qquad (A.19)$$

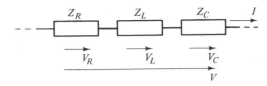

Figure A.8

The total impedance Z for the series circuit is, according to Eq. (A.18),

$$Z = R + j\omega L + \frac{1}{j\omega C}. \tag{A.20}$$

In symbolic form, the relationships between impedances and current and voltage phasors are depicted in Fig. A.8.

Example A.6 Find the impedance of the $R\text{-}L\text{-}C$ circuit assuming the numerical parameter values given in Example A.3.

SOLUTION: We obtain directly

$$Z_R = R = 20\,\Omega,$$
$$Z_L = j\omega L = j377 \cdot 0.040 = j15.08\,\Omega,$$
$$Z_C = \frac{1}{j\omega C} = \frac{1}{j377 \cdot 300 \cdot 10^{-6}} = \frac{8.84}{j}$$
$$= -j8.84\,\Omega.$$

The total impedance thus equals

$$Z = 20 + j15.08 - j8.84 = 20 + j6.24.$$

Example A.7 If a voltage of 100 V rms is applied across the above circuit, find the rms values of the current and voltages across each of the three circuit elements. Also write all variables as time functions.

SOLUTION: We chose the total voltage V as our reference phasor; that is,

$$V = 100e^{j0^\circ} = 100.$$

(Note that the length of the phasor has been given in rms, that is, $v_{max}/\sqrt{2}$. This is more practical as we mostly are interested in rms values of ac variables rather than their peak values.)

From Eq. (A.18) we then get

$$I = \frac{V}{Z} = \frac{100e^{j0^\circ}}{20 + j6.24} = \frac{100e^{j0^\circ}}{20.95e^{j17.33}}$$
$$= 4.773e^{-j17.33^\circ}\,\text{A}. \tag{A.21}$$

(Note that this current value is now in rms because the voltage was expressed in rms.)

For the component voltages we then have

$$V_R = Z_R I = 20 \cdot 4.773 e^{-j17.33°} = 95.47 e^{-j17.33°},$$
$$V_L = Z_L I = j15.08 \cdot 4.773 e^{-j17.33°} = 71.98 e^{j72.67°},$$
$$V_C = Z_C I = -j8.84 \cdot 4.773 e^{-j17.33°} = 42.19 e^{-j107.33°}. \quad (A.22)$$

Note that if we show the above phasors in a phasor diagram, we obtain the same diagram as in Fig. A.5 *but reduced in scale and rotated clockwise by 17.33°.*

If we express the variables as functions of time we obtain

$$
\begin{aligned}
v(t) &= 100\sqrt{2}\ \sin \omega t = 141.4 \sin \omega t \quad \text{V}, \\
i(t) &= 6.750 \sin (\omega t - 17.33°) \quad \text{A}, \\
v_R(t) &= 135.0 \sin (\omega t - 17.33°) \quad \text{V}, \quad (A.23) \\
v_L(t) &= 101.8 \sin (\omega t + 72.67°), \\
v_C(t) &= 59.67 \sin (\omega t - 107.33°).
\end{aligned}
$$

Example A.8 The spring–mass–dashpot system shown in Fig. A.9 is subject to a sinusoidally varying force $f(t)$. Find the displacement x and the velocity s of the mass in steady-state!

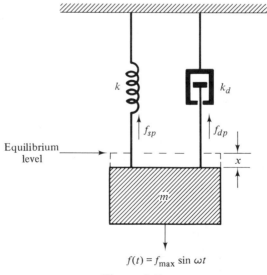

$$f(t) = f_{max} \sin \omega t$$

Figure A.9

SOLUTION: The restraining forces in the spring and dashpot are, respectively

$$f_{sp} = kx \qquad \text{N,} \tag{A.24}$$

$$f_{dp} = k_d \frac{dx}{dt} = k_d s \qquad \text{N,} \tag{A.25}$$

where k and k_d are constants.

According to Newton the acceleration, d^2x/dt^2, of the mass will be proportional to the resulting force acting on the mass:

$$f(t) - kx - k_d \frac{dx}{dt} = m \frac{d^2x}{dt^2}, \tag{A.26}$$

or, in terms of the velocity,

$$f(t) = f_{max} \sin \omega t = k \int s \, dt + k_d s + m \frac{ds}{dt}. \tag{A.27}$$

In steady-state, the velocity will be sinusoidal of the form

$$s = s_{max} \sin (\omega t + \underline{/s}). \tag{A.28}$$

By substitution of (A.28) into Eq. (A.27) we obtain

$$f_{max} \sin \omega t = \frac{k}{\omega} s_{max} \sin (\omega t + \underline{/s} - 90°)$$

$$+ k_d s_{max} \sin (\omega t + \underline{/s}) + m \omega s_{max} \sin (\omega t + \underline{/s} + 90°). \tag{A.29}$$

We now represent the force $f(t)$ by the phasor F and the velocity $s(t)$ by the phasor S in accordance with

$$\left. \begin{array}{l} F = f_{max} e^{j0°} \\ S = s_{max} e^{j/s} \end{array} \right\} \tag{A.30}$$

These phasors are depicted in Fig. A.10, where we have also shown the displacement phasor X. (The latter lags S by 90°. Why?)

In terms of these phasors we can write equation (A.29) as follows:

$$F = \frac{k}{j\omega} S + k_d S + j\omega m S. \tag{A.31}$$

This equation evidently looks very similar to Eq. (A.18) and we therefore are led to define the *mechanical impedance* Z_m:

$$Z_m \equiv k_d + j\omega m + \frac{k}{j\omega}. \tag{A.32}$$

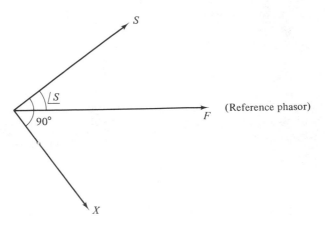

Figure A.10

We then can write Eq. (A.31) in the shorter form:

$$S = \frac{F}{Z_m}. \tag{A.33}$$

....

Example A.9 In the previous example we have

$$f_{max} = 10\,N,$$
$$m = 1\,kg,$$
$$k = 100\,N/m,$$
$$k_d = 1\,N \cdot s/m.$$

Determine the velocity amplitude, s_{max}, as a function of ω.

SOLUTION: We have

$$Z_m = 1 + j\omega + \frac{100}{j\omega}. \tag{A.34}$$

By combining Eqs. (A.30) and (A.33) we thus obtain

$$S = s_{max}e^{j\angle s} = \frac{10e^{j0°}}{1 + j\left(\omega - \dfrac{100}{\omega}\right)}; \tag{A.35}$$

that is,

$$s_{max} = \frac{10}{\sqrt{1 + \left(\omega - \dfrac{100}{\omega}\right)^2}} \quad m/s, \tag{A.36}$$

$$\underline{\angle s} = -\tan^{-1}\left(\omega - \frac{100}{\omega}\right). \tag{A.37}$$

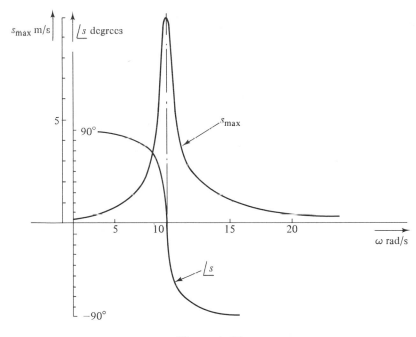

Figure A.11

We have plotted s_{max} and $\underline{/s}$ versus ω in Fig. A.11. Note the velocity peak occurring at $\omega = 10$ rad/s. This is the *resonance frequency*. It is interesting to note that $\underline{/s} = 0$ at resonance; that is, at resonance the velocity and force are in phase:

$$\left.\begin{array}{ll} f_{res} = f_{max} \sin \omega t & N, \\ s_{res} = s_{max} \sin \omega t & m/s. \end{array}\right\} \tag{A.38}$$

A.4 ADMITTANCES

Consider the parallel circuit depicted in Fig. A.12. For the currents I_1 and I_2 through the impedances Z_1 and Z_2 we have

$$I_1 = \frac{V}{Z_1}, \qquad I_2 = \frac{V}{Z_2}. \tag{A.39}$$

The total current I is

$$I = I_1 + I_2 = \frac{V}{Z_1} + \frac{V}{Z_2} = V\left(\frac{1}{Z_1} + \frac{1}{Z_2}\right). \tag{A.40}$$

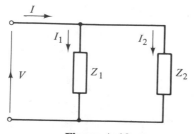

Figure A.12

We introduce now the *admittances*

$$Y_1 \equiv \frac{1}{Z_1}, \qquad Y_2 \equiv \frac{1}{Z_2}; \tag{A.41}$$

and the total admittance

$$Y \equiv Y_1 + Y_2. \tag{A.42}$$

Equation (A.40) can then be written in the form

$$I = V(Y_1 + Y_2) = VY. \tag{A.43}$$

Example A.10   Connect the three impedances in Example A.6 in parallel and feed this circuit from an ac generator delivering 100 V rms, at 60 Hz. Find the current drawn from the generator!

SOLUTION: The three parallel elements represent a total admittance of

$$Y = \frac{1}{20} + \frac{1}{j15.08} + \frac{1}{-j8.84}$$
$$= 0.0500 - j0.06631 + j0.1131$$
$$= 0.0500 + j0.0468.$$

Equation (A.43) then gives directly

$$I = 100(0.0500 + j0.0468) = 5.00 + j4.68$$
$$= 6.85e^{j43.11°}.$$

Note: When the three circuit elements were connected in *series* (Example A.7) across a 100-V source, the current equalled 4.7 A and *lagged* the voltage by 17°—the series circuit is *inductive*.

Now we find that when the same elements are connected in parallel across the same source, the current equals 6.8 A and is *leading* the voltage by 43°—the parallel circuit is *capacitive*. Explain this difference!

....

appendix B

SPECTRAL ANALYSIS

B.1   PERIODICITY

A function or "wave," $f(t)$, is said to be *periodic* if for all values of t it satisfies the equation

$$f(t) = f(T + t) \tag{B.1}$$

where T is the *period time* (Fig. B.1).

Fourier proved that it is possible to express a periodic function as an infinite sum of *harmonic components* in accordance with

$$f(t) = A_0 + \sum_{\nu=1}^{\infty} A_\nu \sin(\nu\omega t + \varphi_\nu), \tag{B.2}$$

where

$$\omega = \frac{2\pi}{T} \quad \text{rad/s}$$

is the *base* or *fundamental radian frequency*.

If use is made of the trigonometric identity

$$\sin(\alpha + \beta) = \sin\alpha\cos\beta + \cos\alpha\sin\beta, \tag{B.3}$$

we can write the series (B.2) in the alternate form:

$$f(t) = A_0 + \sum_{\nu=1}^{\infty} (B_\nu \sin\nu\omega t + C_\nu \cos\nu\omega t). \tag{B.4}$$

The amplitudes A_ν and phase angles φ_ν in the series (B.2) are related to the coefficients B_ν and C_ν in the series (B.4) through the relations

$$\left.\begin{aligned} A_\nu &= \sqrt{B_\nu^2 + C_\nu^2}, \\ \tan\varphi_\nu &= \frac{C_\nu}{B_\nu}. \end{aligned}\right\} \tag{B.5}$$

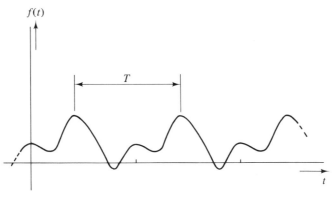

Figure B.1

The constant A_0 represents the *dc component* of the periodic wave. The coefficients, A_1, A_2, ... are the amplitudes of the first, second, ... *harmonics.* The "first" harmonic (of frequency ω) is also referred to as the "fundamental" or "base" component.

B.2 FINDING THE HARMONIC AMPLITUDES

One can readily prove† that the coefficients in Eq. (B.4) may be derived from the following formulas:

$$\left.\begin{array}{l} B_\nu = \dfrac{2}{T} \displaystyle\int_0^T f(t) \sin \nu\omega t \, dt, \\[4mm] C_\nu = \dfrac{2}{T} \displaystyle\int_0^T f(t) \cos \nu\omega t \, dt. \end{array}\right\} \tag{B.6}$$

The dc component is computed from

$$A_0 = \frac{1}{T} \int_0^T f(t) \, dt. \tag{B.7}$$

If the periodic wave has a zero average, it clearly contains no dc component. Certain symmetry features of the wave may greatly simplify the task of finding the harmonics, as will be demonstrated in our first example.

Example B.1   Consider the periodic "triangular" wave depicted in Fig. B.2. Find the fundamental component and all higher harmonics of this wave!

SOLUTION: We explore first the symmetry features of this wave. We note in particular that the wave is characterized by the symmetry equations

$$f(t) = -f(T - t), \tag{B.8}$$

$$f(t) - f\left(\frac{T}{2} - t\right). \tag{B.9}$$

---

† The proof proceeds in the following steps:
1) Multiply both sides of Eq. (B.4) by the factor $\sin (\nu\omega t)$.
2) Integrate both sides over one full time period. In so doing, one finds that all integrals on the right side equal zero except the integral

$$\int_0^T B_\nu \sin^2 \nu\omega t \, dt,$$

which equals $B_\nu T/2$. Thus, the first line of Eq. (B.6) follows.

The second line of Eq. (B.6) can likewise be confirmed by following the above steps, but using $\cos (\nu\omega t)$ as a multiplication factor.

Finally, Eq. (B.7) is obtained by integrating both sides of Eq. (B.4) over one full cycle.

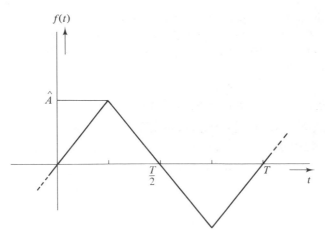

Figure B.2

Based upon these two symmetry features we can now make the following observations:

Observation 1.   Write the integral (B.7) as two parts:

$$A_0 = \frac{1}{T} \int_0^T f(t)\, dt = \frac{1}{T}\left[\int_0^{T/2} f(t)\, dt + \int_{T/2}^T f(t)\, dt \right].$$

Because of Eq. (B.8) the two part integrals are equal but of opposite sign; therefore, $A_0 = 0$—the wave contains no dc component!

Observation 2.   For a cosine wave we have

$$\cos \nu\omega t = \cos \nu\omega (T - t). \tag{B.10}$$

In view of Eq. (B.8), it is therefore obvious that the second set of integrals (B.6) vanishes for all ν;

$$\therefore \ C_\nu = 0 \qquad \text{for } \nu = 1, 2, 3, \ldots, \infty;$$

that is, the wave contains no *cosine* terms.

Observation 3.   For a sine wave we have

$$\left.\begin{aligned}
\sin \nu\omega t &= \sin \nu\omega\left(\frac{T}{2} - t\right) && \text{for } odd \ \nu\text{'s}, \\
\sin \nu\omega t &= -\sin \nu\omega\left(\frac{T}{2} - t\right) && \text{for } even \ \nu\text{'s}.
\end{aligned}\right\} \tag{B.11}$$

In view of Eq. (B.9) it is thus clear that the first set of integrals of (B.6) vanishes for *all even* ν;

$$\therefore \quad B_\nu = 0 \qquad \text{for } \nu = 2, 4, 6, \ldots, \infty;$$

that is, the wave contains only odd sinusoids!

Observation 4.   In view of Eqs. (B.9) and (B.11) the product

$$f(t) \sin \nu\omega t$$

attains identical values for t, $T/2 - t$, $T/2 + t$ and $T - t$ (assuming ν odd). Thus we need only to perform the integration (B.6) over one quarter cycle; that is,

$$B_\nu = \frac{8}{T} \int_0^{T/4} f(t) \sin \nu\omega t \, dt \qquad \text{for } \nu = 1, 3, 5, \ldots, \infty. \tag{B.12}$$

In the interval $0 < t < T/4$ the function $f(t)$ (see Fig. B.2) is a linear sloping line of the form

$$f(t) = 4\frac{\hat{A}}{T}t. \tag{B.13}$$

By substitution into (B.12) we thus have

$$B_\nu = 32\frac{\hat{A}}{T^2} \int_0^{T/4} t \sin \nu\omega t \, dt \qquad \text{for } \nu = 1, 3, 5, \ldots, \infty. \tag{B.14}$$

Integral tables yield the following value for B_ν:

$$B_\nu = \frac{8}{\pi^2} \cdot \frac{\hat{A}}{\nu^2}(-1)^{\frac{\nu+3}{2}} \qquad \text{for } \nu = 1, 3, 5, \ldots, \infty. \tag{B.15}$$

The triangular wave thus can be written as the infinite series

$$f(t) = \frac{8}{\pi^2}\hat{A}\left[\sin \omega t - \frac{1}{9}\sin 3\omega t + \frac{1}{25}\sin 5\omega t - + \cdots\right]. \tag{B.16}$$

In practice one obtains very good accuracy by including only the first few terms in this series. Fig. B.3 shows the close triangular resemblance obtained by including only the first three terms in the above series.

··

B.3   SPECTRAL ANALYSIS BY NUMERICAL INTEGRATION

In many practical cases one cannot perform the analysis as neatly as the previous example lets us believe. Often the periodic wave is obtained experimentally and the function $f(t)$ is available not in analytic form but as a graph.

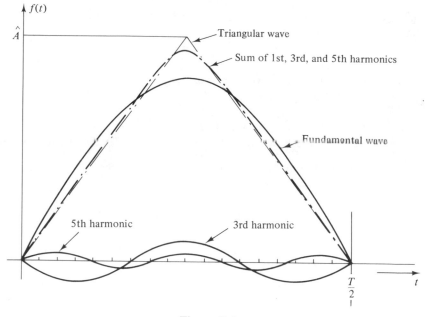

Figure B.3

In such situations, the spectral analysis must proceed numerically using approximations for the integrals B.6 and B.7. Let us exemplify.

Example B.2   Figure B.4 shows the experimentally recorded magnetization current in a power transformer. The base frequency is 50 Hz which means that the period time T equals 20 ms. (Compare also Fig. 5.9!)

Make a spectral analysis of this wave! Specifically, find the amplitude of the 50-Hz component.

SOLUTION: Because the function $f(t)$ is not known analytically, we must perform the computations of B_1 and C_1 using the $f(t)$ values obtained from the graph.

We first write the integrals (B.6) as finite sums:

$$\left.\begin{array}{l} B_1 \approx \dfrac{2\,\Delta t}{T} \sum_{r=1}^{n} i_r \sin \omega t_r, \\[3mm] C_1 \approx \dfrac{2\Delta t}{T} \sum_{r=1}^{n} i_r \cos \omega t_r. \end{array}\right\} \qquad \text{(B.17)}$$

We have divided the time interval $0 < t < T$ into n timeslices of width Δt. In Fig. B.4, we have chosen $n = 20$ which corresponds to $\Delta t = 1$ ms. The current value i_r in the center of each timeslice is read off. We then compute

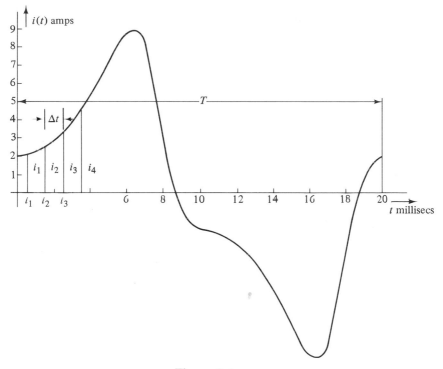

Figure B.4

and tabulate ωt_r, $\sin \omega t_r$, and $\cos \omega t_r$ for each value of t_r in the interval. Finally the products $i_r \sin \omega t_r$ and $i_r \cos \omega t_r$ are computed. The results are tabulated in Table B.1.

Table B.1

r number	i_r amperes	ωt_r degrees	$\sin \omega t_r$	$i_r \sin \omega t_r$	$\cos \omega t_r$	$i_r \cos \omega t_r$
1	2.1	9	0.156	0.329	0.988	2.074
2	2.9	27	0.454	1.317	0.891	2.584
3	3.3	45	0.707	2.333	0.707	2.333
4	4.5	63	0.891	4.010	0.454	2.043
5	6.2	81	0.988	6.124	0.156	0.970
6	8.0	99	0.988	7.902	−0.156	−1.251
7	9.0	117	0.891	8.019	−0.454	−4.086
8	6.0	135	0.707	4.243	−0.707	−4.243
9	0.6	153	0.454	0.272	−0.891	−0.535
10	−1.8	171	0.156	−0.282	−0.988	1.778
Sum				34.267		1.667

Because of the symmetry feature of the wave that can be expressed mathematically as

$$f(t) = -f\left(\frac{T}{2} + t\right),$$
(B.18)

we need only to perform the summation over half the period time. This means that we compute B_1 and C_1 from these expressions:

$$\left.\begin{aligned}
B_1 &\approx \frac{4 \, \Delta t}{T} \sum_{r=1}^{n/2} i_r \sin \omega t_r \\
C_1 &\approx \frac{4 \, \Delta t}{T} \sum_{r=1}^{n/2} i_r \cos \omega t_r
\end{aligned}\right\}$$
(B.19)

By using the computed sums from Table B.1 we have in this case

$$B_1 = \frac{4 \cdot 0.001}{0.020} \cdot 34.267 = 6.853,$$

$$C_1 = \frac{4 \cdot 0.001}{0.020} \cdot 1.667 = 0.333.$$

Using formulas (B.5) we finally compute the amplitude and phase angle of the base harmonic:

$$A_1 = \sqrt{6.853^2 + 0.333^2} = 6.853,$$

$$\varphi_1 = \tan^{-1}\left(\frac{0.333}{6.853}\right) = 2.78°.$$

We can therefore write the fundamental current component in the form

$$i(t) = 6.861 \sin(\omega t + 2.78°) \quad \text{A.}$$

where

$$\omega = \frac{2\pi}{0.020} = 314 \text{ rad/s.}$$

B.4 PERIODICITY IN THE SPACE DOMAIN

In the formulas and examples treated so far, the independent variable was assumed to be time, t. Indeed, this is normally the case as harmonic analysis finds its widest use in analysis of time variables (for example in communication theory).

However, there are important areas of science and technology where the periodic phenomena involve *space* coordinates. For example, the periodic waves of currents, emf, and magnetic flux that are found around the air-gap periphery of an electric machine can be put in this category. Let us exemplify.

Example B.3   Figure B.5 depicts the "current sheet" in one phase of the distributed stator winding of a 3-phase synchronous machine. We remember from Chapter 4 that this current sheet was created by the current in the phase in question and represented a macroscopic (or "smeared out") view of the current distribution in the stator surface.

We also remember that because the current is of ac type, the current sheets will pulsate in time. The picture shown in Fig. B.5 is therefore a snapshot taken at a certain time (for example, when the current reaches its peak).

Find the base wave of the current sheet.

SOLUTION: By placing the origin as is done in Fig. B.5, the wave will have the same symmetry features as the triangular wave in Example B.1. The observations made concerning the harmonics of the triangular wave thus also apply here. The base harmonic must therefore be a sinusoid which has been shown dashed in Fig. B.5. We will find its amplitude B_1.

In our previous theory our formulas were derived in terms of the independent variable t and the period time T. Now the independent variable is x and the period length is $2\pi D/p$.† We can thus use our previous formulas after

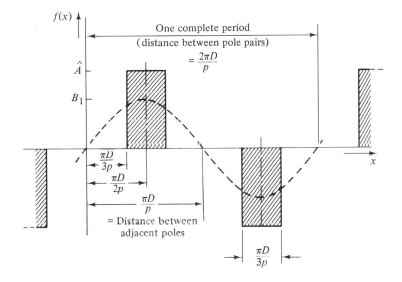

Figure B.5

---

† The expression $2\pi D/p$ is the peripheral width of one pole pair.

making the following variable changes:

$$t \to x,$$

$$T \to \frac{2\pi D}{p}, \tag{B.21}$$

$$\omega = \frac{2\pi}{T} \to \frac{p}{D}.$$

After making these variable substitutions in formula (B.12), it changes to

$$B_\nu = \frac{4p}{\pi D} \int_0^{\pi D/2p} f(x) \sin\left(\nu \frac{p}{D} x\right) dx. \tag{B.22}$$

In our case the function $f(x)$ has the following values (see Fig. B.5):

$$f(x) = 0 \qquad \text{for } 0 < x < \frac{\pi D}{3p},$$

$$f(x) = \hat{A} \qquad \text{for } \frac{\pi D}{3p} < x < \frac{\pi D}{2p}.$$

The integral (B.22) thus yields

$$B_1 = \frac{4p}{\pi D} \left[\int_0^{\pi D/3p} 0 \cdot \sin\left(\frac{px}{D}\right) dx + \int_{\pi D/3p}^{\pi D/2p} \hat{A} \cdot \sin\left(\frac{px}{D}\right) dx \right]$$

$$= \frac{4p}{\pi D} \left[0 + \frac{\hat{A} D}{2p} \right] = \frac{2}{\pi} \hat{A} \qquad \text{A/m.} \tag{B.23}$$

....

Example B.4 Use the result obtained in Example B.3 to prove *analytically* that the stator currents *in all three stator phases* jointly create a rotating "current wave." (In Chapter 4 this fact was "proved" heuristically. Compare Fig. 4.22.)

SOLUTION: The result in Example B.3 can be stated thus:

$$A_{a1}(x) = \frac{2}{\pi} \hat{A} \sin\left(\frac{p}{D} x\right) \qquad \text{A/m.} \tag{B.24}$$

This formula represents the base wave (first harmonic) of the stator current sheet due to the current in phase a as *viewed at a particular instant of time*. If we multiply by $\sin \omega t$ we express the pulsating nature of the wave:

$$A_{a1}(x, t) = \frac{2}{\pi} \hat{A} \sin\left(\frac{p}{D} x\right) \sin \omega t. \tag{B.25}$$

The currents in phases b and c give rise to similar current sheets but they are shifted both in space and time by $2\pi/3$ and $4\pi/3$ radians, respectively: that is,

$$\left.\begin{aligned}
A_{b1}(x, t) &= \frac{2}{\pi} \hat{A} \sin\left(\frac{p}{D} x - \frac{2\pi}{3}\right) \sin\left(\omega t - \frac{2\pi}{3}\right), \\
A_{c1}(x, t) &= \frac{2}{\pi} \hat{A} \sin\left(\frac{p}{D} x - \frac{4\pi}{3}\right) \sin\left(\omega t - \frac{4\pi}{3}\right).
\end{aligned}\right\} \tag{B.26}$$

The total current sheet is obtained by the sum

$$A_{\text{tot } 1}(x, t) = A_{a1} + A_{b1} + A_{c1}. \tag{B.27}$$

By the use of the same trigonometric manipulations as those outlined on p. 162, we determine that the total current base wave assumes the form

$$A_{\text{tot } 1}(x, t) - \frac{3}{\pi} \hat{A} \sin\left(\frac{p}{D} x - \omega t\right). \tag{B.28}$$

This, we know, is the equation for a wave revolving with constant speed and constant amplitude in positive x-direction.

appendix C

THE SI UNIT SYSTEM

C.1 GENERAL

The British Unit System, presently the most favored unit system in the United States, has roots back to Roman times. Internationally, it is fast losing ground to the Metric System conceived by the French Academy of Sciences, in the eighteenth century. On December 23, 1975, the president of the United States of America signed into law the Metric Conversion Act of 1975. Thus, the last major stronghold of the British units had committed itself to "go metric."

The bill stipulates that the metric system "means the International System of Units (SI Unit System) as established by the General Conference of Weights and Measures in 1960 and as interpreted or modified for the United States by the Secretary of Commerce."

C.2 BASIC UNITS

This appendix summarizes the most important features of the SI system as related to electric energy engineering. We tabulate first the basic SI units.

Table C.1

Basic SI Units

Physical Quantity	Name of Unit	Unit Symbol
Length	meter	m
Mass	kilogram	kg
Time	second	s
Electric Current	ampere	A
Temperature	degree Kelvin	K

(Other basic units for radioactivity, luminous intensity, and "amount of substance" exist but will not be needed in our energy story.)

C.3 DERIVED UNITS

Other *derived* units are obtained from the above basic SI units. For example, the unit for force in the SI system is newton (N). It is derived from Newton's Second Law (force = mass × acceleration) and is defined as the force needed to impart an acceleration of 1 m/s^2 to a mass of 1 kg. Force will thus have the dimension "kilogram times acceleration," which is written $\text{kg} \cdot \text{m/s}^2$.

Or consider as another example the SI unit for energy—joule (J). It is derived (see Sect. 2.3) from the basic energy equation (energy = force × distance) and is defined as the work or energy done by a force of 1 N acting over a distance of 1 m. Energy will thus have the dimension "force times distance," which is written $\text{N} \cdot \text{m}$.

The following is a tabulation of the more important derived SI units.

Table C.2

Derived SI Units

Physical Quantity	Name of Unit	Unit Symbol
Force	newton	$N = kg \cdot m/s^2$
Work (or energy)	joule	$J = N \cdot m$
Power	watt	$W = J/s$
Pressure	pascal	$Pa = N/m^2$
Electric charge	coulomb	$C = A \cdot s$
Electric potential	volt	$V = J/C = W/A$
Electric capacitance	farad	$F = C/V$
Electric resistance	ohm	$\Omega = V/A$
Magnetic flux	weber	$Wb - V \cdot s$
Magnetic flux density	tesla	$T = Wb/m^2$
Magnetic inductance	henry	$H = Wb/A$

C.4   MULTIPLICATION FACTORS AND PREFIXES

The SI system is based on decimal arithmetic units of different sizes, all formed by multiplying or dividing a single base unit by powers of 10. In this manner, widely different unit ranges can be easily accommodated. For example, an electric communication engineer is interested in microwatts (μW), whereas his power colleague works with megawatts (MW). These two power units are different by a factor of 10^{12}.

Following is a tabulation of prefixes and letter symbols for the unit multipliers.

Table C.3

Multiplication Factors and Prefixes for Forming Decimal Multiples and Sub-multiples of the SI Units

Multiplication Factors	Prefix	Symbol	Multiplication Factors	Prefix	Symbol
10^{12}	tera	T	10^{-1}	deci	d
10^{9}	giga	G	10^{-2}	centi	c
10^{6}	mega	M	10^{-3}	milli	m
10^{3}	kilo	k	10^{-6}	micro	μ
10^{2}	hecto	h	10^{-9}	nano	n
10	deka	da	10^{-12}	pico	p
			10^{-15}	femto	f
			10^{-18}	atto	a

C.5 CONVERSION BETWEEN UNIT SYSTEMS

Conversion between old units and SI units is possible from a knowledge of the basic conversion constants. The most useful ones are tabulated here.

Table C.4
Some Useful Conversion Constants

1 m = 3.28083990 ft = 39.37007874 inches
1 ft = 0.3048000 m

1 kg = 2.20462262 lbs = 35.27396195 oz
1 lb = 0.45359237 kg

1 m³ = 264.1720 gallons (U.S.)
1 gallon (U.S.) = 0.003785412 m³

degrees Kelvin = $\frac{5}{9}$(degrees Fahrenheit) + 255.37
degrees Fahrenheit = $\frac{9}{5}$(degrees Kelvin) − 459.67

We exemplify unit conversion with the following.

Example C.1   Determine the relationship between pound-force (lbf) and newton (N)!

SOLUTION: One lbf is defined as the gravitational force ("weight") exerted on a 1 lb mass by the "standard gravitational field" g_0 = 9.806650 m/s². Thus, from Newton's Second Law we have

$$1 \text{ lbf} \equiv (1 \text{ lb}) \cdot (9.806650 \text{ m/s}^2)$$
$$= (0.45359237 \text{ kg}) \cdot (9.806650 \text{ m/s}^2)$$
$$= 4.448222 \text{ kg} \cdot \text{m/s}^2$$
$$\therefore \quad 1 \text{ lbf} = 4.448222 \text{ N}$$

(or inversely: 1 N = 0.2248089 lbf).

Example C.2   Express the pressure unit pounds per square inch (lbf/inch² or psi, for short) in pascals (Pa).

SOLUTION: By definition, 1 Pa equals 1 N/m². We thus have

$$1 \text{ Pa} \equiv 1 \text{ N/m}^2 = (0.2248089 \text{ lbf})/(39.37007874 \text{ inch})^2$$
$$= 0.000145038 \text{ psi}$$

(or inversely: 1 psi = 6894.75 Pa = 6.89475 kPa).

Example C.3   Express the energy unit 1 foot-pound force (ft · lbf) in joules (J)!

SOLUTION:

$$1 \text{ ft} \cdot \text{lbf} = (0.30480000 \text{ m}) \cdot (4.448222 \text{ N})$$
$$= 1.355818 \text{ N} \cdot \text{m}$$
$$\therefore \quad 1 \text{ ft} \cdot \text{lbf} = 1.355818 \text{ J}$$

(or inversely: $1 \text{ J} = 0.737562 \text{ ft} \cdot \text{lbf}$).

....

Figures 2.13 and 2.14 in the text give the conversion factors between the most commonly used energy and power units.

REFERENCES

ASTM/IEEE Standard Metric Practice, IEEE Standard STD 268–1976. This Booklet may be obtained from IEEE Service Center, 445 Hoes Lane, Piscataway, N.J. 08854.

ANSWERS TO SELECTED EXERCISES

Chapter 1

1.1 $m = 445.1\,\text{kg}$ **1.2** About 18.31 billion dollars

Chapter 2

2.1 a) 16.5% (or $1.62\,\text{m/s}^2$)   b) 1.62 J/kg

2.2 a) 0   b) 17900 MW (or 24.0 million hp)

2.3 678.6 kW

2.4 The kinetic energy of the elevator must also be included. This energy component amounts to
$$w_{\text{kin}} = \tfrac{1}{2} \cdot 5000 \cdot 5^2 = 62{,}500\,\text{J}.$$

2.5 9.26 miles per gallon   **2.7** $\ddot{x} = \dfrac{(M - m)g_0}{m + M + I/R^2}\ m/s^2$

2.10 $\Delta T = 1.55°\text{C}$

2.12 a) 30.5 MJ/kg   b) 4.8% and 95.2%, respectively

Chapter 3

3.1 a) $3v_0\,\text{V}$ b) $w_0 = \tfrac{3}{2}C\,v_0^2\ \text{J}$

c) A force is required to pull the plates apart (the opposite charges attract). The work done by this force adds to energy of system.

3.3 a) $i(t) = \dfrac{V}{R}\,e^{-2t/RC}\ \text{A}$

b) The two capacitor voltages can be written
$$\frac{V}{2}(1 + e^{-2t/RC}) \qquad \frac{V}{2}(1 - e^{-2t/RC}),$$

respectively

c) $w_\Omega = \tfrac{1}{4}CV^2\ \text{J}$

d) Note that w_Ω is *independent* of the R-value. Thus, charge redistribution via a very small R (short circuit) results in the same heat loss as a large R. (How is this possible?)

3.7 a) $15.6 \cdot 10^{-3}$ N. (The force on each charge is directed outwards in a direction perpendicular to the line between the other two charges.)

b) 11.98 kV/m (directed perpendicular to the base)

c) Each charge contributes a field component vector directed away from the charge. As the three component vectors have equal magnitudes they thus cancel.

3.9 a) $f = 50$Hz   b) $d = 0.00398$ mm

3.10 a) $i = 0.686$ A   b) $p = 8.23$ W   c) $p = 0.00823$ W/m

3.12 a) 1.333 kA   b) 44.8 MW   c) 633.6 kV
d) 844.8 MW   e) 94.7%

3.15 $T = 1000$ N $\cdot$ m

3.18 $L = 62.5$ mH (without air gap)
$L = 2.42$ mH (with 2-mm air gap)

Chapter 4

4.1 $Q = \pm 13.85$ kVAr

4.3 Induced voltage $= 0$. (Note that flux linked with stator coil is zero for all rotor positions.)

4.4 a) 459.3 kW   b) 1377.8 kW   c) 11022 kWh

4.5 a) 15.29 kV   b) 7.647 kV per phase (or 13.24 kV line to line)

4.8 a) 84.8 A   b) 146.9 A   c) $S = 215.9 + j134.7$ (3-phase, kilo-values)

4.10 a) $\delta = 38.68°$ b) $|\text{I}| = 918$ A (leading the voltage by 19.34°)

c) $Q = -6.319$ MVAr. (Generator *absorbs* reactive power, acting like a shunt reactor.)

4.12 a) $\sin \delta = \dfrac{PX_s}{|V| |E|}$

If P, $|V|$, and X_s are constants, it follows that $\sin \delta$ (and thus δ) will decrease if $|E|$ is increased.

b) $Q = \dfrac{|V| |E| \cos \delta - |V|^2}{X_s}$ The generator absorbs reactive power if $Q < 0$; that is, if $|E| \cos \delta < |V|$. If $|E|$ is increased, δ will decrease (see part a) and $\cos \delta$ will increase. Thus the product $|E| \cos \delta$ will increase, meaning that $\| E| \cos \delta - |V\|$ will decrease, thus decreasing $|Q|$.

c) By 17.9%

4.13 a) $\delta = 16.48°$ b) $\varphi = 26.57°$ c) $|E| = 14.69$ kV

4.14 $|E| = 10.76$ kV

Chapter 5

5.1 a) 60 A and 150 A, respectively   b) $|Z| = 1.33 \, \Omega$

5.2 a) 120% of normal flux value   b) 100% (200 V). The high flux may possibly result in core losses (and temperatures) that may damage transformer.

5.3 a) $Z_s = 0.0806 + j0.418$ (on HV side) $Z_s = 0.0129 + j0.0669$ (on LV side) b) 5.11% of normal value

c) Because the core flux is only 5% of normal value, the core losses likewise are very minute (actually less than 5% of normal values, because core losses increase almost by the square of flux).

5.5 a) 63.66 A and 159.2 A, respectively; Secondary voltage = 200 V

b) 61.23 A and 153.1 A, respectively; Secondary voltage = 192.4 V

c) Yes, slightly.

5.7 105 kVA

5.8 a) 1167 V

b) Due to the high flux densities (233% of normal. Why?) the core losses will be very high with overheating as result.

5.11 150 kVA tertiary resistive load in combination with 150 kVA secondary resistive load will result in 300 kVA primary kVA. This is the primary limit.

5.12 a) $R = 538.2\Omega$ b) 74.7 A c) 545.4 A (primary); 129.3 A (secondary)

5.14 By 0.953%

Chapter 6

6.1 The primary winding carries 0.358 A. The secondary winding carries 22.6 A in one section and 13.5 A in the other.

6.3 a) 12.0 MW (3-phase) b) 10.83 MW (3-phase)

6.4 Increase by 0.604%

6.7 a) 149.48 kV

b) Sending end powers: 102.48 MW, 17.85 MVAR.
Receiving end powers: 98.01 MW, 0 MVAR.

6.8 a) 27.1 A/phase b) Line *consumes* 5.04 kW and *generates* 6.58 MVAR

c) $|V_2| = 141.19$ kV

6.10 a) $C = 106.1 \mu$F per phase b) 31.18 kV

6.13 The load flow will appear as follows:

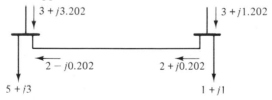

6.15 The load flow will appear as follows:

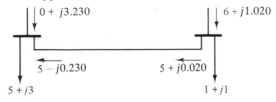

Chapter 7

7.1 a) Maximum force occurs for $s = 0$ b) $f_{m\ max} = \dfrac{BLV}{R}$ N.

7.2 a) Maximum power occurs for $s = \frac{1}{2}s_0$ b) $p_{m\ max} = \dfrac{1}{4}\dfrac{V^2}{R}$

   c) $p_{\Omega} = \dfrac{1}{4}\dfrac{V^2}{R}$ d) $\eta = 50\%$

7.5 $p_{m\ max} = \dfrac{1}{4}\dfrac{V_a^2}{R_a}$

No! It would be overheated. For example, the motor in text Example 7.8 would deliver 13.7 kW and dissipate an equal amount in ohmic heat. (Its normal heat losses are only 256 W according to Example 7.8.)

7.6 $x = 2.193$

7.7 a) $e = 498.0$ V; $i_a = 9.30$ A
    b) $e = 523.0$ V; $i_a = 9.86$ A

7.9 The motor will deliver 86.79 kW (116 hp) to the load. Operating efficiency = 84.67%. Shaft torque = 361 N · m.

7.12 $R_{min} = 2.29\ \Omega$. Generated power = 91.56 kW. Diesel output = 105.55 kW.

Chapter 8

8.2 a) 3.17%   b) 196.83%   c) 156.8 A/phase

8.3 a) $P_1 = 12.01$ kW
       $P_m = 10.64$ kW
       $P_{\Omega} = 1.37$ kW
       $T_m = 88.39$ N · m
   b) $P_1 = -13.98$ kW (delivers power to network)
      $P_m = -15.86$ kW (draws this power from prime mover)
      $P_{\Omega} = 1.88$ kW
      $T_m = -121.2$ N · m (instead of delivering torque to load, will now require torque
         to run at this speed)
   c) $P_1 = 27.33$ kW
      $P_m = -5.54$ kW (needs power from prime mover)
      $P_{\Omega} = 32.87$ kW (With these losses the motor would burn up in a hurry. Note
         that 83.2% of this loss power is drawn from the network and 16.8%
         from prime mover.)
      $T_m = -44.1$ N · m
   d) $P_1 = 23.57$ kW
      $P_m = -11.99$ kW
      $P_{\Omega} = 35.56$ kW (see comment under part c)
      $T_m = 0$

8.5 $P_m = 75.15 \text{ kW} = 100.7 \text{ hp}$
$|I_1| = 107 \text{ A/phase}$
$\cos \varphi = 0.988$
$\eta = 93.3\%$
$T_m = 823 \text{ N} \cdot \text{m}$

8.6 635 A (if IM model is used); 646.5 A (if magnetization current is included)

8.7 Stator current $= 111 \text{ A/phase}$
Motor output power $= 73.30 \text{ kW} = 98 \text{ hp}$
Efficiency $- 89.5\%$
Power factor $= 0.970$

8.9 93.07 kW

8.11 Let the transformer series impedance be

$$Z_t = R_t + jX_T \qquad \Omega/\text{phase}.$$

(We neglect its magnetizing impedance.) Then the formula for maximum torque
will be

$$T_{\text{max}} = \frac{45}{\pi n_s} \cdot \frac{|V|^2}{\sqrt{(R_1 + R_T)^2 + (X_1 + X_2' + X_T)^2} + R_1 + R_T}$$

where $|V|$ is the *primary* transformer voltage (which we consider constant).

INDEX

INDEX